Now Hear This

John Naylor

Now Hear This

A Book About Sound

 Springer

John Naylor
London, UK

ISBN 978-3-030-89876-2   ISBN 978-3-030-89877-9   (eBook)
https://doi.org/10.1007/978-3-030-89877-9

This Springer imprint is published by the registered company Springer Nature Switzerland AG
The registered company address is: Gewerbestrasse 11, 6330 Cham, Switzerland

Sue and her keen ears

Acknowledgements

Like any book that deals with facts and theories that are not the author's own, this one required the help and advice of a large number of people. A fair number of those are the authors of the books and papers I have consulted, and whose names are to be found in the bibliography. Online videos enabled me to pay virtual visits to far away places and listen to the sounds associated with them. I also owe a debt to the those who asked me questions while writing the book, especially the questions that tested my understanding, giving me an opportunity to get things straight.

Particular thanks are owed to Prof. Geoffrey Manley, Steve Tunnicliffe, Antony de Peyer and Joe Naylor who read some of the chapters and offered useful and constructive advice, correcting errors and helping me to express myself more clearly and succinctly. Thanks also to Dr. Gianluca Memoli for organising a visit to the anechoic chamber at the National Physical Laboratory in Teddington. Among those who answered queries or asked searching questions are Dr. Javier Amezcua, Dr. Bruno Fazenda and Les Cowley.

Others who have contributed include Jos Widdershoven (walks and translations), Bob Bery (whispering galleries), Ben Cuddon (whistling jugs), Julian Mumby (Woodstock echo), Mike Benson (lessons in clapping), Francesca Phillips (whistled language of La Gomera), and all the people who have accompanied me on sounds walks in and around London.

Most importantly, I thank Sue, my dear wife, for her company, love, patience and encouragement, who never doubted that this project would see the light of day and whose acute hearing came in useful on many occasions.

Introduction

I have to begin with a confession: I am not by nature a listener. I seldom pay heed to sounds with the same degree of interest and attention as I do to sights. I like to stand and stare, but all too often I fail to linger and listen. At least, such was the situation before I set about writing this book. I now realise that for most of my life I have invariably ignored the vast majority of sounds I have heard in the mistaken belief that they are inconsequential, uninteresting or irritating.

Now, well past middle age, and almost certainly soon in need of hearing aids I am increasingly aware that I no longer hear high frequencies, the very frequencies that give sounds the clarity and bite which are essential to comprehending clearly what is being said to me.[1] So having belatedly resolved to make up for lost time and listen to the world as attentively as I look at it, I now find myself in a race to make the most of what remains of the hearing I took for granted in my youth before my acoustic compass is reduced to murmurs and mumbles.

This book is the product of that decision. It is aimed squarely at anyone with an interest in science and the natural world but who is as unaware of their soundscape as I once was. There are six chapters dealing with the subjective aspects of hearing and listening; the history of some of the key events and discoveries that have led to our present knowledge about the nature of

---

[1] Since writing these words, I have been fitted with hearing aids, much to the relief of my family and friends.

sound; the nature of sound considered both as a physical phenomenon and as a sensation; the workings of the hearing system and how it has evolved in vertebrates; the passage of sound in air and in water; the many ways sounds interact with the environment and the effects of those interactions on what we hear; explanations of a host of sounds and acoustic effects, ranging from the commonplace to the unusual. The footnotes that pepper the text supply annotations to the core text—for no book is entire of itself. My research has been based on primary sources, expert advice and, wherever possible, first hand personal experience of the sounds I write about.

Compared to sight hearing is a hugely underrated sense, not least because we unconsciously assume that the world discloses itself to us most fully by means of light. We seldom pay heed to sounds with the same degree of interest and attention as we do to sights. In almost every situation we are usually far more aware of what we see than what we hear. By habitually ignoring or paying scant attention to sounds we pass up the opportunity to engage more closely with our soundscape, something that will increase our knowledge and awareness of the world at large, not to mention missing out on the unexpected sensuous and intellectual delights that are the true reward of active listening.

A book about sound must also be book about listening. It is only as one learns what to listen for that one begins to engage more fully with one's soundscape because hearing goes well beyond mere sensation. It is a creative act that melds raw auditory sensations with expectation, motive, memory and knowledge to forge a rounded experience of one's sonic environment. Descriptions and explanations of how particular sounds are produced, propagated and altered in a given situation greatly helps here, but is never the whole story, for sounds have psychological and physiological aspects that physical science does not address. The sensations that characterise sound, such as loudness and pitch, are not replicas of the physical vibrations that stimulate them, though that is something very few of us are aware of.

In thrall to the stark and reassuring certainties of physics, during my thirty years of teaching the subject I blithely assumed that there is little more to any sound than complex vibrations within a medium and their interactions with the objects they encounter. I considered that hearing was merely a useful though minor adjunct to those facts. What need for ears when there are mathematical formulas and scientific instruments with which to analyse and make sense of the vibrations we experience as sound? I took it for granted that physics could, and did, tell me everything worth knowing about sound and hearing. How wrong I was. I now realise that although there can be no sound without vibrations to stimulate the hearing system, sounds and the vibrations that give rise to them are worlds apart. One is a visceral, flesh and blood sensation, the other a noiseless physical event.

It was only when I realised the difference that I freed myself from the mistaken assumption that sound is a special form of acoustic vibration. In fact, there is no such thing as a sound wave, there are only vibrations of the medium in which we find ourselves. *All* vibrations of air and water are in principle audible to humans as long as they satisfy two conditions: first that they are sufficiently energetic to cause the eardrum to vibrate and second that they fall between 20 Hz and 20,000 Hz.

And so one of the least expected and most salutary lessons I learned while writing this book is that sound is a product of the auditory cortex: it's all in the mind. In the absence of ears the world would be utterly silent. It hadn't occurred to me that there were no sounds before there were ears, merely feeble, fleeting vibrations that pulsed within the primeval seas that were home to the earliest forms of animal life. The evolution of the ear in its various forms was driven by improving the chances of survival brought about by being able to detect those vibrations because they proved to be an invaluable and often irreplaceable source of information about events in a creature's environment that are out of sight, events that to this day can be a matter of life or death. Within the murky, muddy waters of lakes and rivers and in the sunless depths of seas within which the ear first evolved, hearing was, and still is, a far more useful sense than vision. But to be aware of these vibrations it is necessary to convert them into something palpable, namely sensations that we experience as sound.

Nor did it occur to me to wonder why human hearing is limited to a narrow range of vibrations that lie between 20 Hz and 20,000 Hz, or why the majority of mammals hear vibrations far beyond the upper limit of that range. In fact, the range of vibrations that a given species hears has been determined as much by biology and the environment that a creature inhabits as it has by the physical properties of waves. You will find an explanation for these things in Chap. 3, in which the science of vibrations audible to the human ear is dealt with, and where several pages are devoted to the evolution and the biological purpose of the hearing system. In that chapter you will also learn that there are several illuminating similarities and differences between hearing and vision, hardly surprising given that they both rely on waves that have many properties in common.

Something else dawned on me as I was writing this book: all of us have a vast store of aural knowledge about the world and its workings that we not aware of having made any effort to acquire. We are able to locate the source of a sound with our eyes closed, estimate the relative size of an empty space just from its acoustic qualities, identify a host of events merely from their sounds, and distinguish a hard surface from a soft one simply from the sound of the impact when tapping them lightly. Nor can we recall ever having learned to make sense of the most complex of all sounds, those of speech. Our ears have been beavering away in the background since the day we were born, learning to identify and interpret sounds, seemingly with little active participation on our part. As long as one's ears are in working order and one is exposed to sounds and voices from birth, they perform the task for which nature has fashioned them: to provide information about one's environment that is both useful and reliable. But as you will discover as you read this book, despite the vast trove of acoustically acquired knowledge you already possess, there remains an equally large number of sounds and acoustic effects that you encounter daily of which you will not be aware. All it takes to expand your acoustic horizon is a little knowledge and a willingness to pause and listen attentively whenever you hear a sound.

The ability to acquire reliable knowledge of the world through our ears alone is all the more remarkable given that the acoustic world lacks the permanence of the visual world. By and large a landscape doesn't change significantly from one moment to the next. You can return days or weeks later and find it is more of less as you last left it. But given the transient nature of most sounds, if you don't make a point of listening when the opportunity arises, you may not get a second chance. That transience may be why we often find it difficult to describe a sound in terms of its acoustic qualities alone. In any case, we are usually uninterested in its acoustic qualities; it's the source we find relevant because that tells us something about what is going on in our surroundings. So when we talk about sounds we invariably refer to their sources rather than to their acoustic qualities.

We seldom hear sounds in their pristine state, however. They are almost always altered by their passage through the medium in which they are heard and their interactions with the environment in which we hear them. Indeed, these modifications are sometimes far more interesting than the original sound.[2] So if you are to fully understand why you hear sounds in the form in which they reach your ears, knowing the effect of these interactions on a given sound is as important as knowing how that sound was produced in the

---

[2] This is even truer of light, by the way: sunlight is invisible until it interacts with matter, making environments visible in all their variety and complexity and creating eye catching sky colours, rainbows and mirages. In fact, considered purely as sources, sounds are infinitely varied in comparison to lights. Every sound is due to a unique event whereas during daylight there is but one source of light: the sun. At night the only natural sources of light are stars, lightning, fire, luminous creatures such as fireflies and phosphorescent organisms. Moonlight is reflected sunlight, as is the light of planets.

first place. The nature and effects of those interactions on what we hear are the subjects of Chaps. 4 and 5.

Yet when all is said and done, information about specific sounds and acoustic effects is of little value if you don't listen to them, which is why you will find suggestions and advice in every chapter of how to improve your chances of hearing them. Knowing which features of a particular sound to listen for will influence and alter the way you listen and, consequently, what you hear, a virtuous circle of aurality. If this book achieves anything I hope it will be to encourage you to listen to all sounds, no matter how unwelcome and unpleasant, with greater attention, interest and understanding than hitherto—and thereby be enthralled and entertained and enlightened by the experience.

I don't, however, have much to say about the undesirable effects of sound beyond sketching out a brief history of noise to show that unwanted sounds have posed a problem ever since humans made an appearance on Earth. Interesting and important though the physiological and psychological effects of noise on people and animals are, they are subjects that are well catered for in numerous books, journals and blogs. In any case, I want to encourage you to listen attentively to sounds however intrusive or disagreeable they may be. Not that I advocate that you should tolerate such sounds any longer than is necessary to make sense of them. It's that one of the consequences of living in a noisy environment, as so many of us do these days is that we allow ourselves to become functionally deaf. We don't go out of our way to listen to ambient sounds unless we have a particular reason to do so.

To avoid overwhelming you with dry, matter-of-fact descriptions and explanations, necessary though these sometimes are, I have provided accounts of how some of the unusual sounds and acoustic effects that you will be reading about were discovered. It will put you in the shoes of the men and women who made these discoveries, enabling you to understand the challenges of hearing and making sense of unfamiliar acoustic phenomena.

The history of these discoveries serves another purpose. We usually ignore the distinctive acoustic qualities of most sounds even when they are in principle audible. John Ruskin, the influential 19th century art critic, pointed out "The first great mistake that people make ... is the supposition that they must see a thing if it be before their eyes."[3] Substitute 'hear' for 'see' and 'within earshot' for 'before their eyes' and you have the reason why most of the unusual sounds and their modifications that you will come across in this book were either unknown, ignored or misunderstood until someone

---

[3] Ruskin, J. (1908) Modern Painters, vol 1. George Allen & Sons, p 54.

realised that what they had heard was in some way out of the ordinary or unexpected and required explanation. If nothing else, finding out how and why they made their discoveries teaches us that if we don't make a point of listening, we will hear only noise.

Although Chap. 2 is specifically devoted to the history of acoustics, in every chapter you will find accounts of how we have come to our present knowledge of sound in general and of specific sounds in particular. Arguably, the science of acoustics began with Pythagoras' attempt to discover the basis of musical harmony, an issue that is yet to be settled, if indeed such a thing is possible. In the following centuries the natural philosophers of ancient Greece established several facts about sound, such as that it travels much more slowly than light and that all sounds, whatever their pitch, must travel at the same speed. They also surmised that sound requires a medium, though proof of this was delayed until the scientific revolution of the 17th century during which air pumps capable of producing a vacuum were invented. That century also saw the development of precise measurements and mathematical theories about the nature of sound and how it propagates, which in turn paved the way to the first satisfactory account of light as a wave in the early decades of the 19th century.

The scientific investigation of sound was never conducted in isolation from the wider scientific investigation of nature. The important role that sound played in the development of modern science is all too often ignored in books about the history of science. It took a couple of thousand years before the innermost secrets of light were finally laid bare (i.e. the wave-particle model c.1900 AD), with considerable help along the way from what was known about sound. But the essential nature of sound—that it involves matter in motion—was known to the ancient Greeks c.300 BC.

But until the turn of the 20th century, almost all scientific research into sound and hearing was based on the assumption that only simple sounds that have an identifiable frequency are worth investigating. Complex, dissonant sounds composed of a broad range of unrelated frequencies, such as those due to natural or mechanical events, were lumped together as noise and, on the whole, ignored by scientists in the belief that nothing very useful could be discovered from such sounds. This view owed as much to cultural preconceptions as it did to the limitations of the scientific apparatus and laboratory techniques of the day. Symphonic music was considered to be the highest expression of European culture, so scientific research into sound focused on investigating those aspects of sound that underpin music. It was only with the invention of electrical devices such as the loudspeaker and microphone towards the end of the 19th century that noise began to interest scientists.

Indeed, making scientific sense of noise became a matter of life and death during the two World Wars, as we shall see in Chaps. 4 and 5.

So much for an overview of what you will find between the covers of this book. Now down to business. If you've picked up the book because its subject matter is of interest to you but, like me, you've hitherto taken your ears for granted, you may find it worthwhile priming the pump and, rather than diving straight into Chap. 1, spend some time listening to the soundscape you find yourself in and even making a few noises of your own.

So here is a suggestion: put the book down and take a break. Make yourself a cup of tea (or coffee) and listen to the attendant sounds as if you were hearing them for the first time.

Finished? What did you hear as you filled the kettle from the tap and as it heated up until it began to boil? And what of the sound as you poured water from the kettle into the cup? Did it change as the cup filled up? What of the sound as you stirred the drink with a spoon? If you made instant coffee you might have noticed a change in pitch as you stirred. And how would you describe these sounds to someone else: in terms of their acoustic qualities or the events or things that are their source?

There will, of course, have been many other sounds as you made the cup of tea, ambient sounds that had nothing to do with tea-making. How aware of them were you and can you recall them? I ask only to make the point that to consciously experience sounds you have to make a point of listening attentively, which is the overriding theme of this book.

Alternatively go for a walk and make a point of listening as attentively as you can. Sound walks, as such walks are known, are very popular these days and I shall have more to say about them in Chap. 1.

A final word of advice. The descriptions and explanations of sounds that you will here is only a first step. The real goal is to listen to them. This is not just a matter of being in the right place at the right time, it also helps to know what it is that one should be listening for. Written descriptions of the acoustic qualities of a particular sound is helpful, but these usually have to be heard several times before one can be confident of identifying them without difficulty. Persistence is the key, as I have learned. And when listening be prepared to hear out every sound from the moment you become aware of it until it finally fades away: linger and listen.

Contents

1

Just Listening

Abstract The chapter deals with the subjective experience of hearing and introduces themes and topics that are taken up in later chapters. It opens with a brief history of the universal human aversion to sounds that are loud, intrusive or unpleasant and explores their psychological and physiological consequences before going on to argue that if one is willing to listen attentively one discovers not only that every sound is interesting, but also that they are frequently an irreplaceable source of information about the world.

A Brief History of Noise

In the realm of the senses we unthinkingly consider sight to be first among equals, so much so that we give little thought to the broader picture, that we have a multitude of senses on which we rely for an awareness of our body and the surroundings in which it finds itself. Though the exact number remains an open question, the idea that we have just five is a legacy from ancient Greece, and a gross underestimate.[1] But however many senses we may have,

---

[1] Quite apart from vision, hearing, smell, taste and touch, there is the vestibular system that is essential to keep track of head movements and prevent one from falling over (and is located in a cavity within the skull adjacent to the system responsible for hearing). And touch consists of several distinct senses that are separately responsible for sensations of warmth, hardness and pain. In total we may have as many as 21 senses. Interestingly, the Greek quintet of five senses (sight, hearing, taste, smell and touch) is also found in Chinese and Indian natural philosophy.

J. Naylor, *Now Hear This*,
https://doi.org/10.1007/978-3-030-89877-9_1

only two of them are capable of perceiving the world in detail well beyond arms' reach: sight and hearing.

On the face of it, however, we appear to have far less control over what we hear than what we see. You can avert your gaze from anything that distracts, disgusts or dismays you and direct it at what is of interest or importance. You can close your eyes or turn your back on what you don't want to see, but short of blocking your ears with your fingers there is no equivalent action that will enable you to avoid sounds you don't want to hear.[2] Unlike eyes, ears are more or less permanently open to the world. Indeed, if eyes are scouts that probe and search our surroundings, ears are watchful sentinels that warn of dangers and opportunities that are out of sight Friedrich Nietzsche, the German philosopher and a master of the mordant aphorism, considered the ear to be the organ of fear that evolved to warn us of dangers in the darkness of night.[3]

The price we pay for having such a vigilant sense is that all too often we must endure sounds over which we have no control, sounds that are unpleasant or intrusive and which may be loud enough to damage our hearing. Perhaps this is why there is a specific word for sounds that we would rather not hear—noise—but no comparable collective noun for their visual equivalents, for sights that we find distracting or distressing. And you don't need to be told that we live in a noisy world. Nor that it's becoming noisier, that there are fewer places where one can get away from the host of loud, disagreeable, unwanted sounds due to overflying aircraft and road traffic, not to mention sirens, chainsaws, mowers, strimmers, leaf-blowers and the like that contribute to the all too often aggravating soundscapes of our towns and cities, and which have long since encroached on all but the most remote rural landscapes.

Not that there ever was a time when the world was entirely free of noise, especially noises of which other people were, and continue to be, the source. For noise is more often than not shorthand for any sound that we would rather not hear, either because it is loud or unpleasant or because we find it intrusive. Unwanted sounds can interrupt one's train of thought, cause distress because the source is beyond one's control or prevent one from hearing clearly sounds one wishes to hear. Even sounds that in one context one finds enjoyable or soothing, say a favourite piece of music or a gentle tinkling of a fountain, can be distracting or aggravating in another.

---

[2] Low frequency sound can reach the inner ear by bone conduction through the skull, so sticking your fingers in your ears won't prevent you hearing the bass notes emitted by a source of sound.
[3] Nietzsche F. (1997) Daybreak, Thoughts On The Prejudices Of Morality (trans: Hollingdale R.J.). CUP, aphorism 250, p. 143.

Indeed, to judge from the ancient Mesopotamian tale of Atra-hasis, humans have always had a deep-rooted aversion to unwanted sounds because, long ago, people believed that even the gods detested noise above all else. In this story, which almost certainly predates by several centuries or more the 2nd millennium BC clay tablets on which it first was inscribed, we learn that the storm god and chief deity, Enlil, found the din made by the first humans so unbearable that he decreed that their number should be drastically curtailed. First he visited a drought on them, followed by a plague and finally a famine. But after each disaster the survivors quickly restored their number and became as noisy as before. Exasperated, Enlil concluded that the only solution was to do away with humans altogether by drowning them all in a great flood. One of the gods, Ea, considered the punishment to be unjust and secretly warned Atra-hasis of the impending catastrophe, advising him to build a huge vessel and fill it with artisans and animals, which he did.[4]

Enlil was enraged when he discovered that he had been betrayed, but was eventually persuaded by Ea to spare Atra-hasis and his companions on condition that, to keep their numbers down, henceforth man born of woman shall be mortal. The story of this great flood and its survivors reappears in the Epic of Gilgamesh, a rip-roaring, eponymous saga which dates from the turn of the 2nd millennium BC, though the name of the ark-builder and survivor is now Uta-napishti. After the flood has abated, Uta-napishti and his wife are granted immortality on condition that henceforth they live apart from their fellow men and women. It is the account of the flood in that epic that the Jewish authors of the book of Genesis drew upon for the story of Noah and the flood, though they attributed the punishment to mankind's wickedness rather than to disturbing God's peace of mind.

The Epic of Gilgamesh has been patiently pieced together by scholars over several decades from hundreds of clay tablets, the first of which were recovered in 1850 by an English archaeologist and his Assyrian assistant from the ruins of the royal library of the last great king of Assyria, Ashurbanipal (668–627 B.C.), at Nineveh, in what is now northern Iraq.[5] Nineveh, which was founded several thousand years before the reign of Ashurbanipal, was perhaps the first true city. And of all environments, the hustle and bustle of commerce, manufacture and street life makes the soundscape of cities potentially the most varied, and to sensitive ears, the most gruelling of any. Cities are also

---

[4] Finkel, I. (2014) The Ark Before Noah: Decoding the Story of the Flood. Hodder & Stoughton.
[5] The Epic of Gilgamesh (trans: George, A.). Penguin Books, 1999. See the introduction for an account of the discovery and translation of the cuneiform tablets.

home to poets, philosophers, historians and diarists who have often chronicled life lived within earshot of inconsiderate neighbours, bustling streets and busy workshops.

Writing about life in Rome at the turn of the second century, AD, the Roman poet, Juvenal claimed that the city was so noisy that "Tis frequent here, for want of sleep, to die …What house secure from noise the poor can keep, When even the rich can scarce afford to sleep?"[6] Even the Stoic philosopher, Seneca the Younger, writing a generation before Juvenal, admitted in a letter to a friend that the noises from the gym above which he had rooms (grunting weightlifters, swimmers splashing about in the pool, cries of hawkers peddling their wares), and from the street (trundling carriages, artisans at work, more hawkers), to which, as a Stoic, he claimed he should be indifferent, eventually got the better of him. He moved lodgings.[7]

To judge from "The Statutes of the Streets of this City, against Noysances" drawn up to control inconsiderate neighbours and busy workshops, intrusive noises were an inescapable fact of life in late sixteenthcentury London. Rule 30 required that "No man shall after the hour of nine at the Night, keep any rule whereby any such suddaine outcry be made in the still of the Night, as making any affray, or beating hys Wife, or servant, or singing, or reviling in his house, to the Disturbance of his neighbours." And trades that employed hammers were covered by rule 25, which ordered that "No hammar man, (such) as a Smith, a Pewter, a Founder, and all Artificers making sound, shall not worke after the houre of nyne in the night, nor afore of four in the Morninge." Enforcing these rules was another matter, because the authorities seldom acted upon complaints.[8]

London's "hammar men" paid a heavy price for practicing their trade, one that all metal workers since antiquity have paid, because repeated exposure to loud sounds progressively kills off the tiny hair cells that convert external vibrations into auditory sensations within the inner ear. Hair cells do not regenerate, so their destruction leads to irreversible hearing loss and even to profound deafness. This was particularly true of blacksmiths and coppersmiths because iron and copper are harder to work than softer metals such as silver or gold and so require more forceful blows to fashion them into shape. And it isn't just loudness that harms hearing. The high frequency sounds that are produced when a stiff metal sheet is struck are particularly

---

[6] Juvenal, Satires III, 375–380. In: The Satires of Decimus Junius Juvenalis (1693) Translated into English Verse. By Mr. Dryden, and Several Other Eminent Hands. Printed for Jacob Tonson, London.

[7] Seneca the Younger, Epistles to Luculius, LVI. https://en.wikisource.org/wiki/Moral_letters_to_Lucilius/Letter_56 (accessed 06/08/2021).

[8] City of London (1677) The Laws Of The Market. Printed by Andrew Clark, Printer to the Honourable City of London.

damaging, so hearing impairment in metal workers typically begins with the loss of sensitivity to high frequencies.

Worse was to come. During the early stages of the Industrial revolution, which began during the second half of the eighteenth century, small, muscle-fueled workshops and forges were replaced by relentless and impersonal steam-driven machines housed in large, regimented factories. The noise of the machinery within these factories was often so loud that workers could not hear themselves speak, and occupational deafness was no longer confined to the "hammar men". Weaving mills, in particular, were notoriously noisy, and the level of sound due to the machinery was such that workers were very unlikely to survive life on the shop floor with their hearing intact.

Surprisingly, the link between loud sounds and deafness was not fully recognised until well into the nineteenth century, possibly because workers usually died before the full effects of their occupation on their hearing reached its inevitable denouement. Even as late as 1831, while acknowledging that blacksmiths eventually lose their hearing due to the sound of constant hammering, a physician might be just as likely to ascribe his patient's deafness to "cold air, variable climate, nasal polyps, tonsillitis, fever, bladder infections, measles, catarrh."[9] Nevertheless, the ancient Greeks may have been aware of the link between loud sounds and deafness. Aristotle implies as much in a passing reference to metal workers of his day. Commenting on the claim that mortals are unable to hear the Music of the Spheres—the sound that Pythagoreans claimed was made by the planets as they move through the sky—he wrote that it "is just what happens to coppersmiths, who are so accustomed to the noise of the smithy that it makes no difference to them."[10] Knowing what we now know, it is not unreasonable to conclude that the reason Greek coppersmiths tolerated the sound of constant hammering is that it had rendered them partially deaf and so were unable to hear the worst of it.

And in an early eighteenth century treatise on industrial injuries by an Italian physician, Bernardino Ramazzini wrote that Venetian coppersmiths become increasingly deaf during their working life "To begin with, the ears are injured by that perpetual din … so that workers of this class become hard of hearing and, if they grow old at this work, completely deaf."[11] He made

---

[9] Fosbroke, J. (1831) "Practical observations on the pathology and treatment of deafness. The Lancet, Vol. 16, No. 398, p 69–72.

[10] Aristotle, 350BC, On the Heavens, Book 2, Sect. 9. http://classics.mit.edu/Aristotle/heavens.2.ii. html (accessed 15/03/2020).

[11] Ramazzini, B. (1940) Diseases of Workers Translated from the Latin text De morbis artificum of 1713 by Wilmer Cave Wright. Chicago University Press, p 438.

no mention of other trades in which hearing is damaged by loud noise such as stonemasons, blacksmiths and military gunners.

The loudest and most injurious industrial sounds were those produced in the manufacture of boilers, the *sine qua non* of the steam engines of the nineteenth and twentieth century. Indeed, hearing loss was so common among boilermakers that occupational deafness of industrial workers became known as "boilermaker's disease". The worst affected were the riveters' mates who worked inside the boiler. Although boilermakers knew that their hearing was impaired, many of them insisted that they could hear perfectly well as long as they were in a noisy environment. They seemed not to realise that this was because they had to shout loudly to be heard above the din. When their hearing was tested, they were invariable found to be partially deaf.[12]

But even if you haven't spent your working life bashing bits of metal, your hearing will almost certainly deteriorate. Hearing loss is, alas, an inevitable consequence of aging. The condition is known as presbycusis and is due to irreversible damage of the hair cells within the inner ear responsible for hearing high frequencies.[13] But presbycousis usually occurs so gradually that one becomes aware of the problem only when one finds oneself constantly having to ask people to speak up or repeat what they have just said, adding irritably: "don't mumble". The inability to hear high frequencies makes it difficult for a listener to distinguish clearly sibilant consonants such as /s/ from /z/, in which high frequencies are prominent.[14] This can result in confusing one word for another, e.g. 'fifty' and 'sixty'. Presbycusis makes it difficult for English speakers to distinguish the singular from the plural form of English words because the sibilant /s/ at the end of the plural form can't be heard distinctly.[15] As we shall see in chapter three, the link between presbycusis and the loss of sensitivity to high frequency sounds was not clearly established until the early years of the nineteenth century.

It is also possible to lose one's hearing through illness, injury or poison. Hearing loss in an adult can result from illnesses such as scarlet fever and measles, and, occasionally, head injuries. And as if that were not enough, some people become partially deaf because the tiny bones in their middle

---

[12] Schartz, H. (2011) Making Noise: From Babel to the Big Bang & Beyond. Zone Books – MIT, p 366–67.

[13] The deterioration of vision with age due to loss of elasticity of the lens within the eye, which leads to long sightedness, is known as presbyopia.

[14] See this for yourself with an audio spectrometer such as SpectumView, an app for iOS. Alternately voice each of the consonants into the app and notice that their spectra are (a) similar and (b) higher frequencies are prominent. The spectra of consonants /v/ and /f/ are even more similar. To see the audio spectrum clearly, prolong the utterance of the consonant to allow the spectrometer to reveal the constituent frequencies clearly.

[15] Foley, H.J., Matlin, M.W. (2010) Sensation and Perception. Routledge, p 394.

ear fuse together, which prevents them efficiently transmitting vibrations of the eardrum to the inner ear. The condition is known as otosclerosis and it is the commonest cause of deafness in young adults. Alarmingly, some of the drugs that have been used to treat these conditions, such as quinine or the antibiotic streptomycin, are now known to damage the inner ear and the auditory nerves.

Otosclerosis is probably what put paid to Ludwig van Beethoven's hearing. He became aware of problems with his hearing when he was 30 years old and living in Vienna. He had already been suffering from ringing and buzzing in his ears (i.e. tinnitus) for a few years. Most of the doctors he consulted concluded that his ears were blocked up with wax and that in time they would clear. Given the state of medical science at the time, it is not surprising that the remedies they suggested proved useless. Among these was bathing in the Danube. Eventually he was advised to leave Vienna for a quiet village not far from the city. He was joined there by a friend, Ferdinand Ries. During a walk in the countryside, Ries drew Beethoven's attention to a shepherd playing his pipe, but Beethoven, to his great dismay, couldn't hear a thing.[16] Perhaps this was the moment when he first realised that his hearing was not going to improve. It continued to deteriorate and in 1816 he resorted an ear trumpet. A year later he was reduced to using pencil and paper to conduct conversations. And by 1823 he was completely deaf. Only someone with most determined and uncompromising character could have composed his greatest symphonies while being unable to hear clearly, or indeed at all in the case of his final and possibly his greatest symphony, the 9th. The autopsy following his death was inconclusive regarding the cause or causes of his deafness. A later analysis of his hair revealed high levels of lead, a metal now known to be otoxic and which can cause otosclerosis.

Beethoven lost his hearing over several years, but a serious injury to the head can sometimes deafen one in the blink of an eye. Disconcertingly, a sudden and total loss of hearing is not always immediately evident to the sufferer. John Kitto, a nineteenth century British missionary, lost his hearing aged 12 following a fall from a roof. He was concussed and didn't regain consciousness for 2 weeks. "I was very slow in learning that my hearing was entirely gone. The unusual stillness of all things was grateful to me in my utter exhaustion [which] I ascribed to the unusual care and success of my friends in preserving silence around me. I saw them talking indeed to one another, and thought that, out of regard to my feeble condition, they spoke in whispers, because I heard them not. The truth was revealed to me when

---

[16] Wegeler, F. and Ries, F. (1848/1988) Beethoven Remembered: The Biographical Notes of Franz Wegeler and Ferdinand Ries, (trans: Noonan, F.). Andre Deutsch Ltd.

[I asked for a book I wished to read]. And I was answered by signs which I could not comprehend. 'Why do you not speak', I cried, 'pray let me have the book.' [But instead of speaking to him, someone wrote an answer on a slate.] But, I said in great astonishment, 'Why do you write to me, why not speak? Speak, speak.' The answer was written on the slate: 'You are deaf'."[17]

Another occupation where hearing was and continues to be at risk is soldiery. The blast that accompanies the exit of a musket ball or bullet from the muzzle of a rifle damages the hearing of the shooter, a condition known a "shooter's ear". The problem would have been evident as long ago as the sixteenth century, when firearms were first introduced to the battlefield. Ambroise Paré, who served with the French army as a battlefield surgeon in the sixteenth century, found that cannon fire often left gunners permanently deaf.[18] But it was only following the American Civil War (1861–65), in which at least one third of the surviving combatants were found to have suffered loss of hearing to a greater or lesser extent, that deafness was recognised to be a service-related injury.

Hearing loss, however, is arguably among the least of a soldier's worries when on active service. Long lasting psychological trauma due to the stress of combat, which in addition to the prospect of being maimed or killed includes the shock and noise of battle, afflicted combatants long before it was properly recognised. Herodotus, the Greek historian, writing about the battle of Marathon which took place in 490BC recounted how "An Athenian, Epizelus, son of Cuphagoras, while fighting in the medley, and behaving valiantly, was deprived of sight, though wounded in no part of his body, nor struck from a distance; and he continued to be blind from that time for the remainder of his life."[19] There are countless similar cases in accounts of battles throughout the ages.[20] But it is only since the advent of the modern artillery shell during the early years of the nineteenth century that the nerve-shattering sound of exploding shells became an inescapable feature of the battlefield soundscape, and which undoubtedly contributed to what became known as "shell shock" during the First World War. It was experienced in all armies during that war, hence "kriegsneurose" in German and "névrose

[17] Kitto J. (1845) The Lost Senses, Series 1, Deafness. Charles Knight & Co, p 11–12.

[18] Schacht, J., (2008) Auditory Pathology: When Hearing Is Out Of Balance. In: Schacht, J., Popper, A. N., Fay, R.R. (eds) Auditory Trauma, Protection, and Repair. Springer, p 1.

[19] Herodotus, (1899) The Histories of Herodotus With a Critical and Biographical Introduction by Basil L. Gildersleeve (trans Cary, H.). D. Appleton and Company, Book VI, p 357.

[20] Crocq, M.-A., Crocq, L. (2000) From Shell shock and war neurosis to posttraumatic stress disorder: a history of psychotraumatology. Dialogues in Clinical Neuroscience, Vol 2, No 1, p 47–55.

de guerre" in French.[21] Following the huge number of cases of traumatised American soldiers during the Vietnam War, which ended in 1975, the condition has been known as "post traumatic stress disorder" or P.T.S.D.

Although the "hammar men" of earlier years had ceased to ply their trade in London's residential districts by the nineteenth century, the streets still rang with intrusive sounds. Alongside the harsh, unpleasant sounds of iron-rimmed carriage wheels on cobbled streets, London's chattering classes objected to the legions of street musicians and the all too often incomprehensible cries of a multitude of itinerant street traders. Charles Babbage, the wealthy and notoriously irascible mathematician who designed the first, albeit entirely mechanical, computer, was prepared to stand his ground, though much good it did him. His particular *bête noire* were street musicians, especially organ grinders. He used his political contacts to lobby against them and as a result of his campaign became a target for people who took an uncharitable delight in provoking him. When he left his house, he wrote, "the crowd of young children, urged on by their parents, and backed at a judicious distance by a set of vagabonds, forms quite a noisy mob, following me as I pass along, and shouting out rather uncomplimentary epithets."[22]

For the best part of a year, Babbage kept a tally of the number of times he had been disturbed by organ grinders—165 occasions from August 1860 to May 1861—and appended it to the letter of support he wrote to Michael Bass, an M.P. who was campaigning to make the parliamentary statute that regulated noise more effective. Bass's amendment required the police to enforce an existing bylaw that allowed householders to ask street musicians to move on, something that the police had hitherto been reluctant to do. The bill was enacted in 1864, much to the relief of several hundreds of "professors and practitioners of one or other of the arts and sciences", Bass claimed.[23]

As the administrative and commercial centre of the wealthiest nation on Earth, nineteenth century London grew to become the most populous conurbation in the world. The city expanded as houses and buildings were erected to accommodate the huge increase in its population and the demands of industry and commerce, adding the sounds of their construction to the city's already stressful soundscape. Charles Dickens, another of Bass's supporters, complained that a "speculative builder, who is running up terraces, crescents, and gardens by the score in the suburb where I dwell, has erected a range of workshops at the bottom of my garden, where all his carpentry and joinery is

---

[21] Hendy D. (2013) Noise: A Human History of Sound And Listening. Profile Books, p 269–81.
[22] Babbage, C. (1864) Passages From The Life Of A Philosopher. Longman, Roberts, & Green, p 349.
[23] Bass, M. (1864) Street Music in the Metropolis. John Murray, p 41.

done…He has set up a circular saw—twenty circular saws, I should say. They are sawing my heart in twain. I shudder at the shrill, ceaseless whirr. I can hear the innocent planks screaming as the merciless teeth eat into their very marrow."[24]

Unable to bear the noise of the saw a moment longer, "I rush to the front of the house, desperate; but there, oddly enough, I experience no nervous discomfort when I hear the costermonger crying his "fine savoys", his turnips and his carrots. I shudder not, when the donkey-man who sells fish expatiates in prolonged bawl on the virtues of his fresh cod and "fine cheap soles." The sweep is rather a melodious person than otherwise, with an excellent baritone voice. The four o'clock muffin-woman, with her tinkling bell, fills me with comfort and joy. I could tolerate the milkman if he cried his wares in an honest and rational fashion; but the man who comes at three o'clock utters a cacophonous cry sounding like "Yahoop;" and the milkwoman, who is due a three-thirty—she is presumably of Welsh extraction, and has a pair of legs like the balustrades in the background of a carte de visite—puts her arms akimbo, and in accents as gruff as those of a corporal-major in the Life Guards, says "Cuckoo!" Now, "yahoo" and "cuckoo!" have nothing, I surmise in common with "Milk O!" I am waiting for "afternoon cresses!" a pretty innocent noise, when I am driven to the back of my residence again by the diabolical screech of the knife-grinder's wheel-as dire an infliction in its way as the circular saw. The wretch with the wheel—he will be Ixion [tied to a rotating, fiery wheel as punishment by Zeus] I hope one day—who infests my neighbourhood, is an orator, forsooth; and instead of succinctly delivering himself of his message to the community launches into a long round running, "Ave you hany knives, scissors, razors, penknives, table himplements to grind, or heven humbrellas to mend O!" and a murrain [i.e. a plague] on him!".[25]

These street sounds, about which anyone who has not had to live with them day in and day out may feel a somewhat misplaced nostalgia, are absent from the cities of post-industrial nations. But in their place we city dwellers must endure the sounds of overflying planes, the occasional hovering police helicopter, assorted vehicles hurtling around neighbourhood roads, the wailing sirens of assorted emergency services, and inconsiderate builders who invariably insist that they cannot work without loud music.[26] And in place of the screech of the knife grinder's wheel, one has all too often to put up the

[24] Dickens, C. (1871) Noises. In: All the Year Round, 159, Dec 16, p 56–57.

[25] Dickens, C. (1871) Noises. In: All the Year Round, 159, Dec 16, p 56–57.

[26] The most egregious example of builders and radios I have come across was a man laying paving stones which he cut with a petrol powered stone-cutting saw while his radio was on full blast. Mindful of his hearing he was, of course, wearing ear defenders.

devilish cacophony of a neighbour's assortment of power tools such as electric lawnmowers, strimmers and leaf-blowers.

In short, it appears that intrusive sounds are an unavoidable feature of any soundscape that involves human activity: *Homo clamosus* seems inseparable from *Homo sapiens*. It has probably always been like that, even before there were settled communities. It seems unlikely that a snoring sleeper, a bawling infant or a barking dog would not have tested the patience of our prehistoric nomadic ancestors just as much as it does ours. We have the same hearing system as they did, and like them we can't avert or close our ears as we can our eyes.

As far as we know, however, it was not until there were settled communities in which people lived cheek-by-jowl that attempts were made to control noise with rules and regulations. These emerged piecemeal and were usually based on circumstances in which particular sounds were considered to be a nuisance, as we can deduce from the "The Statutes of the Streets of this City, against Noysances". But any attempt to formulate an all-encompassing, objective definition of noise that can be used to formulate laws to control or limit all unwanted sounds seems doomed to fail. Consider a 1931 proposal by George Kaye, a British physicist, that noise should be considered as sound out of place due to "[its] excessive loudness, its composition, its persistency or frequency of occurrence (or alternatively, its intermittency), its unexpectedness, untimeliness or unfamiliarity, its redundancy, inappropriateness, or unreasonableness, its suggestion of intimidation, arrogance, malice, or thoughtlessness."[27] Such a broad definition means that apart from loudness and composition, which are measurable qualities and therefore a source of objective evidence, noise remains for most of us what it has it always has been, largely in the ear of the beholder; noise needs a listener.[28] In fact, noise legislation, where exists, usually lays down only the maximum permissible loudness to which workers should be exposed and is intended to prevent their hearing being damaged.[29] All other noise issues are dealt with through laws that limit nuisance, whatever its nature.

Nevertheless, as we all know from personal experience, there are sounds such as the proverbial scraping of fingernails on a blackboard, that make one wince and clamp one's hands over one's ears regardless of circumstances.[30]

---

[27] Kaye, G.W.C. (1931) The Measurement of Noise. Proc. Of the Royal Institution of Great Britain, 26, p 435–88.

[28] Hegarty, P (2007), Noise/Music: A History. Bloomsbury, p 3.

[29] UK noise legislation limits worker's average daily exposure to noise between 80 and 85 dB. The decibel scale is logarithmic, so 85 dB is almost twice as loud as 8 dB.

[30] Has anyone heard fingernails on a blackboard now that pen and paper have replaced slates and chalk and blackboards have given way to whiteboards?

Such sounds are not merely unwanted, they are inherently disagreeable. And what makes then so is that they are particularly rough and jarring, which makes them exceptionally dissonant. In fact, the vast majority of sounds that we hear are dissonant, i.e. a short-lived jumble of unrelated frequencies of irregular and constantly changing energies. The only exceptions are sounds composed of harmonically related frequencies, sounds that are the bedrock of music and speech as well as of bird-song and animal calls. Harmonic frequencies are also present in the hums and whines of the rapidly beating wings of flying insects such as bees, flies and mosquitos, not forgetting the eponymous hum of hovering hummingbirds. The almost complete absence of naturally occurring musical tones is hugely significant because it holds clues to why and how our hearing system evolved, a subject that we shall take up in chapter three. Suffice it to say here that it is probable that ears first evolved to hear brief, dissonant sounds, not musical tones.

Dissonant sounds are not necessarily unpleasant, however. All composers of music routinely employ a degree of dissonance to great effect in their compositions.[31] Depending on one's frame of mind, dissonance can be soothing, pleasurable, evocative and fascinating. Who isn't cheered by the gentle chatter of a babbling brook, or soothed by the spellbinding rhythm of wave-driven pebbles grating on a beach, or diverted by the rustle of leaves on a windy day?Even an undeniably vexatious sound such as the prolonged and often alarming roar of a low flying aircraft is full of interest if you are willing to listen attentively. Indeed, as we shall see in later chapters, the different sounds produced by their engines are arguably among the most interesting and instructive of all the mechanical noises that we encounter daily.

But there is no getting away from the fact that dissonance is the major reason why people find some sounds unbearably unpleasant. Dickens claimed that "There can be very little difference of opinion, I should say, as to the repulsiveness of the sounds made by the tearing of calico, the creaking of doors, the passing of a wet finger over silk, the endeavour to remove a glass stopper from a bottle, or the scraping of slate pencil. Concerning sounds the bare thought of which is enough to set your teeth on edge, it is not necessary to say much more."[32]

Dickens may have been afflicted by misophonia or "hatred of sound". This auditory condition makes it difficult for a person to tolerate sounds that the most of us either hardly notice or don't find particularly disagreeable or irritating. Noises such as chewing, slurping, constant sniffing or throat clearing

[31] Ball, P (2010) The Music Instinct: How Music Works And Why We Can't Do Without It. Vintage, p 165–170.
[32] Dickens, C. (1871) Noises. In: All the Year Round, 159, Dec 16, p 56–7.

will trigger a range of emotions from intense annoyance to extreme rage in someone who is misophonic.

Until recently misophonia was not considered to be a medical condition. But a recent study appears to have established that there is a link between an extreme intolerance of particular sounds and an abnormal activation of the limbic system, that part of the brain that deals with the emotional response to sensations. And if there is such a link, then there will always have been people whose excessive reaction to certain sounds has left their acquaintances and families nonplussed.

The nineteenth century appears to have particularly well supplied with misophobics, though perhaps this owes much to the fact that many of them were prepared to record in detail the auditory torments they claimed to suffer daily. Charles Babbage was certainly amongst their number, as were the German philosopher Arthur Schopenhauer ("noise is the true murdered of thought", he wrote, and his least favourite sound was the crack of a drover's whip[33]) and Thomas Carlyle, the notoriously churlish Scottish historian who had an extra floor added to his home to accommodate a sound proof study. It proved to be even more susceptible to unwanted sounds than the room it replaced. Other famous writers known to have found particular sounds distressing include Charles Darwin, Edgar Allen Poe, Anton Chekhov, Marcel Proust, Joseph Pulitzer and Franz Kafka.

Clinical research into misophonia is in its infancy, but there is another form of auditory discomfort known as hyperacousis that is a medically recognised condition and to which misophonia may be related. Hyperacousis causes one to perceive sounds to be louder than they really are. In some sufferers the condition is linked to problems with the vestibular system, the sense organ responsible for orientation and balance, which is located within the bone of the skull next to the inner ear. We shall take a closer look at the vestibular system in chapter three.

To judge from the experience of Jane Carlyle, Thomas's long-suffering wife, in some circumstances an aversion to sound can be contagious. Thomas found the noise of construction of the sound proof study so unbearable that he moved out, leaving Jane in charge of the works. She confided to a friend that "The tumult has been even greater since *Mr C* went than it was before…But now I feel that the noise and dirt and discord with my own senses only and not thro *his* as well, it is amazing how little I care about

---

[33] Schopenhauer, A. (1893) On Noise. In: Studies in Pessimism (trans: Saunders, T.B.). Swan Sonnenschein & Co, London, p 127–133.

it."[34] If you are neither misophonic nor hyperacoustic and live with someone who is, her words will ring a bell.

But what is it about those sounds that almost everyone finds inherently unpleasant, the ones that set one's teeth on edge?[35] A recent study appears to have found the answer.[36] Volunteers were asked to rank several sounds in terms of how unpleasant they found them. The most disagreeable was found to be that made when the blade of a metal knife is scraped against a glass bottle, a variant on fingernails scraped across a blackboard. Indeed, the study found that with few exceptions, of all the sounds that the volunteers listened to the most disagreeable were due to scraping or grinding. Somewhat unexpectedly, a woman's scream and a baby's cry were rated almost as unpleasant as scraping.

To the selection of unpleasant sounds used in this study I would add a couple of my own. One is a bird's distress call. Interestingly, these distress calls do not vary greatly from one species of bird to another. A recorded version of these calls is played loudly through outdoor speakers at a supermarket near where I live and has succeeded in driving away the starlings that once lingered around the entrance in search of morsels to eat. The other is the shriek of a fox, which sounds like the cry of an infant in extreme agony and which puts me on edge. What all these sounds have in common is that they are largely composed of frequencies to which the human ear happens to be most sensitive, namely those between 2000 and 5000 Hz. Equally important is that their loudness varies too rapidly to be heard distinctly, variations which contribute audible frequencies to the original sound, making the overall sound even more dissonant than it would otherwise be.

As for the most agreeable sounds, the study found that most involve flowing water, either cascading or bubbling gently. And, as if to compensate for its distressing cry, a baby's laughter was considered to be the most pleasurable of all the sounds used in the study. All the agreeable sounds used in the study are dominated by low frequencies and variations in their loudness that can be clearly heard, the very opposite of the characteristics of unpleasant sounds.

---

[34] Carlyle, J.W. (1852) letter to John A. Carlyle 27th July 1852.

[35] There is no single word in English for this reaction to unpleasant sounds, though there does seem to be one in Spanish: "grima", which is translated into English as "the creeps".

[36] Kumar, S., Kriegstein, K.v., Friston, K., Griffiths, T.D. (2012) Features versus Feelings: Dissociable Representations of the Acoustic Features and Valence of Aversive Sounds. The Journal of Neuroscience, 32 (41): 14184–14192.

These Ears Were Made for Listening

Yet, it's an ill wind that blows no good. Loud sounds can permanently impair our hearing and unwanted ones drive us to distraction. But there is no escaping the fact that the purpose of ears is to hear sounds, not to avoid them. Sounds are an important and, more often than we realise, an irreplaceable source of information about the world. A knock at the door, an unexpected rattle from a faulty engine, drips from a leaking tap and birdsong from within a bush make us aware of things and events that are out of sight. And while there's no denying that the sound of traffic is an unpleasant and hideously invasive racket, if you don't hear it you run the risk of being run over should you cross a road without looking.

Moreover, by focusing on the unwelcome physiological and psychological effects of sounds, we risk throwing the baby out with the bathwater. By dwelling on their nuisance value, we come to regard the majority of sounds to be noise, i.e. as something to be avoided or ignored, whereas, as we have just noted, they are as often as not an indispensable source of information about what is going on around us, not to say a potential source of interest, pleasure and amusement. Preconceived notions that any sound out of place is insignificant or undesirable may not prevent one hearing sounds, but it almost certainly inhibits one's willingness to listen to them.

And because we tend not to listen attentively to most sounds, hearing is a vastly underrated and all too often an underused sense. It's not that we don't value it; everyone would concede that deafness is a huge handicap. Even mild hearing loss, which sooner or later afflicts us all as we age, makes life difficult, tiresome and dull, impeding free-flowing conversation, increasing the likelihood of accidents because warning sounds are not heard, and hindering one's enjoyment of music and nature's sounds. And, arguably, the drawbacks of profound deafness surpass those of loss of sight. We are such a social species that the inability to communicate verbally can be exasperating and hugely isolating, sign language and lip reading notwithstanding.[37]

In fact, until the latter half of the eighteenth century, when the first systematic sign language was devised and developed in France by Charles-Michel de l'Épée, to be born deaf was a calamity that consigned one to the margins of society. Unable to hear, it was all but impossible to learn language, rendering one mute, which was invariably taken as a sign of a simple mind. Without language there can be no sustained symbolic thought, so unless early remedial action is taken, the inner life of someone born deaf may well be limited to

---

[37] Bathurst, B. (2017) Sound, Stories of Hearing Lost and Found. Profile Books, see chapter 4.

wordless thoughts, feelings and sensations that can't be fully comprehended or readily communicated to others.[38] Even these days, when methods (sign language and Braille) and devices (hearing aids and cochlear implants) are available to overcome the problems of deafness, to be born with impaired hearing can be as much of a handicap as it ever was if the condition is not diagnosed within the first 2–3 years of birth. The most pressing problem is that the physiological development of the necessary auditory pathways within the brain requires early exposure to sounds in general and the sound of speech in particular if an infant is to acquire spoken language. Fortunately, the language centres of the brain operate independently of the hearing system, so profoundly deaf children can learn to think and communicate through sign language, though this too depends on an early diagnosis of hearing impairment so that the necessary training can begin as soon as possible.

Spoken language notwithstanding, few of us exploit our hearing to its full potential because we don't realise what a keen and versatile sense it is. In part this is because we assume that sight is the sense through which the world discloses itself to us most fully and unambiguously, and unwittingly consign hearing to a supporting role. "Seeing is believing", we smugly admonish the doubter, conveniently ignoring the multitude of instances in which our *only* source of information is auditory. Indeed, were it not that two of the traits that distinguish humans from all other creatures, namely spoken language and music, depend wholly on hearing, a cynic might well wonder if deafness is as much of a disadvantage as blindness.

Helen Keller, the remarkable American author and lecturer who overcame the seemingly insurmountable handicap of the loss of both her sight and hearing following a severe illness when she was a year and a half old, knew from first hand experience that "the problems of deafness are deeper and more complex, if not more important, than those of blindness. Deafness is a much worse misfortune. For it means the loss of the most vital stimulus the sound of the voice that brings language, sets thoughts astir, and keeps us in the intellectual company of man."[39]

Speech and music aside, however, most sighted people probably assume that making sense of one's surroundings by means of sound alone must require superhuman effort and concentration, and, at best, yield a some-what sketchy and ambiguous impression of objects and events. But blind people manage to get about without constantly bumping into things by

---

[38] Sacks, O. (1989) Seeing Voices, A Journey Into The World Of The Deaf. See: "Thinking in Sign" for an account of the intellectual difficulties faced by congenitally deaf children.

[39] Keller, H. (1933) Helen Keller in Scotland: A Personal Record Written By Herself, edited with and introduction by J. Kerr Love, M.D., LL.D. Methuen & Co Ltd, p 68.

relying on their hearing. Forced to listen attentively in order to navigate their surroundings they learn to be acutely aware of subtle acoustic effects due to interactions between ambient sounds and objects of which sighted people are usually oblivious.

That is not to say that sighted people rely exclusively on vision. While vision supplies a detailed picture of our physical surroundings, hearing enables us to be aware of events and situations that are out of sight, either behind one's back, around a corner or hidden from view by or within objects. Indeed, our overall experience of the world at any given moment is a fusion of information from several senses, not just eyes and ears, though this is something we take for granted and seldom give a moment's thought to.

Nevertheless, even though most of us are unlikely to listen as attentively to our surroundings as someone who is blind, with practise anyone with good hearing can discern some of the subtle acoustic effects that blind people rely on, such as the slight alteration in the timbre of ambient sounds that can occur as one approaches a solid vertical surface such as a wall. Sighted people are often unaware of these acoustic effects because they almost never need pay them heed: they avoid bumping into walls because they can see them.

Compared to vision, which seems such a purposeful, directed activity over which we exercise considerable control, hearing can seem a somewhat hit and miss affair. By and large we don't listen actively unless there is a particular reason for doing so, and as a result we become accustomed to hearing sounds by chance. In most circumstances our eyes serve us so well that we unwittingly consider sight to be the only sense we need, an attitude that makes us passive and somewhat lackadaisical listeners.

John Hull, a theologian and university lecturer, who had poor eyesight as a youth and eventually became totally blind at the age of 45, knew why.[40] In addition to his academic work, he became widely known for his perceptive accounts of life lived without sight. He found that sighted people are at a disadvantage when they try to understand blindness because all their senses are to a greater or lesser degree dominated by vision. Sounds are invariably associated with objects and events that can in principle be seen, and as a result sighted people don't tend to pay much attention to ambient sounds.[41] As a result, most of us with good eyesight are hard of listening rather than hard of hearing. Usually, it is only when we find ourselves in unfamiliar surroundings

---

[40] He recorded his thoughts on audio cassettes over 3 years after he became blind in 1983, and some of these were used in Notes on Blindness, a 2014 film about coming to terms with his loss of sight.
[41] Hull, J.M. (2001) Sound: An Enrichment or State Soundscape. The Journal of Acoustic Ecology, Volume 2, Number 1, July, p 10.

or situations that we pay particular attention to what we hear and, indeed, to what we see.

Something else that tips the scales in favour of sight is that the majority of sounds are transient bursts of energy that usually reach one's ears without warning, catching one unawares, and disappear without trace. In comparison, the visual world is infinitely more permanent and stable. By and large, objects don't vanish into thin air moments after they have been perceived. Even when they do, we are not surprised when they reappear. Drop a marble onto the floor, it rolls out of sight under a table and seconds later emerges on the far side. But the sound of the impact when the marble strikes the floor lasts a mere fraction of a second.

The relative permanence of the visual world means that our eyes don't have to respond as rapidly to stimuli as do our ears. In fact, our hearing system responds tens of times more rapidly to stimuli than our visual system. If our hearing operated at the same rate as our vision some sounds might be unheard because the auditory pathways in the brain would not act quickly enough for one to become conscious of them, and many of those that were heard would be reduced to an acoustic version of visual blur. As it is, the hearing system is often able to make out the finer details of sounds that last no more than the blink of an eye. And there is no better example of this than our ability to cope with the complex vibrations that buffet our eardrums when we are listening to normal speech, which averages 150 words a minute.

But even though the hearing system responds very rapidly, the conscious brain, that part of it in which thought and perception occurs, can't always keep pace with the torrent of signals reaching it from the ears. And so evolution has furnished us with an auditory memory that in some respects is superior to our visual memory, enabling one to "replay" sounds so that one can often recall a sound long after first hearing it. We can often whistle or hum a simple tune having heard it only once or twice. Try matching a particular colour, say, or a shape if you don't have the original to hand.

Seeing, it should be said, is no less demanding than hearing, even though all that seems necessary is to open one's eyes and direct one's gaze at a scene to have everything revealed as clearly as it is in a photograph. There is, indeed, a superficial similarity between the eye and the camera: in both cases a lens forms an inverted image on a light-sensitive surface within a darkened chamber. But a photograph only preserves a moment and captures every detail indiscriminately. Vision, on the other hand, especially when being actively employed, i.e. when one is looking, is highly selective and focuses narrowly on what is of immediate interest. Unlike a camera, which in principle faithfully captures whatever it is pointed at, we are unable to see clearly everything within our visual field at the same time, let alone attend to it. We must shift our gaze around a scene to take it all in. Nor are we are born with an innate ability to make sense of whatever is before our eyes. As Ruskin pointed out, seeing is a skill that takes time and effort to acquire, and it is only when fresh demands are made on our eyes, such as learning to draw or paint a figure or a scene, or to take a photograph that is something more than an unexceptional snapshot, that we realise that seeing involves much more than merely directing one's gaze at a particular object or situation.

The same is true of hearing. The ear is like a microphone in that it detects aerial vibrations. But unlike a microphone, which as anyone who has recorded a conversation in a noisy environment has found to their dismay picks up every sound, our hearing can single out particular sounds such as someone's voice from a hubbub while ignoring all others.[42] The ability to do so is known as the *cocktail party effect*, a phenomenon we'll return to in chapter three. It could be said that the microphone hears but cannot listen. Listening entails paying attention and that is something only sentient creatures can do.

Unlike sight, however, our hearing is always on, even when we are asleep. We can close our eyes, but unless we don ear protectors or insert plugs into our ear canals we can't do the same with our ears; we don't have earlids. Even with blocked ears, very loud sounds can reach the inner ear through the bone of the skull. But, as if compensate for the lack of earlids, hearing is far more selective than vision, i.e. it is always on, but we are not always listening. Unless a sound is loud, unusual, unexpected or unpleasant, we tend to ignore those that we judge to be of little or no interest or significance to us. Indeed we are usually completely unaware of them. This is particularly true of prolonged, monotonous sounds such as the drone of traffic or the background babble of voices in a public place—we simply tune them out. With sight we usually pay a degree of attention to what we see, if only because we rely on our eyes to navigate our way around our environment, for example looking ahead to choose a route and avoid obstacles. But in a noisy environment such as a busy street we all but stop listening. When you are next out and about, pause for a moment to consider how aware you are of ambient sounds. And should you later recollect your outing you may be surprised that your memory of it unfolds like a silent film: images with little or no soundtrack. In fact, if you can recall ambient sounds, it is likely that it is the events that caused them that first come to mind, not their acoustic qualities. And should you think you are you alone in turning a deaf ear to ambient sounds,

---

[42] There are many types of microphone, each designed to overcome the problems of recording sound in particular conditions. See: Goldsmith, M. (2015) Sound: A Very Short Introduction. OUP, p 69–74.

ask others to tell you about the sounds they have heard that day. You will probably be met with a blank stare.

As we shall see below, turning a collective deaf ear to ambient sounds has potentially detrimental consequences because as a society we have become increasingly tolerant of and apathetic about noisy environments, environments in which both our peace of mind and the health of our hearing system are at risk.

Making Sense of Sounds

When next you hear a distinct sound ask yourself what you are aware of. Usually you won't be particularly conscious of the acoustic qualities of the sound such as its pitch and timbre unless the situation calls for it or you are unable to identify the source. Instead it is almost always the event or object that is the source of the sound that first comes to mind. Events that produce sounds involve physical interactions, so we "hear" the wind in a pine tree rather than a monotonous hiss, an airplane flying overhead rather than a prolonged roar, a bee in flight rather than a faint buzz, a door slamming shut rather than a loud crash, and so on.

On some occasions, what might be called the "auditory texture" of an event can reveal material conditions that can't be directly perceived by the eye. The sound of a glass marble rolling across a wooden floor reveals both the shape and composition of the marble—a smooth, hard sphere—as well as the firmness and texture of the floor, things we can't always discern visually, though we can sometimes do so by touch. And it is those physical qualities rather their acoustic qualities that usually first come to mind in such circumstances.

There is a very good reason why we are aware of objects and events rather than their associated sounds: the primary goal of perception is to extract information that is necessary to our needs from what we see, hear, feel, taste and smell. Not that perception recreates a facsimile of our surroundings. Rather, it organises and repackages a torrent of sensations to create a conscious experience that provides a synopsis of salient features in our environment on which we can act. Thus we see people, houses and trees rather than unrelated coloured patches of different brightness, shape and size, though we can discern these qualities in what we see should we wish to do so. And where hearing is concerned, in most situations it is the event that is the cause of a sound that is usually more relevant to the needs of the listener than the acoustic qualities of sound itself.

And because it is events that are of greatest importance to us, we seldom need to consciously identify the acoustic qualities of sounds. Indeed, we often find ourselves at a loss when asked to describe what we hear in terms of acoustic qualities alone. How would you describe the sound of walking on gravel, or that of a knife scrapping against a glass bowl, or of a plate smashing on tiled floor, or of sheet of paper being crumpled up? With great difficulty, if at all, I would suggest.

That's not to say that we lack words to describe the acoustic qualities of sounds, qualities that are not tied to a specific event or situation. Rusty hinges, loose floorboards and toy ducks *squeak*; refrigerators, electric fans and bees in flight *hum*; distant thunder, avalanches and cart wheels on cobbles *rumble*; fireworks, electric sparks and burning logs *crackle*; a bunch of keys, ice in a glass of water and bicycle bells *tinkle*. And like every other word in the English language, these came into use at different times and were often adaptations of existing words.

According to the Oxford English Dictionary or OED, the earliest written use of *squeak* to describe an inanimate high pitched noise appears in one of John Donne's sermons in which he mentions the "squeaks and whines" of an overloaded cart. *Hum*, as the sound made by a bee in flight, first appears in 1601 in a translation of Pliny the Elder's History. *Rumble* as the description of a noise was first used in 1530.[43] *Crackle* as the sound of something burning made its debut in a 1656 poem by Abraham Cowley. *Tinkle* to mean a ringing sound of a small bell was coined in 1617 by Fynes Moryson in an account of his travels in Europe.

The OED is not confined to definitions, it is also an historical record of the English language and a perusal of entries for these words shows that they were in use long before they acquired their modern usage. For example, the earliest documented use of *squeak* was in 1387 and was used to describe the cry of a child. But some words have retained their meaning over several centuries. Among the oldest words for specific sounds in written English that are still in use in their original sense are "thunder" (c.750 AD), "whisper" (c.950 AD) and "belch", "din" and "whistle" (all c.1000 AD.)[44]

---

[43] Used by John Palsgrave in what is the first book about French grammar: "I romble, I make noyse in a house with remevyng of heavy thynges, je charpente." Despite its French title, "Lesclarcissement de la langue francoys" (1530), the book was written in English.

[44] The history of words for colours provides one of the clearest examples of how and why specific vocabularies evolve. Ancient Greek, for example, had few words for colour, and those it did have were applied in ways that seem haphazard to the modern reader. The fact is that people in the ancient world were less interested in colour than we are, and so did not feel the need to name them. The optical qualities that interested the ancient Greeks were brightness, darkness and lustre. See Deutscher, G. (2010) Through the Language Glass: How Words Colour Your World. William Heinemann.

Name that Sound

Fortunately, our senses can function perfectly well without the need to put into words the perceptions they give rise to. We can *see* colours and shapes, *taste* flavours, *smell* odours, *feel* textures and *hear* sounds without having specific words or phrases for them. The startle reflex, which is triggered by loud, unexpected sounds, causing one to flinch, muscles to tense and eyes to blink, does not require the hearer to identify the sound (i.e. to name or describe it) because all that is required to elicit the reflex is that we hear a sudden unexpected noise.[45] We react without having to think, which is, of course, the essence of a reflex. In fact, the startle reflex is of such importance to survival is that it is common to all vertebrates.

But if we want to go beyond a mere visceral awareness of our sensations and make sense of our experiences, a vocabulary with which they can be described is essential. "We think only through the medium of words", claimed the eighteenth century Enlightenment French philosopher, Étienne Bonnot de Condillac.[46] This doesn't mean that there must be a unique word to describe or identify our every sensation. Indeed, that is undesirable and, almost certainly, impossible. The power of language lies in its flexibility, in being able to express more or less the same thing in several ways, each of which captures or emphasises a different aspect of one's thoughts and experiences.

And where necessary, words are coined, borrowed from other languages and existing ones assigned a new meaning. Where there's a need, there's often an apt word or phrase. We do this with all our senses, though it is often claimed that sound sensations are usually more difficult to describe in terms of their acoustic qualities than visual sensations are to describe in terms of optical qualities.[47] One reason for this is that hearing is primarily concerned with identifying events, so that it is not always necessary (or, indeed, relevant) to refer to specific acoustic qualities when identifying a sound. That, and the fact that we often resort to a visual lexicon when talking about sounds, means that, on the face of it, the vocabulary of words that describe specific sound sensations does not have to be as large as that necessary to describe visual sensations.

Bernie Krause, an American musician and ecologist, noticed that the adjectives that people often resort to when describing music are not particularly enlightening because they are ocular rather than acoustic; he was often

---

[45] The sound need not be loud: a creaking floorboard in the dead of night can be alarming.

[46] Condillac, É. B. de (1780) La logique, ou, Les premiers développemens de l'art de pense. Paris.

[47] Beament, J. (2001) How We Hear Music: The Relationship Between Music And The Hearing Mechanism. The Boydell Press, p 2.

instructed by film directors for whom he was working to create scores that were dark or light or bright or murky.[48]

Michel Chion, a French composer who is also a filmmaker, takes a different view. He disputes the claim that languages have few words to describe the acoustic qualities of sounds and insists that there are far more such words in every language than most people suppose.[49] The real source of the perceived poverty of an acoustic vocabulary, he believes, is due to lack of usage, not a dearth of apposite words. We all know a wide range of words that describe specific acoustic qualities, he says, it's just that we tend not to use them in ordinary discourse.[50] As a consequence we are often unsure which words to use when describing the acoustic qualities of a particular sound. The solution, says Chion, is to flex one's lexical muscles. Whether he is right on this point is something that you can gauge for yourself by trying to describe the sounds that you hear in acoustic terms. It's something that you could do during a listening walk, an activity that we shall explore later in this chapter.

In fact, we have a potentially inexhaustible source of words with which to describe sounds, namely onomatopoeias. These are words that imitate or recreate the sounds of events or the vocalisations of animals by exploiting the phonetic qualities of the range of the distinct sounds known as phonemes that are employed in speech. But onomatopoeias are not mimicry, though they sometimes seem to come close. Moreover, they reflect both cultural differences and the fact that onomatopoeias depend on whatever phonemes are available within a particular language, which explains why onomatopoeias for the same animal noise frequently differ from one language to another. In English, pigs *oink-oink*, but in French they *groin-groin* and in Japanese they *boo-boo*.[51] In Britain frogs *croak* while on the other side of the Atlantic in the United States they *ribbit*.

Indeed, there is a good case to be made that the majority of words in common use—i.e. those found in dictionaries—that describe the acoustic qualities of sounds in the English language are onomatopoeias. Consider bang, buzz, clang, click, fizz, glug, hum, plop, sizzle, swish, thrum, toot, twang, wheeze, whoosh, zing, zoom. If you have any doubt that these are

---

[48] Krause, B. (2013) The Great Animal Orchestra: Finding The Origins Of Music In The World's Wild Places. Profile Books, p 19.

[49] Chion, M. (2016) Sound, An Acoulogical Treatise. Translated with an introduction by James A. Steintrager. Duke University Press, p 224.

[50] Chion has been compiling "Le Livre des Sons, une célébration" (The Book Of Sounds, A Celebration) which he described in 2015 as a "dictionary of sonic evocations and words to describe sound throughout history and in different languages (a big book that I started 20 years ago and should be finished in a year.)" It has yet to be published.

[51] For examples of onomatopoeias in different languages see Derek Abott's Animal Noise: http://www.eleceng.adelaide.edu.au/personal/dabbott/animal.html (accessed 20/07/2021).

onomatopoetic, just read them aloud while recalling (or, better still, while listening to) the sounds they refer to.

It would be very surprising if languages did not make use of the sensuous qualities of sounds to add a further dimension to words over and above their dictionary definitions. Because when voiced words are sounds, the purely acoustic qualities of phonemes can and do contribute something to their meaning. Such words are known as *ideophones*. For example, for English speakers words that begin with "*sl*", such as *slimy*, *slither* and *slithy*,[52] are often associated with unpleasant sensations. And words that include closed vowels, such as *teeny*, *weeny* and *itty-bitty*, are associated with smallness, while those with open vowels, such as *vast* and *large*, are associated with spaciousness. There is no hard and fast rule about this, however: consider *small* and *big*.

Some languages exploit the mimetic features of spoken words more than others. Japanese has more than three times the number of onomatopoeias than English and employs them to convey not only characteristic noises associated with their acoustic qualities, a category known in Japanese as *giseigo*, it also uses them to evoke qualities in which sound plays no part, known as *gitaigo*. For example, *gorogoro*, an onomatopoeia for the purr of a cat, the rumble of thunder or the sound of something rolling on a hard surface, is also used to describe either an upset stomach or being lazy, idle or relaxed.

This suggests that language may have begun as imitations of natural sounds: i.e. deliberate mimicry rather than incoherent grunts and growls. Speculation about the origin of language has a long history. It is the subject of "Cratylus", one Plato's dialogues which was probably written in 399 BC. In this work, two protagonists, Cratylus and Hermogenes, appeal to Socrates to resolve a dispute between them over whether names for things are the result of convention or nature.[53] Hermogenes maintains that words are a matter of convention, a view that two millennia later was memorably declaimed by Juliet Capulet, who mused "What's in a name? That which we call a rose/By any other name would smell as sweet".[54] Cratylus takes the opposite view: names should inform us about the nature of the thing they refer to. In assessing Cratylus' thesis, Socrates comes up with several examples of what we would now class as ideophones. He suggests that the letter *R* requires the speaker to roll his tongue, which adds the idea of motion to certain words

---

[52] Jabberwocky: "'Twas brillig, and the slithy toves/Did gyre and gimble in the wabe: /All mimsy were the borogroves,/And the mome raths outgrabe." (Slithy = lithe and slimy, mimsy = flimsy and miserable). In: Carroll, L. (1872) Through the Looking-Glass and What Alice Found There. Macmillan.

[53] Plato, c.399BC, Cratylus.

[54] Shakespeare, W. (1597) *Romeo and Juliet*, Act ii, scene ii.

in Ancient Greek such as *rheîn* (to flow) and *thraúein* (to shatter). But on examining the merits of the opposing views, he concludes that the relationship between words and their meaning is largely a matter of convention, a view that is widely accepted by modern linguists.[55]

The assumption that the phonetic qualities of words are completely arbitrary has, however, been challenged in a number of recent studies. In one survey of the phonetics of certain categories of words of more than half the world's current languages the authors concluded that words for parts of the body tend to employ similar phonemes. For example, in many languages, words for *tongue* will use either "l" and "u" and those for *nose* use "n" and "oo".[56] This supports other research into the issue. In 2001 a neuroscientist, V.S. Ramachandran, asked native English speakers and native Tamil speakers to choose which of two nonsense words, "kiki" and "bouba", they would associate with a couple of irregular shapes, one spiky and other rounded. Almost everyone in the study associated "kiki" with the spiky shape and "bouba" with the rounded one.[57] One possible explanation why we sometimes associate words with shapes is the way in which lips, tongue and breath are employed to create sounds when speaking. When voicing *kiki*, air is expelled twice in rapid succession from the mouth, mirroring the abrupt change in outline at the pointed edges of a spiky shape, whereas the flow of air when voicing *bouba* is smooth and prolonged, much like the visual appearance of the shape that it the word was associated with (Fig. 1.1).[58]

The origin of language remains an open question, though we now know far more about the issue than did Plato in 400BC.[59] And we also know a great deal more about the next major step: the relatively recent innovation of writing. Writing, at least for languages that use an alphabet, is first and foremost a method of encoding phonetic sounds by means of symbols.[60] Moreover, phonetic sounds are unquestionably the most complex in nature, and are precisely fashioned within the cavities of the pharynx and the mouth,

---

[55] Condillac was of the opinion that language had started as gestures, which he called "the language of action", rather than as sounds, so he was delighted by de l'Epée's invention of sign language for the deaf. See: Rée, J. (1999) I See A Voice: Deafness Language And The Senses, A Philosophical Study. Henry Holt and Company, p 170.

[56] Blasi, D.E., Wichmann, S., Hammarström, H., Stadler, P.F., Christiansen, M.H. (2016) Sound–meaning association biases evidenced across thousands of languages. Proceedings of the National Academy of Sciences, Vol 113, No.39, p 10818–10823.

[57] Meteorological symbols for warm and cold fronts seem to draw on this distinction: warm fronts are represented by semicircles and cold fronts by triangles.

[58] Ramachandran, VS, Hubbard, EM (2001) Synaesthesia—A Window Into Perception, Thought and Language. Journal of Consciousness Studies, 8, No. 12, p 3–34.

[59] Hurford, JR (2014) The Origins of Language: A Slim Guide. OUP.

[60] Alphabetical writing did not begin like this: symbols that stood for things (pictographs or glyphs) evolved into representations of phonemes present in their vocalised forms.

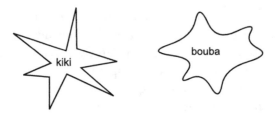

Fig. 1.1 The name fits the shape

collectively known as the vocal tract, that are altered in shape and volume by movements of the tongue to create speech sounds from a monotonous drone due to the movements of the vocal folds in the larynx. St Augustine, the Bishop of Hippo and one of the most influential theologians in the early years of Western Christianity, held that the letters of the alphabet were both the "signs of sounds" and "signs of things we think", which enables the reader to listen to a voice without having the speaker present.[61] In fact, learning to read an alphabetic language is a matter of learning that groups of letters on a page correspond to particular phonemes, which is why we learn to read as children by reading aloud in the presence of someone who already knows how the squiggles on a page are each associated with distinct sounds. In Augustine's day people did not read silently and he was taken aback when he saw his teacher, St Ambrose, the Bishop of Milan doing so.[62]

But the silence that St Augustine witnessed is more profound than he could have imagined because when we read silently we do not laboriously mentally vocalise the constituent phonemes of each word, as we did when learning to read. Silent reading is thus a misnomer because no silent vocalisation is involved. Instead we directly access the meaning of written words and phrases as our eyes scan across a page.[63] It is only when we come across unfamiliar words that we resort to (silently) sounding out its phonemes.

The reason we are able to access meaning directly when reading, whether silently or aloud, is that the part of the brain devoted to understanding language is independent of the hearing system. Hence we can communicate without having to vocalise. Profoundly deaf people who employ sign language are using their eyes in place of their ears—as, of course, sighted people do

[61] Augustine of Hippo Of The Origin And Nature Of The Soul. In: Oats, W.J. (ed) (1948) Basic Writings of Saint Augustine. Random House, iv: 7:9.

[62] Fisher, S.R. (2003) A History of Reading. Reaktion Books, p 90–91.

[63] Once you have learned to read, "it deosn't mttaer in waht oredr the ltteers in a wrod are, the olny iprmoetnt tihng is taht the frist and lsat ltteer be at the rghit pclae... it doesn't matter in what order the letters in a word are, the only important thing is that the first and last letter be at the right place." https://www.mrc-cbu.cam.ac.uk/people/matt.davis/cmabridge/ (accessed 23/07/2021).

when they read silently. Blind people use their sense of touch to read Braille. Helen Keller relied entirely on touch to converse with people. And recently David Eagleman, an American neuroscientist, has been working on a vest that converts speech into a pattern of tactile sensations through an array of small vibrators built into the garment. He calls the idea of encoding sound as a pattern of vibration that is felt through the skin as "sensory substitution."[64]

Seeing with Sound

The ability to perceive the physical world in detail through hearing alone is usually considered to be the preserve of echolocating creatures such as bats and dolphins. As we shall see in chapter three, these creatures emit a rapid succession of powerful bursts of acoustic energy, each consisting of a brief pulse of very high frequency sound that is well beyond our range of hearing, and listen to the reflection. Echolocation not only enables them to perceiving their immediate surroundings in considerable detail, it also enables them to hunt in darkness. What is much less well known is that humans can also perceive their surroundings by listening for reflected sounds, though they can't match the precision or acuity achieved by echolocating creatures.

Nevertheless, blind people, particularly those born blind or who have lost their sight when very young, are remarkably adept at navigating their surroundings and identifying its salient features by listening attentively to ambient sounds, both direct and reflected. As we noted previously, ambient sounds are an vital source of information for sighted people, but because they rely primarily on vision to get around, they are far less aware of the many ways in which sounds are modified by their environment than someone who is blind and so do not realise that these modifications can reveal specific details of their surroundings. Blind people can also echolocate by taping a metal-tipped cane against the ground and listening for the reflection of the impact. Some blind people have taken this a step further and listen for reflections of loud clicks made by flicking their tongue against the roof of their mouth, in effect mimicking bats and dolphins. A leading exponent of this technique is Daniel Kish, an American who lost his eyesight to cancer when he was 13 months old. He is so skilled at this technique that he can ride a bike around his neighbourhood or walk through a wood with ease.[65]

---

[64] See David Eagleman's website for information on what he calls "sensory substitution. https://eagleman.com/?s=sensory (accessed 06/08/2021).

[65] There is a great deal of information online about Daniel Kish.

Human echolocation cannot rely entirely on listening for distinct reflections, in part because the human hearing system is unable to distinguish clearly between even the briefest of sounds such as a click and its reflection if they occur less than one hundredth of a second apart. If the interval is less than this, the reflection is ignored by the hearing system and only the direct sound is heard. The speed of sound in air is 340 m/s, so sound travels 3.4 m in this brief interval, which means that its reflection cannot be clearly distinguished from the original sound if the reflecting surface is less than 1.7 m from one's ears. In fact, the actual time interval varies depending on the nature of the sound, being least for brief clicks. It is considerably greater for speech and music because these sounds are prolonged in comparison to a click and, as the explanation of echoes in chapter five makes clear, their reflection is usually masked by the original sound. The inability of our hearing system to detect reflected sounds from nearby surfaces is known as the "precedence effect", and is believed to help us locate the direction from which a sound reaches the ears. As we shall see in chapter three, locating the direction of the source of a sound is as important as hearing and identifying the sound itself.

Even where it is not possible to hear a distinct reflection, another acoustic effect that depends on reflection can sometimes be very noticeable once one learns what to listen for. And this you can easily do this without even having to put this book down. With your eyes closed, hold the book some 40 cm from your face and make a continuous *shh* sound with the tips of your front teeth clenched and your lips well parted. Move the book alternately towards and away from your face and listen to the effect this has on the sound. You should hear a whooshing that rises and falls in pitch. How does the direction in which the book moves affect the pitch you hear? Now repeat the exercise in front of a wall. Can you detect the wall as you move closer to it simply from changes in the sound of the hiss? What difference, if any, does an open doorway make?[66] Blind people learn how to interpret this change in a reflected sound to perceive objects, surfaces and openings when they are too close to echolocate. In some circumstances, they can also hear ambient sounds being modified in this way.

In fact, as you'll find out in chapter five, you will have often experienced this effect without being fully aware of it. It occurs when a sound and its reflection are combined in such a way that the overall timbre of the sound is altered. Part of the reason why it goes unnoticed it is that, as a sighted person,

---

[66] Cobble together a "sonic torch" consisting of a mono Bluetooth speaker tethered to an app that produces white noise on a tablet or smartphone. Hold the speaker 25 cm from a hard surface, turn on the white noise and listen as you move the speaker either towards or away from the surface. What do you hear? Does the same thing happen when you direct the sound at a soft surface, say curtains or a pillow?

you assume that all the necessary and useful acoustic information is present in the sound that reaches our ears directly from the event that is its cause. You usually don't pay much attention to subtle modifications such as those due to the interaction between the original sound and its reflection. Your lack of awareness of this effect is also likely to be due to lack of detailed knowledge and understanding of physical nature of sound, something that this book is intended to redress.

In addition to "hearing" the presence of walls with your eyes closed, you can learn to hear a difference between surfaces (e.g. soft and absorbent v. hard and smooth). By the same token, the absence of reflections from a wall reveals open doorways and windows. It is even possible to perceive individual objects, though doing so is akin to being short sighted, even for the most proficient human echolocator: only objects that are very large or very close can be unambiguously perceived aurally.

It is also possible to perceive aurally a great deal about the nature of the space in which one finds oneself. Every enclosed space, from a small room to a huge concert hall, has a distinct acoustic signature because it modifies sounds in two important ways. Firstly, we have all experienced reverberation in an empty room due to the rapid succession of overlapping reflections from walls and ceilings of ambient sounds. As we shall see in chapter five, both the nature of the reflecting surfaces and the volume of the room play a part in the duration and timbre of reverberation. Secondly, you will also have noticed that within an enclosed space some frequencies of ambient sounds are emphasised more than others. This happens when the wavelength of some of the frequencies present in a sound match the dimensions of the enclosed space, a phenomenon known as resonance. A striking example of resonance is the apparent improvement in the quality of one's voice when singing in a shower cubicle, though, as we shall see in chapter six, this is by no means the most interesting or important instance of this phenomenon.

We also make use of sound to investigate the interior composition or condition of objects; in many situations this is our only source of information. Tap an object and you will usually be able to tell whether it is solid or hollow, and even determine something about its composition, as we shall shortly see.

Another acoustic modification of which we are all aware is the decrease in loudness that occurs as a sound spreads out from its source. This is always accompanied by a loss of its high frequencies because they are absorbed as a sound travels through air, which is why the peal of thunder is heard at a distance as a rumble. We unconsciously employ these two effects to estimate how far away we are from the source of a sound.

Soundworlds

Sighted people inhabit a world largely defined by objects and their spatial relationships. But according to John Hull, if you are blind you are acutely aware that acoustic space lacks the boundaries and limits that are evident in visual space. As a result, a blind person inhabits an infinite space.[67]

But even though it lacks boundaries, an infinite space is nevertheless a space within which objects are located and events take place. So hearing has spatial attributes, despite a claim by the Oxford philosopher, Peter Strawson, that sounds in themselves have no intrinsic spatial characteristics such as their position relative to the listener. He concluded that a concept of space can't be conveyed in purely acoustic terms.[68] But although it is not always possible to estimate unequivocally how far one is from the source of a sound by listening alone, with few exceptions sounds are invariably experienced as being outside oneself. Moreover, as long as both ears are in good working order, both sighted and blind people directly perceive the direction from which they hear sounds and therefore can determine the direction of the source relative to themselves: footsteps approaching from behind, bird calls from within a bush to one's right or the roar of an overflying aircraft.

Indeed, among the many differences in how eyes and ears perceive space, among the most significant is that hearing is omnidirectional. But such is the primacy of vision that sighted people don't give the omnidirectional nature of hearing a second thought. Even allowing for peripheral vision, our attention is usually narrowly focussed on whatever lies before eyes whereas you can hear sounds from all directions. And as we shall see in chapter three, this helps us to direct our gaze at noisy events that lie outside our narrow field of view.

Nevertheless, perceiving three-dimensional space is primarily a visual matter. In the first place, light from objects within one's field of view form images on different parts of the back of the eye, so their separation in the physical world is directly replicated on the retina. At the same time each eye has a slightly different view of whatever one is looking at, producing a sense of depth for objects that are closer than 30 m. For anything further away we rely on the relative movement between objects that occurs as we change the position from which we view them. Objects in the foreground change position with respect to their background much more than do those that are further away from us. Our visual system also infers that objects are located at different distances from the eye, either because nearby objects partially cover

---

[67] Hull, J.M. (2001) Sound: An Enrichment or State. The Journal of Acoustic Ecology, Volume 2, Number 1, July, p 10.

[68] Strawson, P.F. (1959) Individuals: An Essay in Descriptive Metaphysics. Routledge, p 65–66.

those further away or because distant objects appear to be smaller than those near at hand.[69]

But these ocular clues have to be learned, though the process of doing so is largely unconscious because it takes place during early infancy. We are not born able to perceive visual depth directly, as is evident in cases of people who have had their eyesight restored after a lifetime of blindness, either by having corneal transplants or the by the removal of congenital cataracts. Among the most vexing of many perceptual problems experienced following the operation is that everything appears to be equally distant, so that perceiving that some objects are further away than others is almost impossible. This has been known since the very first published account of the restoration of sight by the excision of congenital cataracts, which appeared in the *Philosophical Transactions of the Royal Society* in 1728. The author of the account, a pioneering English eye surgeon William Cheselden, wrote that "When [the anonymous 13 year old boy—Cheselden did not name him] first saw, he was so far from making any judgment of distances, that he thought all objects whatever touched his eyes (as he expressed it)".[70] Nor did this difficulty abate as the boy grew older, Cheselden noted. However, the perceptual problems following the removal of cataracts experienced by people who are blind from birth was not systematically investigated until the middle of the twentieth century. Richard Gregory, an English neuropsychologist, made several studies of people whose sight was partially restored. These confirmed that visual depth perception is *not* regained following the removal of cataracts to restore sight.[71]

In fact, in the absence of the necessary optical clues everyone finds it difficult to perceive depth. Nowhere is this more evident than with the night sky. Although stars differ in brightness, it is impossible to overcome the impression that they are all equidistant from the Earth because a starry sky offers none of the perceptual clues such as parallax or perspective on which depth perception normally relies upon. This is almost certainly why the ancient

---

[69] There are exceptions: the Moon Illusion whereby a full or nearly full moon appears to be unexpectedly larger when seen near to the horizon than it does when seen high above it. Rainbows are also subject to this illusion: the arc of a distant rainbow appears broader than one near at hand, though the angular dimensions of a rainbow are constant however far you may be from the rain in which it is seen.

[70] Cheselden, W. (1728) An Account of Some Observations Made by a Young Gentleman, Who Was Born Blind, or Lost His Sight so Early, That He Had no Remembrance of Ever Having Seen, and Was Couch'd between 13 and 14 Years of Age. Philosophical Transactions, Vol. 35, Issue 402, p 447–450.

[71] Gregory, R.L., Wallace, J.G. (1963) Recovery from Early Blindness: A Case Study. Experimental Psychology Society Monograph No. 2. The paper can be downloaded from Richard Gregory's website http://www.richardgregory.org (accessed 1/03/2121).

Greeks were prepared to accept that all stars lie on a surface of a vast sphere that has the Earth at its centre, a notion that seems indisputable until one has to deal with the finer points of planetary motion. And although most astronomers had abandoned the idea of a geocentric universe by the end of the sixteenth century, it wasn't until 1838 that improvements in telescopic design and techniques enabled the German astronomer F. W. Bessel to measure the distance to a star using parallax due to the almost imperceptible apparent motion of the star against the backdrop of much more distant stars as the Earth orbits the Sun.[72] But the knowledge that stars lie at different distances from the Earth can't overcome the lack of parallax that would be necessary to enable us to see the night sky in three dimensions.

As we shall see in chapter three, hearing can be restored to people born deaf, though, as we noted in connection with learning to speak, this too is not without perceptual complications similar to those that accompany the restoration of sight.

The Business of Listening

These days, when automated diagnostic devices are widely available in hospitals, the cliché of a doctor with a stethoscope draped around his or her neck seems an anachronism. Surely, ultrasound scanners, X-ray machines, electro-cardiograms and the like have made listening to bodily sounds through a length of rubber tubing about as useful as consulting a patient's horoscope before deciding how and when to carry out treatment, as was once the case? Not so; the simplicity and ready availability of an acoustic stethoscope makes it the first choice for a preliminary diagnosis of a range of conditions of the heart, lungs and bowels.

Perhaps the stethoscope's days are numbered now that pocket-sized diagnostic gadgets are available, such as a hand-held echocardiogram that pairs with a smartphone to monitor heart problems. And with the introduction of portable automated devices that employ artificial intelligence, the ability to recognise the tell-tale sounds of a damaged organ will surely become unnecessary. In any case, as an aid to diagnosis, a stethoscope is only as good the person using it. Reliably identifying problems by ear alone demands hours of training, and years of practise, something that is no longer considered a priority in many medical schools.[73]

---

[72] The star was 61 Cygni, and Bessel's measurement of how far it is from Earth was about 10% less than the most recent value (11.4ly).

[73] Websites to help doctors hone their skills in auscultation abound.

Listening to bodily sounds as a means of diagnosing ailments was used by physicians in Ancient Rome and Greece. The practice fell into disuse and wasn't revived in Europe until the second half of the eighteenth century when an Austrian physician, Leopold Auenbrugger realised that it is possible to tell if the lungs of a patient are diseased by tapping the chest with a finger, a procedure now known as *percussion*. Apparently, when he was a boy he worked in his father's cellars where he learned how to estimate how full wine barrels were by tapping them and listening to the resulting sound, and that this gave him the idea of using the same technique to investigate the internal condition of bodily organs. A healthy lung sounds hollow, he wrote, but "if it yields only a sound like that of a fleshy limb when struck—disease exists in that region."[74]

But it wasn't until a Frenchman, René Laennec, invented the stethoscope in 1816 that it became possible to use bodily sounds to diagnose a range of diseases beyond that of congested lungs. The technique is known as *auscultation*, a word coined by Laennec from the Latin verb "auscultare", to listen. His stethoscope was a short, hollow wooden tube that was pressed against a patient's body while the doctor put his ear against the other end. Surprisingly, this makes it possible to hear bodily sounds that are inaudible even when the ear is pressed directly against a body because they cause the air within the tube to resonate, making them louder.

The idea for the device came to him during a consultation.

In 1816, I was consulted by a young woman labouring under severe symptoms of diseased heart, and in whose case percussion and the application of the hand were of little avail on account of the great degree of fatness. ... I happened to recollect a simple and well-known fact in acoustics and fancied, at the same time, that it might be turned to some use on the present occasion. The fact I allude to is the augmented impression of sound when conveyed through certain solid bodies-as when we hear the scratch of a pin at one end of piece of wood, on applying our ear to the other. Immediately, on this suggestion, I rolled a quire of paper into a sort of cylinder and applied one end of it to the region of the heart and the other to my ear, and was not a little surprised and pleased, to find that I could thereby perceive the action of the heart in a manner much more clear and distinct than I had ever been able to do by the immediate application of the ear.[75]

---

[74] Auenbrugger, L. (1761) Inventum novum ex percussione thoracis humani ut signo abstrusos interni pectoris morbos detegendi [New Invention by Means of Percussing the Human Thorax for Detecting Signs of Obscure Disease of the Interior of the Chest]. Vienna: Johann Thomas Trattner.

[75] Laennec, R. (1836) A Treatise On The Diseases Of The Chest And On Mediate Auscultation (trans by Forbes, J.). Thomas and George Underwood, p 4.

For doctors in Laennec's day his device also had the added advantage that it made it unnecessary to place the ear in contact with the body of an unwashed and possibly unhygienic patient and did not invade a woman's privacy.

Of course, sounds heard through a stethoscope are of no value unless they can be correlated with specific conditions. Laennec achieved this by listening for the distinguishing features of sounds he heard through the instrument and, whenever a patient died, following up his auditory diagnosis with an autopsy. Over several years he compiled a vast catalogue of sounds associated with various conditions. Faced with the need to identify the huge variety of noises made by both diseased and healthy organs, Laennec had to resort to similes. Some of these were straightforward: lungs of someone in the initial stages of pneumonia produced what he described as a *rale crépitant* (wheezing crackle). But others were more difficult to pin down: obstructed bronchi produced sounds that reminded him of "the cry of small birds, the kind of noise made by two marble slabs coated with oil which are abruptly pulled apart."[76] The ringing quality of the cough of a patient with lung disease sounded like "a fly buzzing in a porcelain vase."[77]

Laennec's nomenclature was excessively convoluted and may well have confused some of the doctors who adopted his invention. In any case, successful auscultation requires a good ear and a thorough knowledge of anatomy. Oliver Wendell Holmes, better known as a poet, was by profession a doctor and penned a wry poem that warned of the dire consequences of the mistakes a physician might make by placing too much faith in the sounds heard through a stethoscope. In the poem, the sounds of two flies trapped within the stethoscope lead to all sorts of misdiagnosis and as a consequence the patient eventually dies.[78]

Medicine is by no means the only occupation that routinely employs sound to make a diagnosis. Automobiles are immeasurably simpler than flesh and bone, and, arguably, the range of their characteristic sounds are correspondingly less numerous and more easily traced to their source. An experienced mechanic can improve the performance of a car's engine by ear; it's known by a suitably musical term: tuning. Moreover, the first indication of a mechanical fault is often an unusual and distinctly audible noise. The throaty rumble of a faulty exhaust, the rattle of a loose exhaust pipe and the squeal of a loose fan belt or worn brake pads are all disconcerting sounds that require the services

---

[76] Laennec, R. T. H. (1819) De L'auscultation Médiate, Ou Traité Du Diagnostic Des Maladies Des. Poumons Et Du Coeur, Fondé Principalement Sur Ce Nouveau Moyen D'exploration. Tomé Second. Paris, p 5.

[77] Laennec, R. T. H. (1819), p 95.

[78] Holmes, O.W. (1848) The Stethoscope Song; A Professional Ballad.

of a mechanic. And should the noise be unfamiliar, there are any number of websites that provide examples of the tell-tale sounds of a huge range of engine faults. Experienced car mechanics, whether amateur or professional, are likely to recognise the significance of such sounds without recourse to such a resource. And to pinpoint the source of the problem, mechanics can make use of a stethoscope that has a thin metal shaft in place of the diaphragm of the medical version. The mechanic presses the tip of the shaft against the running engine as close to the fault as possible and listens through the earpieces. This makes it possible to pick out the ticks, squeals, rattles and hisses due to faulty components from the overall thrum of the engine.

Sound can also be used to track down the location of a leak in an underground high-pressure water supply because water escaping from a crack in a buried pipe makes an audible sound. The noise will travel some distance along the pipe and can be heard from the surface above the pipe either as a hiss or a whoosh by someone equipped with the necessary instrument, as long as the pipe can be directly accessed from the air—the high frequency sounds that give this hiss its particular acoustic signature are rapidly absorbed by soil. Despite the availability of sensitive microphones and the necessary ancillary apparatus to make these sounds audible, the favoured method of detection remains a long, thin, solid iron rod known, aptly enough, as a listening stick. One end of this rod is placed firmly against the pipe and the ear pressed against a small hollow chamber at the other end, which is designed to resonate at frequencies that are typical of leaks and thus make them easier to hear distinctly.[79] As we shall see in Chap. 4, listening for underground sounds has often been a matter of life and death during military campaigns. It was routinely used in the trenches of the Western Front during WW1 to detect enemy tunnelling.

But one occupation that has been consigned to history by the microphone is the wheeltapper. His task was to test the state of the wheels of a stationary train by striking each one with a long handled hammer and listen to the resulting sound. In the In the absence of flaws the wheel would ring, but cracks within the steel, even those invisible to the eye, prevent the wheel from vibrating freely. In place of a ring there would be a dull thud that wheeltappers were trained to recognise. These days the job is done using an ultrasound probe.

---

[79] Auto mechanics sometimes resort to a long screwdriver if they don't have a automotive stethoscope to hand. The tip of the screwdriver is placed against the body of the running engine and the ear is pressed against the end of the handle. In the hands of an experienced mechanic this is almost as effective as a stethoscope.

Tapping is a surprisingly effective method for discovering the composition or the internal state of an object. It is also one that has been exploited since antiquity. Marcus Vitruvius Polio, the Roman architect, advised that prior to firing a ballista their " strings must not be clamped and made fast until they give the same correct [musical] note to the ear of the skilled workman. For the arms thrust through those stretched strings must, on being let go, strike their blow together at the same moment; but if they are not in [musical] unison, they will prevent the course of projectiles from being straight."[80]

Masons, sculptors, potters and bricklayers have long known that a stone, a fired pot or brick that does not ring when struck a light blow with a hammer probably has internal faults that weakens it.[81] And you will probably have tapped a wall to discover whether it is solid, useful in determining if it is a supporting wall or locating a wooden stud when hanging a picture. A couple of taps with a fingernail is usually all it takes to tell whether a window pane or a vase is made of glass or plastic and, indeed, its relative thickness.

One of the tests carried out to determine the maturity of Parmensan cheese is to tap it lightly with a small hammer. Home cooks are advised that a freshly baked loaf of bread should sound hollow when rapped with one's knuckles if it is fully cooked, though many professional cooks say that this is not always an infallible test.[82] And as for advice about how to test if a watermelon is ripe, should we heed Henry David Thoreau, the nineteenth century American poet, philosopher and nature writer and author of 'Walden, or, Life in the Woods'? "Of two otherwise similar [melons], take that which yields the lowest tone when struck with your knuckles, i. e., which is hollowest. The old or ripe ones sing base; the young, tenor or falsetto. Some use the violent method of pressing to hear if they crack within, but this is not to be allowed. Above all no tapping on the vine is to be tolerated, suggestive of a greediness which defeats its own purpose."[83] We shall return to cooked loaves and ripe melons in chapter six, where the secrets of their sounds will be revealed.

---

[80] Vitruvius (1914) The Ten Books On Architecture (trans Morgan, MH). Harvard University Press. Book 1, p 8–9.

[81] When I was a child it seemed to me that my father, a civil engineer, could never walk past a stack of bricks on a building site without picking a couple up and banging them together. I realised that this was a rough and ready test of quality, but gave no thought to the role that sound played. I must have assumed that faulty bricks would break, as they sometimes did. But he was also probably listening to the sound they made.

[82] As an occasional home baker, I find that "hollow sound" a reliable test that a loaf is done, though the colour of the crust is just as important. A stiff crust vibrates when tapped and the air pockets within the cooked loaf determine the resonance of its interior.

[83] Thoreau, H.D. Journals, August 27, 1859.

Sound also plays an unexpectedly important role when we eat. We asses the freshness or otherwise of many foods from the sounds they make when we bite into them. A tell tale sign that a biscuit is stale is that it is not crunchy. Nutritionally it may be identical with the fresh article, but a *silent* biscuit is unappetising in comparison with one that emits a crunch.[84] Other foods that are more appetising when they emit an audible crunch when bitten or chewed include apples, carrots, celery, some lettuces, cucumbers, crackers, crisps and baked nuts. Indeed, along with variously flavoured potato crisps, there is a huge range of snacks whose appeal depends in large part to being crispy that include Pringle chips, tortilla chips, popcorn, croutons, pretzels and breadsticks.

In a recent investigation into the effects of sound on one's evaluation of food freshness, researchers discovered that the presence of a broad range of frequencies above 2000 Hz when biting food is closely associated with freshness. Each person in the study was asked to take a single bite at a time out of dozens of Pringle chips and assess their freshness merely from the resulting sound. The sound of the bite, however, was heard by participants through earphones, which allowed the researchers to alter the balance of frequencies in the sound. Whenever loudness of the higher frequencies present in the sounds produced when the chip was bitten into was reduced, the chips were invariably judged to be less fresh than those where the sound was not altered even through all the chips were from the same packet and so equally fresh.[85]

We can all agree that crunchy snacks are tasty and moreish—unless you are consuming them in a cinema or theatre, where the noise is likely to annoy those around you and make you wish they weren't quite so loud. And as if to add insult to injury, all these snacks (with possible the exception of popcorn) are sold in packets made of crinkly materials. Go the snacks section in any supermarket and squeeze the packets: they all crackle and rustle loudly. The reason why the material used in snack packaging is so noisy is that it is fairly stiff. Changing the shape a sheet of stiff material by crumpling it involves a series of sudden, rapid deformations. Even a sheet of writing paper crackles when crumpled. The energy released as sound during crumpling is louder than expected because it is concentrated in brief bursts during the multiple deformations that occur as the material is crushed. And the stiffer the material, the louder the sounds. Cloth, in comparison, isn't stiff and so doesn't crinkle when crumpled, though it may rustle as surfaces rub against one another.

---

[84] Apropos of soft biscuits, in the light of the link between sound anf food freshness, what is one to make of that peculiarly British custom of dunking a biscuit in hot tea before eating it?

[85] Spence, C. (2017) Gastrophysics. Penguin. See chapter 4 for details of this and similar experiments.

The record for the loudest bag of snacks, incidentally, was that used for Sun Chips. It was made from a biodegradable plastic that produced sounds as loud as 95 dB when crumpled. The sound was so loud that manufacturer of the chips changed the packaging.[86] The average snack packaging can reach 75 dB when handled, so the sound produced by those Sun Chips bags was 100 times more powerful, hence twice as loud.[87] Even though noisy packaging is not necessary to keep snacks fresh, it is probably considered desirable by manufacturers because we consumers associate crinkly, crunchy sounds with freshness.

Although some of the sounds described in this section may be new to you, you will have often relied on sound to discover the composition or internal state of things and doubtless have your own examples to add to those above. Moreover, as we have noted, sounds are often the only source of information we have about events in our immediate environment. And, of course, we rely on sounds to warn of danger and, just as importantly, to indicate the direction of the source of the danger relative to ourselves.

Darkness and fog are circumstances in which the ear comes into its own. In pitch darkness, as if to compensate for enforced sightlessness, it can seem that even the faintest sounds are audible. And in a dense wood or forest, where vision is limited to one's immediate surroundings, the presence of people, animals, birds, flowing water, wind and even rain is revealed principally through their sounds.

Bernie Krause has recorded natural soundscapes all around the globe. On one of his expeditions he spent time with the Jivaro, an Amazonian jungle-dwelling tribe. He was amazed by how easily and unhesitatingly they could find their way through the dense, dark forests through which they travelled simply by listening to the soundscape and the creatures that inhabited it.[88]

Arctic explorers have noted that the Eskimos they lived with often relied upon sounds when travelling by kayak. The British soldier and explorer, F. Spencer Chapman, was once kayaking off the eastern coast of Greenland with a party of Eskimo hunters when they became enveloped in a dense fog. Although waves breaking on the shore were audible, the shore itself was no longer visible, and he began to worry that they would miss the narrow entrance of their home fjord. But the hunters were undismayed, "indeed they beguiled the time by singing verse after verse of their traditional songs and

---

[86] The packaging was changed in 2010. "Sun Chips loud bag" https://www.youtube.com/watch?v=kki32mt8p6w (accessed 23/07/2021).

[87] Heller, E.J. (2013) Why You Hear What You Hear: An Experimental Approach To Sound, Music And Psychoacoustics. Princeton University Press, p 162–63.

[88] Krause, B. (2012) The Great Animal Orchestra, Finding the origins of Music in the World's Wild Places. Profile Books, p 67.

occasionally they threw their harpoons from sheer *joie de vivre*." But after a further hour of paddling, the lead kayak swung abruptly inshore and entered the narrow fjord without difficulty. It was only much later that he learned how the hunters had achieved this remarkable feat of navigation: "All along this coast there were snow-buntings nesting, and each male bird…used to proclaim the ownership of his territory by singing his sweet little song from a conspicuous boulder. Now each cock snow-bunting has a slightly different song, and the Eskimos had learned to recognise each individual songster so that as soon as they picked out the notes of the bird who was nesting on the headland of their fjord, they knew it was time to turn inshore."[89]

Edmund Carpenter, an American anthropologist, had a similar experience while living with Eskimos. He recalled travelling by kayak with a party of Eskimos along a dangerous coast in dense fog. But although visibility was zero, they forged ahead without hesitation. The sound of the surf and the cries of nesting birds was all they needed to keep them on track and avoid ending up on the rocks.[90]

In fact, anyone can use sounds to find their way in fog on land as long as they are within hearing range of a distinct source of sound such as a waterfall, a noisy cataract or waves breaking on a shore. A few years ago, finding myself disorientated by sea fog in unfamiliar countryside near the Seven Sisters, the well known chalk cliff on the South East coast of Britain, I was able to find my way across several fields to the path along the cliff top principally by listening to the sound of waves breaking on the rocky shore below. Contrary to popular opinion, sound travels well in fog. Indeed, given that the conditions that favour fog, a calm, stable atmosphere, sound will often be heard a greater distances in fog than in clear weather. We'll come back to the effect of fog on sound in Chap. 4.

And in the days before radar, in poor visibility due to fog or darkness, ships sailing close to a rocky shore would routinely sound the ship's whistle and time its echo with a stopwatch. Given that the speed of sound in air is 340 m/s, every second that elapses between the whistle and its echo corresponds to a distance of 170 m between ship and shore.[91] We shall see in Chap. 4 that this method was not always reliable because atmospheric conditions strongly affect the propagation and absorption of sound.

While most of us can never hope to match the necessary lifetime of experience required to interpret indirect clues as an aid to navigation as expertly or

[89] Spencer Chapman, F. (1932) Northern Lights: The Official Account of the British Arctic Air-Route Expedition. Chatto and Windus.

[90] Carpenter, E. (1973) Eskimo Realities. Holt, Rinehart and Winston, p 36.

[91] Windsor, H.H. (1927) Echo Sailing In Dangerous Waters. Popular Mechanics, 47, p 794–97.

instinctively as the Jivaro and Eskimo can, it is possible to learn the rudiments of how to find one's way in unfamiliar territory without map and compass, should one wish to do so.[92]

Soundscapes

These examples of the many ways in which sounds inform us about our environment, together with others that we shall come across in later chapters, prove that hearing is a useful, not to say a vital sense. But because we usually consider sounds to be a source of information about events, I suspect most of us take a decidedly utilitarian approach to hearing. Unless circumstances call for it, say when using sound diagnostically or heeding an unfamiliar noise, we take little interest in the acoustic qualities of a particular sound or in its physical cause. Have you ever made a point of listening attentively to seemingly inconsequential sounds such as the crunch of footsteps on a gravel path or the tinkle of a metal spoon as you stir a cup of tea or coffee? In most situations, I would hazard, our listening is perfunctory at best. Add to that a perfectly reasonable desire to avoid, or at least ignore, sounds we find unpleasant or intrusive, and the result is that we habitually turn a deaf ear to the wider soundscape, that random and seemingly purposeless sonic backdrop of daily life.

According to Raymond Murray Schafer, the Canadian musician and a pioneer of acoustic ecology, who popularised the term in the 1970's, a soundscape is the sum of all the individual sounds that one hears (or would hear if one makes the effort to listen) at any given place and moment, together with the preconceptions and expectations that affect how we experience it.[93]

Several decades later, Bernie Krause coined three terms that have proved very useful for classifying soundscapes: anthropophony, biophony and geophony.[94] Anthropophony, as its etymology suggests, is the soundscape due to human activity, which all too often drowns out the sounds of the animal kingdom, the biophony, and those of inanimate nature, the geophony. In fact, the distinction between the anthropophony and biophony is not clear-cut because if we take the biophony to be the vocalisations

---

[92] Huth, J.E. (2013) The Lost Art of Finding Our Way. The Bellnap Press of Harvard University Press. See p 10–11 for an example of navigating in fog using sound.

[93] Acoustic ecology is the study of the relationship between living creatures and their environment in terms of sound.

[94] Krause, B. (2012) The Great Animal Orchestra, Finding The Origins Of Music In The World's Wild Places. Profile Books, p 80.

of animals, then that soundscape must include humans talking, laughing, crying, shouting and singing.

Inevitably, soundscapes, like landscapes, are not all equally appealing or interesting. Moreover, a sound that one person finds appealing or interesting is not necessarily so to others. For Schafer, the urban soundscapes of the modern world are the least desirable because the clamorous anthropophonic hubbub of those environments frequently prevents us from hearing any sounds clearly. In his opinion the ideal soundscape should be free of prolonged noises, allowing individual sounds to be heard distinctly. And as things stand, he says, only environments devoid of permanent human presence provide such conditions; and there are not very many of those these days.

But Schafer doesn't claim that nature ought to be silent. He acknowledges that the sounds made by the creatures that inhabit natural environments and of the events that shape those environments are an inescapable and, indeed, a desirable aspect of nature. He would doubtless agree with Richard Jefferies, the nineteenth century English naturalist and writer, that sounds animate the world.

Describing a visit to Kew Gardens, which in his day had not yet been swallowed up by the expansion of London and its attendant hubbub, Jefferies paused to "recline upon the grass and with half-closed eyes gaze", listen and ruminate.

"The delicious silence [of the park] is not the silence of the night, of lifelessness; it is the lack of jarring mechanical noise; it is not silence but the sound of leaf and grass gently stroked by the soft and tender touch of summer air. It is the sound of happy finches, of the slow buzz of bumble-bees, of the occasional splash of a fish, or the call of a moorhen. Invisible in the brilliant beams above, vast legions of insects crown the sky, but the product of their restless motion is a slumberous hum.

These are the real silence; just as a tiny ripple of the water and the swinging of the s as the boughs stoop are the real stillness. If they were absent, if it was the soundlessness and stillness of the stone, the mind would crave for something. But these fill it and content it."[95]

The "soundlessness and stillness of the stone" is attainable in an anechoic chamber, which as we shall see in chapter five is a dedicated acoustic laboratory carefully engineered to suppress all extraneous sounds. But whether the funereal quiet of these laboratories—they are dark as well as silent—is to one's liking only first hand experience can resolve.

---

[95] Jefferies, R. (1887) Nature Near London. Chatto and Windus, p 193.

Out in the open air, and in the absence of the most intrusive mechanically generated anthropophonic noises—those due to motorcars, planes, helicopters and assorted machinery—we become aware of the both the biophony and geophony, as well as potentially pleasurable anthropophonic sounds such as church bells, the clickety clack of train wheels and tinkling fountains. And when not drowned out by anthropophonic noise, not only do individual sounds become audible, giving them an intimacy and appeal that they lose in an urban hubbub, one's hearing is no longer confined to one's immediate surroundings as one's ears, unbidden, open up to the world at large.

But biophonies differs from geophonies in one very important respect: biophonies consist of sounds that are meant to be heard *and* understood. Whether it lives on land or in water, a huge number of living creatures —humans included—make sounds to communicate, sounds that must be carefully fashioned to ensure that they convey specific meanings that can be identified as unambiguously as possible by others of their kind. And, as Bernie Krause discovered, to help with this that animals that had occupied the same ecosystem for a long time tended to pitch their vocalisations so that they do not intrude on one another's acoustic turf.[96] All of which makes it easier to pick out and identify the calls of particular creatures, none more so than those of birds and whales, which of all animal groups have evolved the most varied, complex and enchanting vocalisations in the animal kingdom.

Unsurprisingly, the anthropophony intrudes on the acoustic turf of both birds and whales, not to mention every other creature that relies on sound to communicate. There is widespread evidence that birdsong is affected in urban environments; birds in towns and cities have been found to sing shorter, louder, higher pitched songs than their country cousins. The effect of man-made mechanical sounds on whales is not yet fully understood, though given that whales and dolphins depend on hearing far more than they do on vision, can there be any doubt that the sound of rotating ship's propellers and the widespread use of powerful sonar by naval vessels must have an adverse effect on the health and behaviour of all cetaceans?

Geophonic sounds, on the other hand, have no intrinsic meaning despite being an important source of information about physical events that are out of sight. And one of the consequences of this lack of intrinsic meaning is that a great many geophonic sounds are sufficiently similar acoustically that in some circumstances it can be difficult to distinguish one from another. The reason is that geophonic sounds are often composed of a broad band of frequencies. At a distance, the roar of an avalanche can be mistaken for rolling

96 Krause, B. (2012) The Great Animal Orchestra, Finding the origins of Music in the World's Wild Places. Profile Books, p 98.

thunder, and wind blowing through a wood can sound like a distant cataract. Many anthropophonic sounds are similarly meaningless, being an unwanted but unavoidable by-product of interactions between inanimate objects or the moving parts of machinery. Hence the considerable effort made by manufacturers to reduce the sounds made by cars and aeroplanes.[97] And in some situations anthropophonic sounds can be mistaken for geophonic ones and vice versa: heard from a distance, heavy traffic and fast flowing cataracts are indistinguishable.

Unfortunately, according to Schafer, even far from anthropophonic soundscapes, silence alone no longer guarantees that we will hear ambient sounds distinctly because as a culture we appear to have forgotten what listening should entail and what it can achieve, namely to forge a closer bond with the world by actively and patiently attending to its sounds whether or not we judge them to be relevant to our immediate needs.

To overcome our functional deafness, Schafer proposed a program of auditory exercises that he called "ear cleaning", an ugly phrase for what is an instructive and hugely pleasurable and rewarding activity—not that anyone has come up with a more euphonious alternative phrase. Mercifully, ear cleaning does not involve having one's ears syringed—though that might help if one's ear canals are clogged up with wax—it is simply a matter of listening to ambient sounds attentively and in silence.[98] To achieve this, he recommended that one should cease making sounds (including not speaking) and instead listen to one's surroundings for an entire day.[99]

If a full day of listening attentively seems excessive, as it assuredly would be even for the most committed listener, let alone the novice auditor, Schafer suggests less demanding alternatives. One of these is to go for a stroll and note all the sounds you hear, whatever their source. He calls this a listening walk. Nor need it be solitary. Several people at a time can take part, the only condition being that they spread out so that they are just out of earshot of the footsteps of their companions. At the end of the walk the participants can compare notes.

When you first set out on a listening walk you will probably find that you have to deliberately avert your eyes to avoid being distracted by your surroundings. Your mind wanders, eyes take over and you fail to hear some

---

[97] Ironically, the intrinsic silence of electric cars poses a danger to the heedless pedestrian. Very soon, manufacturers will have to come up with a suitable sound that can be played through external loudspeakers from the car. What sound or sounds will they choose?

[98] Schafer, R.M. (1992) A Sound Education, 100 Exercises in Listening and Sound-Making. Arkana Editions, Canada.

[99] Schafer, R.M. (1994) The Soundscape, Our Sonic Environment And The Tuning Of The World. Destiny Books, p 208.

sounds until you remind yourself that you are supposed to be listening, not looking. Having got a grip of yourself, you become absorbed in the soundscape, something that puts you very much in the moment as you strain to hear sounds, which more often than not are likely to be solitary, sporadic and fleeting. Even the most mundane sound is enthralling when listened to attentively. Indeed, you may find that the walk is recalled more vividly and in greater detail through what you have heard than what you have seen.

There is also the question of how to listen. Given that our interest in sounds is usually limited to what they tell us about things and events, we are not in the habit of paying much attention to their acoustic qualities. A listening walk offers an opportunity to listen acoustically, i.e. attend to the acoustic qualities of sounds. As a bonus, listening acoustically forces one to pay closer attention to what one hears.

You might also want to test the claim that there are relatively few words with which to describe the acoustic qualities of sounds. So in addition to listening to sounds you might try to come up with words or phrases that describe their acoustic qualities. You'll be surprised how difficult this can be. But even if you fail to describe a sound, the attempt to do so will make you listen much more attentively than you might otherwise do and think about the acoustic qualities of the sounds that reach your ears.

Nor need you limit yourself to listening. Are you able to identify the physical process responsible for the sound you hear? And how is the resulting sound altered by its surroundings as it makes its way to your ears, i.e. by being reflected, refracted, absorbed and so on, effects that are the subjects of Chaps. 4 and 5. Indeed, as was noted in the introduction, the modifications can often be far more interesting than the original sound.[100]

Nor is it necessary to listen passively. You can interact with your surroundings by tapping or striking objects and noting their particular acoustic characteristics and how they differ among themselves, or clap, hum or sing and listen to the resulting echoes or reverberations. Or you could, instead, confine yourself to a particular class of sound, say those of flying insects (bees, mosquitos, flies), or the noise of flowing water (babbling brooks, rapids, cascades) and note their particular acoustic characteristics and how they differ among themselves.

A listening walk is passive in the sense that one is straining to hear any and every sound whatever their source. Alternatively, you might seek out the keynote sounds of a particular neighbourhood or landscape, i.e. listen to its

---

[100] This is also true of light: direct sunlight is just an unpleasant glare until it is transformed into something eye catching when it is reflected, refracted and so on, to make objects visible, produce sky colours, rainbows, mirages, etc.

soundscape. By noting the relative contributions of geophonic, biophonic or anthropophonic sources you can build up a detailed picture of the sound-scape you find yourself in. Schafer calls this a soundwalk.[101] A soundwalk is more structured than a listening walk, and requires greater planning and organisation.

In the years since Schafer first proposed soundwalking it has been employed by environmentalists, ecologists, planners and sound artists to explore or assess the state of particular environments, both urban and rural. Organised soundwalks frequently feature in the cultural calendar of many cities.[102]

The purpose of both a listening walk and a soundwalk is to make one more aware of sounds, particularly those one would ordinarily ignore because we judge them to be unimportant, and to take an interest in what we hear. It's a way to make listening second nature, so that every outing becomes an opportunity to expand and increase one's engagement with the world at large with all the insights, rewards and pleasures that brings.

The Gentle Art of Listening

"Plant pine to hear the sound of the wind; plant banana to hear the sound of the rain", an ancient Chinese poem advises. An enticing suggestion, but why stop there? Pine needles and banana leaves are not the only things that give voice to wind and rain, as anyone who delights in nature's sounds will know. Moreover, on a breezy day it soon becomes apparent that the sound of wind depends upon the circumstances in which it is heard, for there is a reciprocal relationship between the elements and the physical world. Without the latter, a torrential downpour would be as silent as Scottish smirr and a hurricane would be mute.[103] As things stand, the sounds of wind and rain are as varied as are the things with which they interact.

And there's a challenge. Can you identify something just from the sound it makes when it is lashed by rain or blasted by wind? It's not impossible, though it takes not a little practice. Thomas Hardy learned to do so, as is obvious from the opening paragraph of his novel, Under The Greenwood Tree.

---

[101] Schafer, R.M. (1994) The Soundscape, Our Sonic Environment And The Tuning Of The World. Destiny Books, p 213.

[102] An online search will provide details of times and places of soundwalks near you.

[103] There is a specific word for the sound of leaves rustling in the wind: *psithurism* (derived from the Greek for whispering: psithuros).

To dwellers in a wood almost every species of tree has its voice as well as its feature. At the passing of a breeze the fir trees sob and moan no less distinctly than they rock; the holly whistles as it battles with itself; the ash hisses amid its quiverings; and beech rustles while its flat boughs rise and fall. And winter, which modifies the note of such trees as shed their leaves, does not destroy its individuality.[104]

John Hull noted similar seasonal variations in Birmingham, the city where he lived. Wind caused trees to whistle and groan in winter, sound fluffy in spring when leaves first appeared, resemble the quiet roar of ocean waves in summer and tinkle metallically in autumn. Leaf fall in autumn proved useful to him because when blown along the ground their sound revealed the location and direction of roads.[105]

Hull claimed that of all nature's sounds, it is rain that best reveals the topography of one's surroundings, the sound of countless drops striking roofs, pavements or trees in leaf being readily distinguishable from one another.[106] And so one might pick out "the surging hiss of the flying rain on the sod, its louder beating on the cabbage-leaves of the garden, on the eight or ten beehives just discernible by the path, and its dripping from the eaves into a row of buckets and pans that had been placed under the walls of the cottage."[107]

But a very heavy shower can so dominate a soundscape that it drowns out all other sounds and can become a source of phantom sounds as one's mind struggles to identify what one's ear hear. One day during a week-long walk in 1913 to explore the Icknield Way, reputed to be the oldest surviving Neolithic trackway in England, Edward Thomas, the English poet and writer, was caught in a deluge:

For half an hour … Every sound was the rain. For example, I thought I heard bacon frying in a room near by, with a noise almost as loud as the pig made when it was stuck; but it was the rain pouring steadily off the inn roof.[108]

What might Thomas have made of the sounds of the intense rain showers that occur in dense tropical forests? According to Bernie Krause, the thick

---

[104] Hardy, T. (1920) Under The Greenwood Tree. E.P. Dutton and Company, p 3.

[105] Hull, J.M. (2001) Sound: "An Enrichment or State Soundscape". The Journal of Acoustic Ecology, Volume 2, Number 1, July, p 11.

[106] Hull, J.M. (1997) On Sight and Insight: A Journey into the World of Blindness. Oneworld Publication, p 26–7.

[107] Hardy, T. (1912) Wessex Tales. Macmillan Company Ltd, p 8.

[108] Thomas, E. (1913) The Icknield Way. Constable and Co. Ltd., p 278.

upper canopy of vegetation intercepts the rain before it reaches the ground so that the first sign of the shower is a roar of sound from far above one's head. Most of the rain doesn't reach the ground directly so what you hear are drops striking leaves and puddles.[109]

Water is an inexhaustible source of some of the most intriguing and delightful sounds in nature. You are bound to have noticed that flowing water, ranging from gentle streams to waterfalls, is a source of a huge variety of sounds. If you have ever paused to listen to a fast flowing mountain stream you will recognise Tennyson's description:

> I chatter over stony ways,
>
> In little sharps and trebles,
>
> I bubble into eddying bays,
>
> I babble on the pebbles.[110]

At the other end of the scale, we have this account of a river in full flood during a storm by John Muir, the Scottish-born American explorer and pioneering environmentalist, who tramped across the Sierra Nevada in 1874 and seems always to have been acutely aware of the soundscapes he encountered in his journeys.

> Here I was glad to linger, gazing and listening, while the storm was in its richest mood—the gray rain-flood above, the brown river-flood beneath. The language of the river was scarcely less enchanting than that of the wind and rain; the sublime overboom of the main bouncing, exulting current, the swash and gurgle of the eddies, the keen dash and clash of heavy waves breaking against rocks, and the smooth, downy hush of shallow currents feeling their way through the willow thickets of the margin. And amid all this varied throng of sounds I heard the smothered bumping and rumbling of boulders on the bottom as they were shoving and rolling forward against one another in a wild rush, after having lain still for probably 100 years or more.[111]

And one of the undoubted pleasures of being by the sea, or indeed by any large body of water, is the sound of waves breaking on or lapping against the shore, as the evocative description of waves washing back and forth on a

---

[109] Krause, B. (2013) *The Great Animal Orchestra, Finding the origins of Music in the World's Wild Places*. Profile Books, p 46–7.

[110] Tennyson, A., (1886) *The Song of the Brook*.

[111] Muir, J. (1894) *Mountains of California*, vol 1. Houghton Mifflin Company, p 263.

pebbly beach in the opening stanza of Matthew Arnold's poem, Dover Beach, so effectively conjures up.

Only, from the long line of spray

Where the sea meets the moon-blanched land,

Listen! you hear the grating roar

Of pebbles which the waves draw back, and fling,

At their return, up the high strand,

Begin, and cease, and then again begin,

With tremulous cadence slow, and bring

The eternal note of sadness in.[112]

As the final line implies, we don't invariably link a sound with its source or attend exclusively to its acoustic qualities. Sounds, like sights and smells, can stir up a succession of memories emotions and associations.

According to the following account, a Japanese tea ceremony deliberately creates conditions that take the possibilities of exploiting sounds imaginatively a stage further.

Quiet reigns with nothing to break the silence save the note of the boiling water in the iron kettle. The kettle sings well, for pieces of iron are so arranged in the bottom as to produce a peculiar melody in which one may hear the echoes of a cataract muffled by clouds, of a distant sea breaking among the rocks, a rainstorm sweeping through a bamboo forest, or of the soughing of pines on some faraway hill.[113]

That contemplative approach to sound is a key feature of traditional Japanese gardens, which often employ devices designed to create sounds for purely aesthetic purposes. While the principle elements of these gardens are visual, i.e. the plants themselves and the way they are laid out, as you wander its paths you may encounter a *suikinkutsu* or a *shishi-odoshi*. A *suikinkutsu* consists of a large ceramic pot that has been buried upside down so that only the drainage hole in its base is open to the air. The other end of the pot is partially immersed in a shallow subterranean pool. Water from a small stone reservoir directly above the pot drips into this pool through the drainage hole

---

[112] Arnold, M., (1867) Dover Beach. In: New poems. London: Macmillan and Co. p 112–14.
[113] Okakura-Kakuzo (1919) The Book of Tea. T.N.Foulis, p 75.

producing an enchanting bell-like chime as each drop hits the surface of the water and causes the air within the pot to resonate. A *shishi-odoshi* creates a loud tap by means of hollow length of bamboo pivoted about its centre. One half of the bamboo is slowly filled with water that trickles from a spout positioned just above it. At regular intervals enough water has filled the bamboo to cause it to tip over. It empties and then falls back against a rock, producing its characteristic hollow tap. The original purpose of the device is said to have been to frighten away deer, but these days it is appreciated for its aesthetics rather than its utility.[114]

The sound of gently trickling water of a *shishi-odoshi* has a boisterous parallel in the fountains of European gardens. The sound of jets of water cascading into a pool is not only soothing, it is also a form of acoustic camouflage because it can mask extraneous noises such as traffic. Fountains first appeared in Europe in the gardens of Italy in the latter half of the sixteenth century, i.e. in the late Renaissance. The grandest and most famous these Renaissance gardens are those at the Villa d'Este at Tivoli, which is near Rome. There are several dozen fountains and cascades in the gardens, and one of these powers an organ and is called, suitably enough, *Fontana dell'Organo* or Fountain of the Organ. The Italian fashion for garden fountains soon spread to other European countries.

Renaissance fountains played an unexpected role in the history of science. One of hydraulic problems that had to be solved by the engineers who designed and built these fountains inadvertently undermined the Aristotelian claim that nature abhors a vacuum. Why, they wondered, is it impossible to raise water vertically more than 10 m with a pump? As we shall see in the next chapter, the sounds of tiny tinkling bells and ticking clocks played a crucial role in settling these important issues.

Many of these fountains were designed to produce a spray of tiny drops in which it was possible to see spray bows on a sunny day. They were known as rainbow fountains and René Descartes claimed that it was seeing a spray bow in such a fountain in France that inspired him to come up with the first correct mathematical explanation of the shape and size of natural rainbows. Moreover, Descartes' explanation of the rainbow was arguably the first instance of the successful application of mathematical physics to a natural phenomenon.[115]

---

[114] Videos of *suikinkutsu* and *shishi-odoshi* can be found on YouTube, as can instructions on how to make your own.

[115] Werrett, S. (2001) Wonders Never Cease: Descartes's *Météores* And The Rainbow Fountain. BJHS, 34, p 129–47.

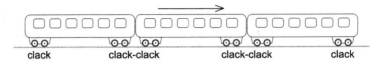

Fig. 1.2 The sequence of sounds from the wheels of a train travelling at speed over points

The rewards of listening to seemingly inconsequential sounds are not limited to their associations or their aesthetics. Many sounds are confusing when first heard. It is only when one has listened attentively to a particular sound on several occasions and made an effort to analyse it that one can make sense of what one has heard. Living approximately half a kilometre from a railway line, I often hear the delightful syncopated clatter as the wheels of fast moving trains race over gaps between the rails. The sounds are paired: clack-clack, clack-clack, clack-clack. But on listening more attentively I became aware that the final sound is always a solitary "clack" and realised that the repetitive "clack-clack" comes from the combination of the rear set of wheels of the leading coach and the front set of wheels of the next coach. The interval that separates one paired sound from the next is due to the time it takes a coach to travel its length across the points. The solitary clack is from the final set of wheels of the last coach, which, of course, is unpaired. There should also be a solitary clack from the front wheels of the first coach, though it is difficult to anticipate the moment the train reaches the gap unless one can see the train, so it catches one unprepared (Fig. 1.2).

Of course, even when one sets out with the sole intention of listening, as one would on a listening walk, the sounds that one is most likely to hear are those with which one is already familiar. Hearing new sounds is challenging, something that a listening walk will go some way to overcoming because it requires one to listen with an open mind. It also helps to be prepared, which is where this book should help because it not only provides examples of unusual sounds, sounds that it is worth seeking out, it also explains their causes and some of the many ways they are modified by their surroundings.

Above all, heed the advice of John Cage the avant-garde American composer: every sound is interesting when you take the trouble to listen.[116] Nor is it necessary to plant pine to hear the sound of the wind or plant a banana tree to hear the sound of the rain because the world is full of things that give the elements a voice. All you have to do is listen.

---

[116] Cage, J. (1961) "The Future Of Music: Credo". In: Silence: Lectures and Writings. Wesleyan University Press, p 3.

2

Pythagoras' Hammers

Abstract An account of some of the key discoveries about the nature of sound and how they influenced developments in other branches of science, notably ideas about the nature of light, the vacuum and atomism. Among the topics covered are the works of Marin Mersenne, Issac Newton, Ernst Chaladni, Daniel Colladon, A. Kircher, O. v.Guericke and R. Boyle.

Music and the Origins of Science

If science is the marriage of experiment and mathematics then acoustics may well be the oldest science. The union was supposedly first forged by Pythagoras, the semi-legendary Greek natural philosopher said to have founded a secretive brotherhood in Crotone in Southern Italy sometime during the 6th century BC. You will doubtless know him as the author of the famous mathematical theorem about right angled triangles that bears his name and to which, in the words of W.S. Gilbert, we owe "many cheerful facts about the square on the hypotenuse."[1] In fact, the numerical relationship between the three sides of a right angled triangle was already well known in his day, though only as a rule of thumb. These days Pythagoras is credited with supplying the first of the more than sixty proofs of the theorem

---

[1] Gilbert, W.S., Sullivan, A. (1879) Pirates of Penzance, Act 1.

© The Author(s), under exclusive license to Springer Nature Switzerland AG 2021
J. Naylor, *Now Hear This*,
https://doi.org/10.1007/978-3-030-89877-9_2

that mathematicians have come up with over the succeeding two and a half millennia.[2]

The facts of Pythagoras' life and works have never been established with any certainty. We know almost nothing about him except that he was born c.570 BC on Samos, an island in the Aegean Sea. He left no writings and we know about him only through fragmentary accounts written by men who were born long after he is supposed to have died, by which time he had come to be regarded as a Promethean genius responsible for a host of discoveries and ideas. Indeed, it is more than likely that some, if not most, of these ideas were developed at a later date by his followers. The uncertainties that surround the historical Pythagoras, however, are arguably of less importance than the ideas associated with his name. And one of the most enduring and influential of these is that musical harmony is a matter of arithmetic rather than sensation.

The story of how Pythagoras made this discovery took root early and became one of the founding myths of modern science, the first in a somewhat rose-tinted pageant of scientific discovery with which budding scientists are routinely regaled by their teachers, and which includes Archimedes leaping naked from his bathtub, Galileo dropping cannon balls from the top of the Leaning Tower of Pisa, and Newton observing a falling apple in his mother's orchard. Even though such tales are more often than not fanciful embellishments of the truth, if not downright fabrications, as long as they are taken with large pinch of salt they help enliven the occasional dry lecture. But a good story often makes it difficult to separate fact from fiction, necessary though this is, because the actual event is invariably more interesting and certainly more enlightening than the myth.[3]

The tale of Pythagoras' discovery is a case in point, and is worth recalling in full because it illustrates that plausible stories and uncritical respect for authority are no basis on which to found a science. Although the tale focused attention on an important discovery—a relationship between musical tones and mathematics—the actual details of how it was supposedly made were taken at face value for the next two thousand years, during which no one bothered to repeat the experiments that Pythagoras was said to have performed.

According to one version of events, which was widely and uncritically circulated by philosophers and historians of ancient Greece and Rome, and

---

[2] Crease, R.P. (2008) The Great Equations, The Hunt For Cosmic Beauty In Numbers. Constable and Robinson. See Chapter 1: The Basis of Civilization: The Pythagorean Theorem.

[3] Numbers, R.L. (ed) (2009) Galileo Goes to Jail and Other Myths About Science and Religion. Harvard University Press.

which was not called into question until the last decades of the 16th century, Pythagoras was musing about the relationship between musical tones when he happened to walk past a busy smithy. Musicians had long known that certain tones combine harmoniously and had tuned their instruments accordingly, but why it is that only particular combinations are harmonious seems to have been ignored. The noise of hammers striking anvils interrupted his train of thought, and as he listened he is said to realised that the sound of their blows formed a harmonious sequence.[4] His first thought was that this was due to the force of the impact, so he asked the blacksmiths to exchange hammers. But however hard the blow, the note of the sound produced each of the hammers remained the same. Pythagoras then had the hammers weighed and found that not only does the weight of a hammer determine the note produced, if the pitch of the sound produced by one hammer is twice that of another (i.e. the notes are an octave apart), then the weight of that hammer will be twice that of the other.[5] If their weights are in the ratio 2:3 then the resulting notes are a fifth apart[6]; and if the ratio is 3:4, their notes are a fourth apart.[7] In short, he discovered that the sequence of harmonious tones he had heard is based on the first four whole numbers: 1, 2, 3, 4.

The encounter with the hammers was said to have prompted Pythagoras to perform further experiments. One of these involved suspending weights from cords of catgut of equal length. As with the hammers, he is supposed to have found that when the cords are plucked the ratio of resulting tones to the weights hanging from them correspond to the same harmonious musical intervals as those produced by the hammers. In other words, doubling the weight hanging from a particular cord doubles the pitch it produces when it is plucked. The fact that this is *not* the relationship between tension of a cord and musical intervals wasn't discovered until the end of the 16th century, when Vincenzo Galileo, an eminent musician with an interest in musical theory, and father of his much more famous son, put Pythagoras to the test by repeating this spurious experiment and found him sorely wanting.

Despite these far-fetched claims, Pythagorean musical theory was not built entirely on sand, for one of the experiments attributed to him does yield the results claimed for it. To perform this experiment Pythagoras is said to have invented the monochord, a device that in its modern form consists of a length of steel wire (in Pythagoras' day it would have been catgut) stretched tightly

---

[4] Boethius De Institutione Musica I. In: Cohen, M.R., Drabkin I.E. (eds) (1975) Source Book in Greek Science. Harvard University Press, p 298–99.

[5] As in "Somewhere over a rainbow…"

[6] As in "Twinkle, twinkle, little star…"

[7] As in "Here comes the bride…"

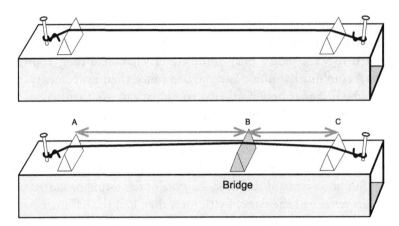

Fig. 2.1 The monochord. To show that halving the length of a string doubles its pitch, a bridge is used to divide the monochord's string so that AB is twice the length of BC. Each length is plucked to compare pitches

between two anchor points attached to an oblong, hollow box that acts as a resonating chamber and amplifies the sound of the wire when it is plucked.[8] With this device, it was claimed, he discovered that the pitch of a stringed instrument is determined by the length of the string, for when the pitch of a pair of otherwise identical wires is compared, one being twice the length of the other, when plucked the shorter one will produce a note that is precisely one octave above the other. To achieve this, a bridge is used to divide the string of the monochord in a ratio of 1:2 (Fig. 2.1). By repositioning the bridge the other Pythagorean musical ratios, 2:3 and 3:4, can be obtained.

This sequence of notes—the octave, the fourth and the fifth—was already in use as a musical scale in Pythagoras's time, so what he discovered was not a musical scale but a mathematical relationship between its notes.[9] Music, of course, predates Pythagoras by several millennia. The remains of several lyres were found by the British archaeologist, Leonard Woolley in 1929, during excavations of the 4300 year-old Royal Cemetery of Ur in Iraq. And more recently several bone flutes that are at least 30,000 years old have been discovered in Germany. A modern replica of one of these Paleolithic flutes was found to produce a range of notes comparable to modern ones: i.e. it

---

[8] In the modern version of this device, the tension is controlled by hanging a weight from one end of the cord.

[9] To the ancient Greeks frequencies in the ratio of 1:2 were known as the DIAPASON (and to us as the octave: e.g. *Some-where*, over the rainbow); frequencies in the ratio 3:2 were known as the DIAPENTE (and to us as the perfect fifth: e.g. *Baa, Baa, Black* Sheep); those in the ratio 4:3 were known as the DIATESSERON (and to us as the perfect fourth: the *mins-*trell boy).

is possible to play a recognisable tune with these flutes—you can listen to examples online.

Pythagoras' discovery not only appeared to offer the possibility of a rational approach to music, it also served to reinforce a faith in number that all Pythagoreans believed is the key to understanding nature. The central tenet of their philosophy was the assumption that the only reliable knowledge is mathematical in nature: the key to understanding the world lies in numbers and the relationship between them. And not just *any* numbers, they must be whole numbers, and small ones at that. They saw particular significance in ratios based on the first four integers: 1, 2, 3 & 4, which came to be known as the *sonorous* numbers.

In fact, the Pythagoreans were quite right to place number at the heart of our knowledge and understanding of the physical world, but wrong to put their faith in number alone.[10] The physical sciences—i.e. physics and chemistry—made little progress until the 17th Century when natural philosophers began to combine mathematics and quantitive experiments. The single-minded, not to say simple-minded, Pythagorean faith in number to the exclusion of experiment (Pythagoras and his monochord notwithstanding) led to extravagant yet beguiling numerological fantasies such as the Music of the Spheres, a doctrine that held that the arrangement of planetary orbits is determined by the sonorous numbers and which underpinned astronomical thinking for some two millennia until its most fervent advocate, Johannes Kepler (1571–1630), discovered in 1605 that planets orbit the sun in elliptical orbits.[11]

Pythagoreans held that all the planets, which in the ancient world included the sun and moon, orbit a central fire that is hidden from view from the Earth by the sun. Moreover, they claimed that motion of each of these heavenly bodies creates a steady note, the pitch of which depends on a planet's distance from the central fire, a distance that is strictly determined by the ratios between the sonorous numbers. Conveniently, this celestial melody was said to be inaudible to mortals, with the exception of Pythagoras himself. Aristotle, arguably one the most influential thinkers of all time, and certainly among the greatest natural philosophers, dismissed the idea as a poetic fiction on the grounds that the sound produced by the motion of large bodies would be so great that no one would fail to hear it.[12]

---

[10] "Actually," as the Pythagorean philosopher Philolaus of Croton put it, "…everything that can be known has a Number; for it is impossible to grasp anything with the mind or to recognise it without this [i.e. number]."

[11] Kepler announced this discovery in his *Astronomia Nova*, published 1609.

[12] Aristotle, (350 BC) On the Heavens, book 2, part 9. http://classics.mit.edu/Aristotle/heavens.2.ii.html (accessed 15/03/2020).

Kepler was not disheartened when he discovered that the orbits of planets are not perfectly circular and that consequently they do not move with constant speed. He simply recomposed the heavenly melody of the planets based on their changing speeds so that instead of producing a single note, the sound of each planet rises an falls as it circles the sun, producing what one modern scholar has called "a dreary succession of uninteresting scale-passages".[13] The final chapter of *Harmonices Mundi*,[14] the work in which these ideas are presented, includes, almost as an afterthought, an altogether more important insight into the motion of planets. This is the third of Kepler's three laws of planetary motion, which later that century was shown by Isaac Newton to be a direct consequence of the Law of Universal Gravitation.

Pythagorean numerology also exerted its siren allure on Newton when he came to give an account of the number of colours in the spectrum. His assertion that sunlight is composed of seven distinct colours was based on his conviction that Pythagorean harmonies apply equally to both sound and light. It led him to claim that the spectrum of sunlight is necessarily composed of seven fundamental colours that correspond to the seven intervals of the traditional divisions of the musical octave. And to make the facts fit the theory he inserted a fictitious extra colour between blue and violet and named it indigo, after the dark blue dye from India.[15]

Newton's theory of light and colour confused and misled natural philosophers for more than a century and a half, long after Thomas Young, a polymath English physician, had proposed in 1801 that Newton's spectral colours could be adequately explained by assuming that colours are sensations produced by no more than three types of receptor in the retina, each sensitive to a different region of the spectrum. Colour is thus entirely in the eye of the beholder, i.e. a sensation, something that even Newton accepted.[16] Exasperatingly, Newton's seven-colour spectrum is alive and well, routinely taught as fact in school science lessons and given the nod in many contemporary textbooks on light.

---

[13] Drake, S. (1970) Renaissance Music and Experimental Science. Journal of the History of Ideas, Vol 31, Oct-Dec, p 483–500, see p 500.

[14] Kepler, J. (1619) Harmonices Mundi. Linz: Johann Planck for Gottfried Tampach.

[15] Should you look through a spectrometer at sunlight you will not be able to distinguish Newton's indigo as a distinct hue in the blue/violet end of the spectrum.

[16] Newton, I. (1979) Opticks, Or A Treatise Of The Reflections, Inflections & Colours Of Light, Bk 1, Pt 1. Dover Publications, p 124–125: "For the rays to speak properly are not coloured. In them there is nothing else than a certain Power and Disposition to stir up a Sensation of this or that Colour."

Newton's Pythagorean bias was not limited to the spectrum. Even though he famously refused to be drawn on the subject of the nature of gravity, declaring "Hypothesis non fingo",[17] privately Newton believed that his theory of gravitation was known to Pythagoras and that the claim that there is a simple direct whole number relationship between the tension of a string and the note it emits when plucked was a deliberate mistake. The truth, as Newton saw it, is that the "mistake" is in fact a riddle that conceals an important insight into the mathematics of the theory of gravity that only the best minds can decipher.[18]

Although the musical discoveries attributed to Pythagoras could not have been made without someone or other performing experiments, experiment was never the ancient Greeks' strong suit, at least as far as the physical sciences are concerned. Indeed, the possibility that Pythagoras performed experiments that involved precise measurement raises more questions than it answers because there was no tradition of precise, quantitative experimentation in ancient Greece, or indeed anywhere else on Earth at the time. Someone, very possibly a Pythagorean acolyte, did discover a numerical relationship between the length cord and its pitch. But where did this individual get the idea that knowledge about nature can be discovered through measurement alone? And if he or she—the Pythagorean Brotherhood was open to women—did perform qualitative experiments, why were most of the supposed Pythagorean discoveries about the physical factors that affect the pitch of a sound false?

The fact is that, with few exceptions, the observational base of Greek physical science was invariably casual and opportunistic.[19] Greek natural philosophers never developed a shared tradition of deliberate experimentation, let alone quantitative experimentation. In any case, most of them preferred to speculate rather than, as they saw it, demean themselves by getting their hands dirty. As far as we know, no one in the ancient world ever repeated the Pythagorean experiments with hammers or with strings to discover if he was right. Pythagoras himself set aside the experiments that revealed a numerical relationship between string length and pitch in favour

---

[17] See the penultimate paragraph of Book 3 of Principia Mathematica: "But hitherto I have not been able to discover the cause of those properties of gravity from phenomena, and I frame no hypothesis; for whatever is not deduced from the phenomena is to be called an hypothesis; and hypotheses, whether metaphysical or physical, whether of occult qualities or mechanical, have no place in experimental philosophy."

[18] Gouk, P. (1988) The Harmonic Roots of Newtonian Science. In: Fauvel, J, Flood, R., Shortland, M., Wilson, R. (eds) (1988) Let Newton Be: A New Perspective On His Life. OUP, p 101–25.

[19] There are very few unequivocal examples of deliberate quantitative experiments or investigations in the entire 700 year history of Greek science. Apart from Pythagoras the few other examples of deliberate experimentation include Euclid (law of reflection), Heron of Alexandria (steam turbine) Archimedes (specific gravity & law of the lever) and Ptolemy (law of refraction).

of conjecture, always the preferred approach of the majority of the natural philosophers of ancient Greek when it came to giving an account of nature.

Anaxagoras, another of the early *physiologoi* or natural philosophers, spoke for all his kind when he claimed that "through the weakness of the sense-perceptions, we cannot judge truth."[20] In other words, distrust what you see or hear and don't rely on experiments as a source of knowledge. Plato, arguably the most profound and influential thinker of the ancient world, was in no doubt that the senses and sensory experiences were thoroughly unreliable as a source of knowledge about the world, and ridiculed those who sought an empirical basis for musical harmony as "excellent fellows who torment their strings, torturing them, and stretching them on pegs."[21] For Plato, as for Pythagoras mathematics alone was the key to understanding the world.

The Nature of Sound According to the Ancient Greeks

As far as we can tell, Pythagoras had no interest in the physical nature of sound; at any rate, no views on the subject are attributed to him. But over the succeeding centuries, Greek natural philosophers assembled piecemeal an account of the nature and properties of sound that was probably as comprehensive as it is possible to be without recourse to the sort of systematic investigations based on quantitative experiments that didn't became a central feature of natural philosophy until the turn of the 17th century. Indeed, by late antiquity (c. 300 AD) the accumulated literature on the subject of sound provided an account of the phenomenon that is often surprisingly modern in approach and which would probably satisfy the casual inquirer of today as long as they didn't press for too much detail.

Nor was it bettered until the 17th century, not least because Greek science and philosophy formed the basis of European thought until the last decades of the 16th century. However, it would be a mistake to imagine that the Greeks spoke with a single voice on this, or indeed on any other, natural phenomenon. There was no coherent body of knowledge or accepted technical vocabulary that all Greek natural philosophers were agreed on, as scientists are today. Successive schools of natural philosophy either ignored

---

[20] Anaxagoras Frag 21. In: Freeman, K. (1948) Ancilla to the Pre-Socratic Philosophers. Basil Blackwell.

[21] Plato, Republic VII, line xxx. His target was Aristoxenus (fl.335BC).

the ideas of their predecessors, or replaced them with ideas of their own. Very rarely did they build on what had gone before.

The prejudice against systematic experiments, Pythagoras or his doppel-gänger notwithstanding, meant that the natural philosophers of the ancient world relied almost exclusively on chance observations of commonplace acoustic phenomena, which meant that their knowledge of sound was limited to easily noticeable features such as that sound involves motion of some sort and the very obvious difference between the speed of sound in air and that of light. Not the broadest of bases on which to erect a comprehensive account of any phenomenon, particularly if one is unwilling to investigate it with experiments specifically devised to discover more about it.

We now know that sound always involves vibration. Yet on the face of it there is a world of difference between the sound of, say, a hum and that of a clap. The one obviously involves vibrations because you can feel your throat and jaw quivering as you hum; the other is equally obviously a colli-sion between the palms of your hand. But there is no palpable sign of any vibration when you clap; all you feel is a slight sting as your palms collide.[22] And to detect the resulting vibrations as they travel through the air requires instruments and techniques that weren't developed until the 19th century. The ancient Greeks, of course, had to rely entirely on the evidence of their eyes and ears, even though they did not entirely trust their senses as a source of knowledge. Yet, despite this self-imposed handicap, they got surprisingly close to the truth in many cases.

The earliest surviving account that deals with the nature of sound is that of the mathematician Archytas of Tarentum, one of the last of Pythagoras' disciples. He claimed that "there cannot be sound without the striking of bodies against one another".[23] Unless one is prepared to give Archytas the benefit of the doubt in the light of what we now know about sound, this rather vague statement would appear to exclude sounds created by plucking strings and blowing into flutes, the very sounds that are supposed to have been used by his master to discover the mathematical basis for harmony. But Archytas' assertion became the starting point for all subsequent accounts of the nature of sound in the ancient world.

Aristotle, who was not usually favourably disposed to Pythagorean ideas, agreed with Archytas but went further: "What is required for the production of sound is an impact of two solids against one another and against the air.

---

[22] A clap produces a weak shock wave that is very quickly transformed within the surrounding air into a sound wave. See Chapter 6 for more on shock waves.
[23] Archytas, Fragment 1. In: Cohen, M.R., Drabkin I.E. (eds) (1975) Source Book in Greek Science. Harvard University Press, p 286.

The latter condition is satisfied when the air impinged upon does not retreat before the blow, i.e., is not dissipated by it. That is why it must be struck with a sudden sharp blow if it is to sound—the movement of the whip must outrun the dispersion of the air."[24] In other words, sounds are created only when air is compressed rapidly, something that was not fully understood until the middle of the 19th century. Note, however, that there is no mention of vibration: sounds are due to collisions rather than the vibrations of solid objects.

It appears that Aristotle may also have given some thought to how sounds travel through air, though the author of work in which this is explained, De Audibilibus, is probably one of his followers, Strato of Lampsacus. But whoever is the author of that work, he has this to say on the subject:

> For when the breath ... strikes the air with successive blows, the air is at once driven forcibly on, thrusting forward in like manner the adjoining air, so that the sound travels unaltered in quality as far as the disturbance of the air manages to reach. For, though the disturbance originates at a particular point, yet its force is dispersed over an extending area, like breezes which blow from rivers or from the land.[25]

In other words, sound is a succession of pulses that travel through the air, growing weaker as they do so. But a pulse is not a wave, it is a solitary disturbance, such as the puff of air produced by the toy known as a vortex cannon.[26] A wave, on the other hand, propagates through a medium as a sequence of linked vibrations: movement in one part of the medium causes movement in the neighbouring part and so on. So we must conclude that for Aristotelians sound did not involve wave motion in the sense that it is understood today.

But one of Strato's contemporaries, the mathematician Euclid of Alexandria, did make a connection between the rate at which an object vibrates and the pitch of the sound it produces. He wrote that "Some sounds are higher pitched, being composed of more frequent and more numerous motions, ... Sounds lower pitched than what is required reach the required pitch ... by an increase in the amount of motion."[27]

---

[24] Aristotle, 350BC, De Anima, Book 2 part 8: http://classics.mit.edu/Aristotle/soul.2.ii.html (accessed 20 March, 2020).

[25] De Audibilibus (1913) (trans Loveday, T., Forster, E.S.). Oxford, Clarendon Press, p 3.

[26] Instructions on how to make this entertaining device, variously known as an "air bazzoka" or "vortex cannon" are widely available online.

[27] Euclid, Sectio Canonis. In: Cohen, M.R., Drabkin I.E. (eds) (1975) Source Book in Greek Science. Harvard University Press, p 291.

The propagation of sound through air was sometimes compared to ripples on the surface of water. According to the Stoic philosopher Chrysippus "Hearing occurs when the air between that which sounds and that which hears is struck, thus undulating spherically and falling upon the ears, as the water in a reservoir undulates in circles from a stone thrown into it."[28] This analogy also has the merit of explaining why sounds grow weaker with distance: all one has to do is notice that ripples die down as they spread out across the surface of water.

A couple of centuries later, Vitruvius also compared sound to ripples spreading across the surface of water.

"Voice is a flowing breath of air, perceptible to the hearing by contact. It moves in an endless number of circular rounds like the innumerably increasing circular waves which appear when a stone is thrown *into* smooth water, and which keep on spreading indefinitely from the centre…In the same manner the voice executes its movements in concentric circles; but while in the case of water the circles move horizontally …the voice not only proceeds horizontally, but also ascends vertically by regular stages."[29]

These passages are sometimes cited as evidence that Crysippus and Vitruvius, and by extension, their contemporaries, understood that sound propagates through air as a vibration. But this is to give them the benefit of the doubt for although they used ripples as an analogy to suggest that sound spreads out from its source in all directions, there is no evidence that they were aware of the mechanism by which ripples propagate. There is no indication they realised that the propagation of waves across the surface of water is due to the up and down motion of the water or that, unlike sounds, the speed at which ripples propagate depends on their wavelength. And if they didn't know that, how could they have realised that sound propagates though air as a vibration when, indeed, those vibrations aren't directly visible? In fact, as we shall see below, the first satisfactory mathematical account of how a mechanical pulse can propagate through a medium was formulated in 1677 by the Dutch natural philosopher and mathematician, Christiaan Huygens, and published a decade later in his Traité de la Lumière.[30] And a few years later, Isaac Newton derived a formula for the speed of sound based on the

---

[28] This passage is attributed to Crysippus by his biographer, Diogenes Laertius (fl. first half of third century A.D.).

[29] Vitruvius (1914) The Ten Books On Architecture (trans Morgan, M.H.). Harvard University Press. Book 5, p 139.

[30] Huygens, C. (1690) Traité de la Lumière: Huygens, C. (1912) Treatise on Light (trans: Thompson S. P.). Macmillan and Co. Ltd, London.

idea that sound is propagated through air by the vibration of the particles of which air is composed. It was only with Newton's mathematical underpinning that it became possible to state with confidence that sound propagates through a medium as a succession of linked vibrations, a form of motion more commonly known as a wave.

The Dark Ages and Beyond

Greek science and philosophy ran out of steam long before the Latin half of the Roman Empire began its final disintegration during the fifth century A.D. due to a combination of economic collapse, successive barbarian onslaughts and hostility of the early Christian church to what it considered to be pagan philosophy. By then Rome had been divided into a Western and an Eastern Empire in 395AD by Theodosius, the last Roman emperor to rule over both halves. The separation had disastrous consequences for the intellectual life of Western Europe because the centres of learning were in the cities of the Eastern empire such as Athens and Alexandria and because the language of scholarship in the Roman world was predominantly Greek, a language that was not spoken in the West. The decline and eventual collapse of the Western Roman Empire isolated Europe, especially North Western Europe, from the bulk of Greek learning, including all of Aristotle's scientific writings, because few of its works had been translated into Latin. As a result, European scholars remained ignorant of almost all Greek science, mathematics and philosophy from the 7th to the 11th centuries, the so-called Dark Ages. During this period, the Church held a monopoly of learning and literacy, and what scholarship there was lost its independence and became a handmaiden of theology.

All was not quite lost, however, because one branch of Greek natural philosophy to which the scholars of Europe's Dark Ages had access was acoustics. Boethius, a Roman civil servant and a scholar, had set out to translate all the works of Plato and Aristotle into Latin. He succeeded only in translating Aristotle's *Logic* before he fell foul of Theodoric, the first ruler of the short-lived kingdom established in Italy by the Ostrogoths, and was executed for treason in 525AD. Boethius also wrote a treatise on music entitled *De institutione musica* in which he did more than merely preserve the ideas of Greek natural philosophy on the subject for it is in this work that the story of Pythagoras and his hammers is told.

Boethius organised music into three categories, an arrangement that had a huge influence on medieval scholarship and education. These were: *Musica mundana*—the Music of the Spheres, *Musica humana*—the harmony of human body and spiritual harmony, and *Musica instrumentalis*—instrumental music, including human voice. Of these, the noblest and highest form is *Musica Mundana* while *Musica instrumentalis* ranked least. And right up to the 17th century, the study of music was a largely intellectual exercise, focussing on the mathematics of time and the harmonious ratios and proportions that were assumed to govern both the heavens and the Earth. Music itself merely mirrored nature's mathematical harmonies and making music, *Musica instrumentalis*, was considered to be inconsequential in comparison.

The Eastern Roman Empire also came under attack. Following the death of the Prophet Muhammad in 632 AD, Arab armies overran and conquered most of the lands of the Eastern Roman Empire where Greek learning and culture had continued to flourish—though it never matched the originality and vigour of earlier times. When the dust of battle finally settled in the following century, the Arabs found themselves heirs to Greek science, mathematics and philosophy. They commissioned translations of Greek texts from their Christian subjects, which enabled Arabian and Persian scholars to study, expand and build upon the achievements of earlier philosophers, astronomers, mathematicians and physicians.

After the fall of the Western Roman Empire in the 5th century AD, however, there was little progress in acoustics, either in the west or in the east. Despite the many innovations during the centuries when science and mathematics flourished in Arabia, Egypt and Persia following the conquests by Islamic armies, no natural philosopher in those lands added very much to what was already known to the Greeks where the nature of sound and its propagation is concerned. As for scholars in Western Europe, until the Renaissance in the 14th century their focus remained largely theological, rather than secular in the spirit of the natural philosophers of ancient Greece.

The Scientific Revolution

Pythagorean ideas on musical harmony held sway for almost two thousand years until they came under sustained attack from several quarters during the latter half of the 16th century. By then, the insistence that the intervals used in musical composition must be based on the so-called four sonorous numbers had long been at odds with developments in musical practice.

In an attempt to bring musical theory in line with musical practice, Gioseffo Zarlino, an Italian composer and musical scholar, published a scheme in 1588, which he called the *senario*, based on six sonorous numbers, which allowed him to employ musical intervals that were forbidden by the Pythagorean system of four sonorous numbers. Zarlino's contemporaries, however, were already using intervals that lay outside even his *senario*, let alone the Pythagorean scheme.

Zarlino's most vociferous critic was one of his former pupils, Vincenzo Galileo. Vincenzo had been a committed Pythagorean until he learned that even in ancient Greece not everyone agreed that numbers are the sole foundation of musical harmony. Aristoxenus of Tarentum, a follower of Aristotle, took a diametrically opposed view to the Pythagoreans and maintained that musical harmony is a matter for the senses, not for mathematics.[31] His writings were rediscovered during the Renaissance and were an inspiration to those who saw the need to reform the basis of musical scales and they persuaded Vincenzo to repeat one of Pythagoras' experiments, something that no one had ever considered doing before.[32]

The experiment in question was the one where Pythagoras was supposed to have established that the ratio of the tensions in two otherwise identical cords whose sounds when plucked are separated by an octave is 2:1. But when Vincenzo performed the experiment he discovered that in order to increase the pitch of a stretched string by an octave, the tension has to be increased fourfold (i.e. it is 4:1). Moreover, the fifth requires tensions to be in the ratio of 9:4 (not 3:2), and for the fourth it is 16:9 (not 4:3).[33]

Vincenzo concluded that his discovery undermined the entire basis of any musical scale based on sonorous numbers, however many such numbers might be used. Musical harmony, he concluded, is a matter of sensory judgement, not abstract mathematics, and consequently that Zarlino's *senario* was merely papering over cracks in a discredited system.

In the meantime, another scholar, the Italian mathematician Giovanni Battista Benedetti, had suggested that the real basis of musical harmony lies in rates of vibration, not in abstract numerical ratios. So where Pythagoreans claimed that notes separated by a fifth sound harmonious because they are in the ratio of 3:2 on the basis that 2 and 3 are sonorous numbers, Benedetti claimed that they are harmonious because the vibrations responsible for these notes are in step with one another every six vibrations of the higher pitch. Vincenzo agreed with Benedetti. Surprisingly, neither man considered performing experiments to verify this claim. Not that they would have

---

[31] Hunt, F.V. (1978) Origins in Acoustics, The Science of Sound from Antiquity to the Age of Newton. Yale University Press, p 28.

[32] Drake, S. (1970) Renaissance Music and Experimental Science. Journal of the History of Ideas, 32, 4 (Oct-Dec), p 483–500, p 492.

[33] In other words, the pitch of a string is not directly proportional to its tension, rather it is proportional to the *square* of its tension.

succeeded because the vibrations of a string that produces an audible sound are too rapid to be counted by eye.[34]

It took an ingenious French friar, Marin Mersenne, to discover how to measure the frequencies of audible sounds. Mersenne is all too often consigned to the odd paragraph or footnote in histories of science, despite being one of the most enterprising and one of the most affable and intellectually generous figures of the early years of the scientific revolution. Although he was not a innovative theoretician in the mould of his better known contemporaries, René Descartes and Galileo Galilei, who between them were largely responsible for the mechanical philosophy that displaced the Aristotelian world view during the 17th century, Mersenne was one of the most inventive and skilled experimentalists of his day, easily the equal of any of the savants of that era, including Galileo. He also kept in close touch with most of the leading natural philosophers and mathematicians of the time, his address book a who's who of European natural philosophy and mathematics during the first half of the 17th century. He organised regular meetings of savants in his rooms in Paris—meetings that eventually led to the formation of the state-funded *Académie Royale des Science* in 1666 under the somewhat disengaged patronage of Louis XIV—and acted as a clearing-house for the latest scientific ideas and discoveries by circulating copies of letters he received from his numerous correspondents. How he found the time and energy for these extracurricular activities is astonishing because the religious order to which he belonged, the Minims, was one of the most ascetic in France. In addition to the daily round of ritual observances, Minims undertook a perpetual Lenten fast i.e. they avoided eating meat and dairy products and went barefoot.

Mersenne was one of the first to recognise the importance of Galileo's scientific work and wrote to him offering to help with the publication of "the new system of the motion of the earth which you have perfected, but which you cannot publish because of the prohibition of the Inquisition."[35] Galileo never answered this or, indeed, any of Mersenne's letters.[36] Undeterred, Mersenne used his contacts to make Galileo's ideas known throughout Europe. One of these was his close friend, René Descartes, the reclusive French mathematician and natural philosopher who had decamped to the Dutch Republic in 1629 in search of a quiet life that he hoped would allow

---

[34] Drake, S. (1970) Renaissance Music and Experimental Science. Journal of the History of Ideas, 32, 4 (Oct-Dec), p 483–500, p 492–93.

[35] The work was Galileo's "Dialogue Concerning the Two Chief World Systems", which was eventually published in 1632. For details, see entry for Marin Mersenne in Gillispie C.G. (ed) (1970–1980) The Dictionary of Scientific Biography. Charles Scribner's Sons.

[36] Crombie, A. C. (1974) Marin Mersenne. In: Dictionary Of Scientific Biography Vol, IX, ed C. Gillispie, Amer. Council of Learned Societies, Charles Scribner's Sons, New York, NY, p 316–22.

him to develop his ideas and write his books free of state or church interference.[37] Throughout the twenty years he spent there, he and Mersenne maintained a steady correspondence on mathematics and natural philosophy. In fact, their correspondence was usually Descartes' only link with the wider world, not least because he frequently changed address to protect his privacy. On learning of Galileo's run-in with the Inquisition over the publication in 1632 of his *Dialogue Concerning Two Chief World Systems*, in which the Copernican theory was clearly favoured over Aristotle's geocentric system, Descartes took fright and shelved plans to publish his own version of a world system based on the idea that matter is inert and in which he promoted the Copernican system even more overtly than Galileo had done.[38]

Bearing in mind that Mersenne was a devout Catholic, and an ordained priest to boot, he seems an unlikely recruit to the ranks of mechanical philosophers. But he had become convinced that the mechanical philosophy could be used by the Catholic Church to combat the occult sciences such as magic, astrology and alchemy that had flourished in Europe since the early Renaissance and which blurred the line between the natural and supernatural world.[39] Mersenne believed that the inert, lifeless world that underpinned mechanical philosophy allowed for the possibility of miracles, i.e. events that are of out of the ordinary and which can only be explained as the handiwork of God. The practitioners of the occult sciences claimed that they could achieve similar ends whenever they wished to do so because the world is itself a living organism and that they alone understood its deepest secrets and how to manipulate them.

One of Mersenne's interests was music, one that he shared with Descartes and a subject about which they corresponded extensively.[40] But rather than theorise, Mersenne set out to put things on an experimental basis. Like Benedetti and the Galileos (father *and* son), he rejected the Pythagorean doctrine of sonorous numbers and instead sought a physical basis for musical harmony.[41]

---

[37] When Mersenne visited Descartes in Holland he had to remove his habit at the border before being allowed to enter the country.

[38] It was eventually published in Paris in 1664 —fourteen years after his death—under the title *Le Monde de M. Descartes ou le Traité de la Lumierè*.

[39] Grayling, A.C. (2016) The Age Of Genius: The Seventeenth Century & The Birth Of The Modern Mind. Bloomsbury Publishing, p 119.

[40] Shea, W.R. (1991) The Magic Of Numbers And Motion: The Scientific Career Of René Descartes. Science History Publications, Chap 4.

[41] The exact dates of Mersenne's experiments are not known, but they probably took place after 1627.

To overcome the difficulty of counting the vibrations of a cord when those vibrations are too rapid to be seen clearly, Mersenne stretched a long hemp rope between two supports some 30 m apart. The vibrations of a rope of this length are slow enough that they can be followed by eye. Using his pulse to determine the rate at which it vibrated—clocks capable of measuring seconds were not available at the time[42]—he found that each vibration of the rope took place in half a pulse. In those days the interval between pulse beats was assumed to be exactly one second. It's not, of course, so measuring time using one's pulse is a hit and miss affair, which is why Mersenne soon turned to a simple pendulum as a time-keeper. This device really is simple: a small, solid sphere made of brass, iron or lead attached to one end of a long string suspended from a fixed point and allowed to swing freely.

Although Galileo was the first person to realise that a pendulum can be employed to measure small intervals of time, it was Mersenne who pioneered its use in his experiments with vibrating strings. What makes a pendulum suitable as a time-keeper is that, as long as the amplitude of the swing is small, the time it takes to complete a swing, known as its period, depends solely on its length.[43] And although it was Galileo who discovered this fact, it was Mersenne who first attempted to determine the exact length of a pendulum that takes one second to make a single swing, with, it must be said, limited success.[44]

The problem he faced is that the only reliable independent standard for the second available to him was the length of the day, which by definition is 86,400 s long. Mersenne had first to estimate the length of his seconds pendulum and then count the number of swings it makes, say, between the passage of two stars that are a known angle apart as they cross the meridian.[45] This would require counting several thousand swings. If the pendulum was either too short or too long, its length would have to be adjusted and the procedure repeated the following night.[46] Mersenne eventually settled on a pendulum "3 pieds & demy" long, which is approximately 114 cm in

---

[42] The first clock capable of measuring seconds reliably was designed by Christiaan Huygens in 1657 and built by the renowned Dutch clockmaker, Salomon Coster. The best of these clocks was lost only 10 s a day.

[43] The mathematical theory of the simple pendulum assumes that the string is weightless sand that all the mass of the device is concentrated at the centre of the suspended weight.

[44] These days we measure the time it takes a pendulum to return to its starting point, so on that basis Mersenne set out to make a pendulum with a *period* of 2 s.

[45] Due to the Earth's rotation about its axis, which takes 24 h, stars appear to move across the night sky at a rate of 1° every 4 min, so their movement can be used as a clock.

[46] Koyre, A. (1953) An Experiment in Measurement. Proceedings of the American Philosophical Society, Vol. 97, No. 2, p 222–237, see p 234.

modern measure. The period of such a pendulum is 2.14 s, which, neverthe-less, allowed him to time his experiments a great deal more accurately than he could using his pulse.[47]

Indeed, it was Mersenne rather than Galileo who recommended that astronomers should use a pendulum to measure the duration of eclipses of the Sun and Moon and that it could used by musicians to "make known to everyone the time to be given to each measure in singing all kinds of music."[48] A pendulum would also be of use to physicians to measure pulse rates and determine "how the passions and other fevers hasten or retard it."[49] Moreover, as he pointed out, not only was a pendulum more accurate than the mechanical clocks of the time, it was also much cheaper to make: "a chorde can serve all the uses that one makes of ordinary clocks, which it surpasses in accuracy. Besides which, as experience teaches, for two half-sous one can make three or four clocks which will mark second minutes, by attaching a chorde three and a half feet long to a peg, for if one attaches to the other end something heavy which hangs freely toward the center of the Earth, each of its returns will last exactly one second minute."[50]

Mersenne repeated his rope experiment with wires of brass and iron, each 30 to 40 m long. Using wires of different diameters he was able to formulate a mathematical relationship between the length, tension, the mass per unit length of a wire and its rate of vibration (i.e. its natural frequency), whatever material it is made from. The resulting relationship is known as Mersenne's law.[51]

Knowing the factors that determine the frequency of a vibrating wire, he was able to measure the frequency of the notes of musical instruments. He tuned a short bronze wire so that it was in unison with a particular organ pipe, and using the relationship for vibrating wires he had discovered, he could calculate the frequency of vibration of the wire and hence that of the

---

[47] In Mersenne's day, the period of a pendulum was taken to be a single swing whereas these days it's a double swing.

[48] Mersenne, M. (1636) Harmonie Universelle. Sebastien Cramoisy, Paris. Livre Premier des Instrumens, prop, XIX, p 45.

[49] Mersenne, M. (1636) Harmonie Universelle. Sebastien Cramoisy, Paris. Livre Second de mouvemens de toutes sortes des corps, prop XV, p 136.

[50] Mersenne, M. (1636) Harmonie Universelle. Sebastien Cramoisy, Paris. Livre Premier des Instrumens, prop, XIX, p 46. Mersenne's "second minute" is, in fact, a second. Taking a French foot (*pied du roi*) to be 32.5 cm, the period of such a pendulum would be 2.12 s. Half a complete swing of this pendulum, which is what Mersenne used as his standard of time, would be 1.06 s.

[51] Mersenne, M (1636) Harmonie Universelle. Sebastien Cramoisy, Paris. Livre Troisiesme des Instrumens a Chordes, prop, VII, p 123–126. Mersenne's equation is: $f = 1/L\sqrt{T/\mu}$ where L is the length of the string, T its tension and μ the mass per unit length of the material.

organ pipe.[52] This was the very first time anyone had measured the frequency of a musical note. Using this method he was able to confirm that the pitch of a musical note when estimated by a trained ear depends only on its frequency, whatever the musical instrument.[53]

In 1681, Robert Hooke, Curator of Experiments to the Royal Society, devised an alternative method of measuring the frequency of sound. Holding a thin metal sheet against the teeth of a rotating toothed wheel he could produce sounds of any frequency. The frequency of the sound could be calculated by multiplying the number of teeth by the rate of rotation.[54]

Another device for determining frequency was the tuning fork, invented in 1711 by John Shore, a trumpeter in the Queen's band. It was used by musicians to tune their instruments and was later employed by both Ernst Chladni, a natural philosopher and amateur musician from Wittenberg, Germany, and Herman von Helmholtz in their experiments on sound, as we shall see later in this chapter.

The Sound of Music

Pure tones, i.e. sounds that consist of a single frequency, are extremely rare. Everyday sounds invariably consist of a combination of unrelated frequencies of unequal loudness that blend together to create a distinctive sound, the overall quality of which is known as *timbre*. Timbre depends on the frequencies that are present in a sound and how loud they are relative to one another. It is timbre that enables us to distinguish one sound from another. But in most circumstances it is usually very difficult to pick out individual frequencies present in a sound, especially for the untrained ear.

Sounds produced by musical instruments are a notable exception because the frequencies present in their sounds are always multiples of the simplest mode of vibration of the source, known as the fundamental frequency. This is the frequency that is heard when we listen to a musical instrument. But different musical instruments don't sound the same even when they play the same note because the fundamental frequency is always accompanied by several higher harmonically related frequencies known as overtones.[55] Not

---

[52] Mersenne gives the frequency of the organ pipe as 84 Hz, but this value should be taken with a pinch of salt because precise measurement was not his strong suit.

[53] Mersenne, M. (1636) Harmonicorum libri, Liber Secundus De Causis Sonorum, proposition xxxiii.

[54] Gouk, P. (1982) Acoustics in the Early Royal Society 1660–1680. Notes and Records of the Royal Society of London, Vol. 36, No. 2, p 155–175, p 169.

[55] The fundamental plus its overtones are known collectively as harmonics. i.e. the fundamental is the first harmonic, the first overtone is the second harmonic, and so on.

only does each instrument produce a unique combination of overtones, the relative loudness of individual overtones also differs from one instrument to another. Hence each type of instrument has a unique and unmistakable timbre.

Mersenne was probably the first natural philosopher to draw attention to the presence of overtones in musical sounds, although the phenomenon had been noticed, if rather vaguely, centuries earlier by Aristotle. Mersenne wrote that he had noticed that "the string struck freely and sounded freely makes at least five sounds at the same time, the first of which is the natural sound of the string and serves as the foundation for the rest..."[56]

What Mersenne called the natural sound of the string was what we now call its fundamental frequency. The other sounds he heard were higher frequency overtones. But, as Mersenne learned later, craftsmen who made musical organs were well aware of this phenomenon and had been tweaking the design of organ pipes to alter their timbre and create a fuller sound centuries before he began his acoustic investigations, as, indeed, had bell founders. But Mersenne was unable to provide a convincing physical explanation for the phenomenon, saying that the presence of overtones was "impossible to imagine" and "against experience". The problem is that the eye cannot follow the movement of a string that produces an audible sound and so is unable to see how it vibrates (Fig. 2.2).[57]

Mersenne's conjecture was confirmed when indirect physical evidence for overtones in a vibrating string was discovered later that century by a couple of Oxford savants.[58] By balancing several tiny bits of folded paper on the sting of a monochord and making it resonate in response to another monochord that was tuned to a frequency an octave higher, they established the presence of stationary points alternating with regions of maximum vibration along the length of the vibrating catgut. The bits of paper remained in place where the catgut was stationary and were thrown off where the catgut vibrated strongly. The stationary points were later named "nodes" and the points of maximum vibration "loops" in 1701 by Joseph Sauveur.[59]

Sauveur, whose impaired hearing seems not to have been an impediment to his researches into sound, was a tutor at the court of Louis XIV and held

---

[56] Mersenne, M. (1636) Harmonie Universelle. Sebastien Cramoisy, Paris. Livre Quatriesme Des Instruments A Chordes, p 210.

[57] Thee days the vibrations of a string can be "frozen" by illuminating the string with a strobe light flashing at the same rate as the vibrations of the string.

[58] The savants were William Noble & Thomas Pigot. See: Gouk, P. M. (1982) Acoustics in the Early Royal Society 1660–1680. Notes and Records of the Royal Society of London, Vol. 36, No. 2, p 155–75, p 167.

[59] Sauveur's "loops" are known in English as "antinodes".

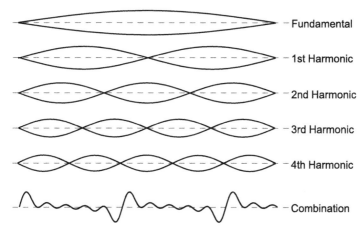

Fig. 2.2 Overtones in a string. These are here shown individually, but they all occur simultaneously in the vibrating string. A snapshot of the combination revealed by freezing the motion of the string with a stroboscope is show at the bottom

the chair of mathematics at the *College Royale*. Like every natural philosopher in the 17th century, he assumed that music held the key to understanding sound. But his research into sound led him to conclude that harmony exists only in the ear of the listener and so music is essentially an aesthetic activity. The idealised Pythagorean notions that harmony is ultimately underpinned by mathematics is untenable and offered no useful insights into sound considered merely as vibration. Sauveur declared that music is merely sound that is agreeable to the listener and maintained that the study of sound encompasses a far broader range of phenomena. He proposed a new study concerned exclusively with vibrating bodies and waves travelling through a medium and called the new science "acoustics", from the Greek word for hearing.[60]

The difficulty of actually hearing individual frequencies present in a musical sound was solved by Herman von Helmholtz. At the time he was professor of anatomy and physiology in Bonn, though later he was appointed professor of physics at the University of Berlin. During his time in Bonn, Helmholtz became interested in the application of physics to music and believed that this would establish the physical basis for the aesthetics of music. To analyse musical sounds he invented a device that enabled him to hear individual frequencies present in a musical note. This was a hollow globe, made of either glass or brass, with openings on opposite sides. Sound enters the sphere through the larger opening, while the smaller opening is pressed against the ear canal. Depending on the volume of the globe, the air within it will vibrate

---

[60] Sauveur, J. (1679) Traité de la theorie de la musique. Paris.

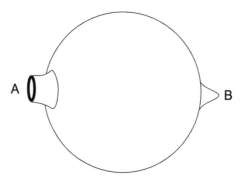

Fig. 2.3 Helmholtz's original resonator. Sound to be sampled enters the sphere though the opening at A and the ear is placed against the nipple at B

in unison—i.e. resonate—with only one of the frequencies present in the musical note. The vibration of the air within the sphere makes that particular frequency noticeably louder than the others present in the note, enabling a musically trained listener to identify its pitch. Using globes of different volumes, Helmholtz showed that a musical note is always composed of a series of harmonically related frequencies, and that the frequencies present in the same note varied from one instrument to another. Hence he was able to provide clear experimental evidence for the cause of timbre (Fig. 2.3).[61]

Resonant Sounds

One way or another, just about every object can be made to vibrate, however briefly. Moreover, an object capable of vibration does so most readily at a frequency known as its natural frequency, which varies from one object to another. There are any number of ways in which this can be achieved, plucking, striking or rubbing being the most common. It is also possible to set up vibrations in an object by shaking it at a rate that matches its natural frequency. The phenomenon is known as resonance. In many situations resonance is a nuisance because it results in unwanted vibrations. Indeed, it can sometimes be more than a nuisance because unless the resonant vibration is constrained, an object can vibrate uncontrollably and even shake itself to bits. A spectacular example of the destructive effect of unconstrained resonant vibration is the shattering of a wine glass by sound alone, a phenomenon that we shall consider in detail in Chap. 6.

---

[61] Helmholtz, H., (1895) On The Sensations Of Tone As A Physiological Basis For Theory Of Music (trans: Ellis A.J.). Longman Green & Co, p 43.

Resonance has its uses, however, because should the natural frequency of an object coincide with that of audible sound, its vibrations may be audible. Indeed, resonance is the *sine qua non* of just about every musical instrument. Without resonance, flutes and trumpets would be all but silent, as would violins and guitars. Resonance also plays a central role in determining the sounds of the human voice, as we shall see in Chap. 6 in connection with how and why inhaling helium alters the timbre one's voice.

You won't be surprised to learn that the first systematic study of resonance was carried out by Mersenne. During his experiments he had noticed that when a tensioned string vibrates it can cause a neighbouring tensioned string that has the same natural frequency to vibrate in unison. Sympathetic vibrations between strings tuned in musical unison was already well know centuries before he became interested in the phenomenon. Some fifteen centuries earlier, Tung Chung-Shu, a Chinese scholar, observed that a particular note emitted by the string of a lute would elicit the same note in a string of a neighbouring lute as long as it too was identically tuned: "the *kung* note or the *shang* note struck upon one lute will be answered by the *kung* note or the *shang* note from the other tuned instruments. They sound by themselves."[62]

And a century before Mersenne had noticed resonance in tensioned strings, that eagle-eye all rounder, Leonardo da Vinci, noted that "The blow given in the bell will cause a slight sound and movement in another bell similar to itself, and the chord of a lute as it sounds produces movement and response in another similar chord of like tone in another lute, and this you will perceive by placing a straw upon the chord similar to that which has sounded."[63]

Galileo, whose interest in resonance predated Mersenne's by several years, provided the explanation in his *Dialogues Concerning Two New Sciences*. Towards the end of the first day of the *Dialogue*, his mouthpiece, Salviati, is asked by Sagredo to explain some aspects of sound such as "…the old problem of two stretched strings in unison; when one of them is sounded, the other begins to vibrate and emit is note."

Salviati answers that "A string which has been struck begins to vibrate and continues the motion as long as one hears the sound; these vibrations cause the immediately surrounding air to vibrate and quiver; then these ripples in the air expand far into space and strike not only all the strings of the same instrument but even those of neighbouring instruments. Since that string

[62] Ronan, C.A. (1981) The Shorter Science & Civilization In China, An abridgement by Colin A. Ronan of Joseph Needhams's original text, Vol 2. CUP, p 364.
[63] The Notebooks Of Leonardo Da Vinci (1938) Arranged, Rendered Into English And Introduced By Edward Maccurdy, Volume 1. Jonathan Cape, p 284.

which is tuned to unison with the one plucked is capable of vibrating with the same frequency, it acquires, at the first impulse, a slight oscillation; after receiving two, three, twenty, or more impulses, delivered at proper intervals, it finally accumulates a vibratory motion equal to that of the plucked string, as is clearly shown by equality of amplitude in their vibrations. This undulation expands through the air and sets into vibration not only strings, but also any other body which happens to have the same period as that of the plucked string."[64]

But resonance had been exploited long before Galileo or Mersenne studied the phenomenon, even though it had not been understood. Musical instruments, of course, have always relied on resonance to create specific notes. Military mining, that is to say digging tunnels under enemy fortifications, has been employed since antiquity, and one of the ways to detect the noise of tunnelling relied on resonance. Vitruvius describes how during a siege of Apollonia in Sicily in 214 BC the defenders were able to discover the location of enemy mines: "But at this time Trypho, the Alexandrine architect, was there. He planned a number of countermines inside the wall, and extending them outside the wall beyond the range of arrows, hung up in all of them brazen vessels. The brazen vessels hanging in one of these mines, which was in front of a mine of the enemy, began to ring from the strokes of their iron tools. So from this it was ascertained where the enemy, pushing their mines, thought to enter."[65]

According to Thomas Young, the technique was still in use in the early 19th century: "…it has sometimes been usual, in military mining, to strew sand on a drum, and to judge, by the form in which it arranges itself, of the quarter from which the tremors produced by the countermining proceed"[66] As we shall see in Chap. 4, during the First World War military mining was widely used, and in the early years of the conflict some of the methods used to detect tunnels as they were dug relied on resonance.

How Fast Does Sound Travel in Air?

During a thunderstorm one can't fail to notice the lag between seeing a flash of lightning and hearing the resulting thunder. It's probably the clearest

---

[64] Galileo Galilei (1914) Dialogues Concerning Two New Sciences, translated from the Italian and Latin by Henry Crew and Alfonso de Salvio. The Macmillan Company, p 99.

[65] Vitruvius (1914) The Ten Books On Architecture (trans: Morgan, M.H.). Harvard University Press. Book 10, p 318.

[66] Young, T. (1845) Lecture XXXII: On the Sources and Effects of Sound. In: Young, T. (1845) Young's Course of Lectures Vol 1. Taylor and Walton, p 300.

evidence that sound travels more slowly than light and was known as such to the natural philosophers of ancient Greece. But given their penchant for speculation over systematic observation, it was inevitable that many of them misinterpreted the evidence of their eyes and ears.

Archytas of Tarentum, the Pythagorean, was the first natural philosopher to consider the issue. He asserted that sounds "that reach us swiftly and violently from the blow seem high pitched, those that reach us slowly and weakly seem to be low pitched."[67] Although this passage is sometimes taken to imply that Archytas realised that the pitch of a sound is due to the frequency of vibration of its source, an alternative reading is that he was claiming that the speed of sound differs according to pitch, something that we now know to be wrong.

At any rate, the latter reading is how his peers interpreted his words. Theophrastus, Aristotle's successor as head of the Lyceum in Athens, pointed out that if Archytas was correct then music heard at a distance would sound very odd because the high and low notes of a piece of music would get out of step "for if it did, it would lay hold on the hearing sooner, so that there would not be concord."[68] Since both the harmony and melody of a piece of music is preserved, however far from the source it is heard, all sounds must travel at the same speed whatever their pitch.

As for the all too noticeable difference between the speed of sound and that of light, it seems that it was Aristotle who first considered this, if in a round about way, when he claimed in his *Meteorologica* that we see lightning before hear thunder because although lightning "comes into existence after the collision and the [resulting] thunder, we see it earlier because sight is quicker than hearing."[69] The unfortunate implication that lightning is caused by thunder, rather than the other way around, persisted for centuries.

Several centuries after Aristotle, a far better account of the phenomenon was offered by the Roman poet and atomist, Titus Lucretius Carus, in his *De Rerum Natura*, a poem that is the most complete account we have of Epicurean materialism and its moral consequences. Epicurus was an atomist who rejected the possibility of a non-material reality and who taught that the aim of life is happiness based on an absence of physical and mental pain rather than the hope of a life after death. Taking his cue from Epicurus, Lucretius

---

[67] Archytas, Fragment 1. In: Cohen, M.R., Drabkin I.E. (eds) (1975) Source Book in Greek Science. Harvard University Press, p 287.

[68] Cohen, M.R., Drabkin I.E. (eds) (1975) Source Book in Greek Science. Harvard University Press, Note 1, p 287.

[69] Aristotle, 350 BC, Meteorologica, Book 2, Sect. 9. http://classics.mit.edu/Aristotle/meteorology.2.ii.html (accessed 25/07/2021).

set out to free his readers from superstition by persuading them that there are no supernatural forces in nature.

In *De Rerum Natura* we are informed that

We see the lightning ere the thunder hear;

For quicker comes the impulse to the eye

Than to the ear; as plain to be perceived

By watch of woodman labouring with axe,

To fell the pride of some wide-spreading tree;

The falling blows are seen before the sound

Comes to the ear; so blinding lightnings come

Before the thunder, though from self-same cause

Springs thunder and its fleet, vaunt courier, light.[70]

A pithier account of the difference between the speed of sound and light is to be found in an encyclopaedia of natural history compiled by Pliny the Elder, the Roman author, naturalist and naval commander who lost his life during the eruption of Mount Vesuvius in 79 AD that destroyed Pompei: "It is certain that the lightning is seen before the thunder is heard, although they both take place at the same time. Nor is this wonderful, since light has a greater velocity than sound."[71]

No natural philosopher or mathematician of the ancient world ever got around to measuring the speed of sound or, indeed, ever considered doing so. Not that they would have succeed because in those days there were no methods or instruments for measuring the small intervals of time that determining the speed of sound accurately by direct measurement entails. In fact, the first attempt to measure the speed of sound did not take place until the early 17th century, using a simple pendulum to measure time. And the person who made that measurement was, of course, Mersenne.[72]

The manner in which the measurement might be made was suggested by Francis Bacon in his *Sylva Sylvarum, or a Natural History in Ten Centuries*, published posthumously in 1627. In this eclectic and influential collection

---

[70] Lucretius (1916) De rerum natura, (trans: Leonard, W.E.). Dent, Book VI lines 164–72.

[71] Pliny the Elder (1855) The Natural History Of Pliny vol 1 (trans: Bostock, J. and Riley, H.T.). Henry G. Bohn, London, chap 55, p 84.

[72] Mersenne's measurements of the speed of sound were carried out between 1628 and 1636. Exact dates are not known.

of observations and conjectures on natural phenomena, Bacon declared that sound is "one of the subtlest pieces of nature." But he complained that "the nature of sound in general hath been superficially observed". Despite the high regard in which Bacon was held by the savants of the 17th century as the advocate of experiment as the only reliable source of knowledge of nature, there is no record that he ever performed any experiments; he was a man of letters, not of science.

This is obvious from his description of how one might go about measuring the speed of sound.

> …let a man stand in a steeple, and have with him a taper; and let some veil be put before the taper; and let another man stand in the field a mile off. Then let him in the steeple strike the bell, and in the same instant withdraw the veil; and so let him in the field tell by his pulse what distance of time there is between the light seen and the sound heard: for it is certain that the delation of light is in an instant. This may be tried in far greater distances, allowing greater lights and sounds.[73]

The potential sources of error in this experiment—the difficulty of striking the bell at the same instant that the light is exposed while feeling one's pulse to measure the elapsed time—would make it impossible to obtain an accurate value for the speed of sound. The altogether more practical Mersenne employed both echoes and gunfire.

The problem, as ever in those days, was measuring small intervals of time because, as we saw earlier, there were no clocks capable of measuring seconds. The method Mersenne employed initially was to utter a short phrase of several syllables and listen for its echo. Unsurprisingly, given that he was a priest, the phrase he chose was "*Benedicam Dominum*"—"Let me bless the Lord"—which he would have been used to enunciating. He claimed that he could utter its seven syllables in one second as measured by the swing of a pendulum. By positioning himself sufficiently far from a wall so that the reflection of the last syllable (…*num*) reached him just as he finished uttering the whole phrase, he would know both the distance the sound had travelled (i.e. twice the distance between himself and the reflecting surface) and the time it takes to do so (the time taken to say "*Benedicam Dominum*", i.e. one second).[74] In modern units he obtained a speed of 319 m/s (10% below its

---

[73] Bacon, F., 1627, Sylva Sylvarum, London. Century III, 209. In: Bacon, F. (1826) The Works of Francis Bacon, Vol 1. London.

[74] Mersenne, M. (1636) Harmonie Universelle. Livre Troisiesme Des Mouvemens, Prop XXI, p 214.

correct value). Not bad considering the difficulty in uttering the phrase in exactly one second and hearing the echo of the final syllable clearly.

Mersenne's use of gunfire involved measuring the time between seeing the smoke and flame when the gun is fired and hearing the sound of the discharge. This should have resulted in a far more accurate value for the speed of sound in air because the distance between the gun and the observer can be far greater than it is with an echo, which should make it possible to measure the time between seeing the discharge of the gun and the arrival of its sound far more accurately (i.e. the percentage error in the measurement of time should be much less than with echoes). Yet these gunfire experiments consistently yielded an average speed of 448 m/s, 30% above its true value. Assuming that Mersenne could measure the distance between himself and the cannon accurately, the source of error in the gunfire experiments must lie in timing, which was done by counting the swings of a pendulum. But even though Mersenne did not realise that the results from his gunfire method were far too high, he knew that in principle they should be more accurate than those involving echoes and so always used the higher value for the speed of sound in his subsequent writings.

Mersenne also took account of the speed of the wind and the prevailing weather when making these measurements, even though he believed that neither has any effect on the speed of sound. In fact, as we shall see in Chap. 4, these do affect the speed of sound, but his measurements were not accurate enough to reveal this. He concluded that "Whoever wishes to measure the velocity of sound under various condition, by night, by day, in valleys, in woods or mountains, either with or against the wind, in fair or rainy weather, in all these circumstances experiment always leads to the same velocity of sound."[75]

In an impractical, not to say surreal twist on the delay between seeing and hearing the woodsman's axe in Lucretius' poem, Mersenne claimed that "…a soldier watching the firing of a gun at 100 six-foot intervals whose fire he has already seen is able to dodge the shot" because the sound of gunfire will reach him one second before the shot giving him time in which to move. How the soldier was supposed to know where the bullet would fall is ignored. In any case, a bullet from a high velocity rifle travels faster than sound, so if you hear the shot, the bullet has either missed you or hasn't killed you outright.

Mersenne also considered how one could estimate distance by timing the interval between seeing an event and hearing the sound it makes. "Suppose then, for example, that the natural well-tempered pulse beats 3 times before

[75] Mersenne, M. On the Velocity of Sound in Air. In: Lindsay, R.B. (1973) Acoustics: Historical and Philosophical Development (Benchmark Papers in Acoustics, V1). Dowden, Hutchinson & Ross, p 64.

one can hear a sound made 500 paces away … and that there are 66 beats of such a pulse in a minute of an hour. I say then that the pulse beats at least 18 times before one hears the sound of a canon, an *arquebuse*, a trumpet, a bell, a *marteau*, a *tonnerre*, or any other instrument at distance of one of our leagues; and consequently that a sound which is strong enough to be heard around the world could only be heard after a time in which the pulse would beat 129,600 times, … but a sound would not last so long nor be strong enough to be heard so far away, unless God wished to produce such a sound; that will perhaps be when the Angels sound the Trumpet on the great day of Judgment to summon those who are about to die."[76] As we shall see in Chap. 4, a mighty natural trumpet worthy of the Day of Judgement, if only for the poor souls who were caught up in the event, sounded in 1883 when Krakatoa, the infamous Indonesian volcano, erupted. But it wasn't the audible sound of the eruption that encircled the Earth (several times, in fact) on that occasion, it was an inaudible infrasonic wave.

Although echoes were occasionally used to determine the speed of sound during the remainder of the 17th century, gunfire became the favoured method of measurement. Gunfire can be heard over a greater distance than any echo, which reduces the percentage error in measuring both the time of travel and the distance between the source of the sound and the observer. It was used in Italy in 1656 by Vincenzo Viviani and Giovanni Borelli, founder members of the world's first scientific think-tank, the short-lived Florentine *Academia del Cimento*, and in France in 1677 by three of the leading astronomers of the *Académie Royale des Sciences*, Giovanni Domenico Cassini, Jean Picard and Olof Römer. The academicians obtained 361 m/s, the astronomers 356 m/s.

During the first decade of the following century, a far more thorough determination of the speed of sound using gunfire was carried out by William Derham, the rector of St Laurence, a small church in Upminster, Essex, and an active member of the Royal Society. On numerous occasions over several years he arranged for guns to be fired at different locations that were in those days within sight from the top of the tower of St Laurence, and used a pendulum clock accurate to half a second to measure the time taken for the sound of the gunfire to reach him. The greatest distance over which he made these measurements was 20 km, using a gun in Blackheath, which lies just south of Greenwich Park. Derham's average value for the speed of sound in air was "1142 feet in a whole second", i.e. 348 m/s which is within 3% of that now accepted for air at 20 °C.

---

[76] Mersenne, M. (1636) Harmonie Universelle. Sebastien Cramoisy, Paris. Livre Premier De La Nature Et Des Proprietez Du Son, prop XXI, p 38.

He also took note of atmospheric conditions when making these measurements and concluded that "...in all weathers, whether fair and clear, or cloudy and lowering; and whether it snow or rain (both which weaken very much the audibility of sound) and whether it thunder or lightening; in hot or cold weather; by day or by night; in summer or in winter; or whether the mercury ascend or descend in the barometer; in short, in all the various states of the atmosphere (excepting the winds) the motion of sound is neither swifter nor slower..."[77] Surprisingly, the one atmospheric condition that Derham did not measure was air temperature. But temperature usually has a much smaller effect on the speed of sound than wind, so perhaps he might not have succeeded in discovering its effect on the speed of sound had he undertaken to do so.[78]

That temperature does affect the speed of sound was discovered some 40 years later by an Italian physician, Giovanni Lodovico Bianconi, who measured the speed of sound on a number of occasions during summer and winter in Bologna.[79] He found that it is slightly greater on warm days than on cold ones. Bianconi's findings were confirmed a few years later by Charles Marie de la Condamine, who compared measurements he made in 1740 when he was in Quito[80] (3000 m above sea level and cool) with those he made later in 1744 in the warmer climate of Cayenne[81] (at sea level and warm).[82]

All these measurements, however, were at odds with the value that Newton had obtained on purely theoretical grounds. In Book II of his *Principia Mathematica*, Newton derived a mathematical formula for the speed of sound by assuming that sound travels through a medium by means of linked vibrations

---

[77] Derham, W.I. (1708) Experiments and Observations on the Motion of Sound, &c. By the Rev. Mr. Derham, Rector of Upminster, and F. R.S. In: Hutton C., Shaw, G. Pearson, R. (eds.) The Philosophical Transactions Of The Royal Society Of London, From Their Commencements, In 1665, To The Year 1800; Abridged With Notes And Biographic Illustrations, Vol V from 1703 to 1712, London 1809, p 380–95, p 384.

[78] Gabrielson, T.B. (2009) Background And Perspective William Derham's De Motu Soni (On The Motion Of Sound). Acoustics Today, Vol 5, Issue 1, p 18.

[79] Zuckerwar, A.J. (ed.) (2002) Handbook of the Speed of Sound in Real Gases, Vol 3. Academic Press, p 266.

[80] Quito is the capital of Ecuador, a country which in 1740 was part of the Spanish viceroyalty of Peru. Condamine was the leading member of the French expedition sent to determine the length of one degree of latitude at the equator, which lies some 25 km north of Quito, in order to determine the exact shape of the Earth. An similar expedition was dispatched to Lapland. When measurements from the two sites were finally compared they confirmed that the Earth is an oblate spheroid as predicted by Newton.

[81] Cayenne is in French Guiana, South America.

[82] Condamine, C. M. de la (1745) Relation Abregee d'un Voyage fait dans l'interieur de l'amerique meridional. Memoires de l'Academie Royals des Sciences, p 391–492, p 488.

of the particles of which the medium is composed.[83] Based on this assumption, Newton established that the speed of sound in any medium—solid, liquid or gas—is determined by the ratio of its elasticity to its density.[84] This was an extraordinary achievement, arguably as ground breaking and important as his mathematical account of gravity, because it was only the second time in history that a natural constant had been derived on purely theoretical grounds.[85]

However, when he substituted the known values of the compressibility[86] and density of the atmosphere into this formula, the calculated result was 288 m/s, some 15% less than the average speed of sound obtained experimentally, which included Newton's own measurements using echoes within the cloisters of Trinity College, Cambridge University, where he was Lucasian Professor of Mathematics from 1669 to 1702.[87] In the second edition of *Principia*, published in 1713, Newton attempted to correct the discrepancy by claiming that he had overlooked the fact that the size of the particles of the medium through which sound travels is not negligible, and made several arbitrary assumptions to allow for this. The real source of error, however, was that he had not taken into consideration that when air is compressed it warms up, even though he was aware that sound waves in air consist in successive condensations and rarefactions.

When a gas is rapidly compressed it warms up, something you may have noticed when pumping up a bicycle tyre with a hand pump.[88] In the case of sound, the successive compressions of the gas through which it travels occur so rapidly (even at low frequencies) that the resulting heat does not have time

---

[83] Newton, I. (1687) Philosophiæ Naturalis Principia Mathematica, Book II, 1687, Proposition 48, Theorem 38.

[84] Elasticity is the measure of the amount by which a substance will change shape when a given force acts on it. Gases are very elastic (small force cause a large change in volume), liquids and solids are far less elastic (large force causes a small change in volume).

[85] By 1657, Christiann Huygens had derived a value for the acceleration of the Earth's gravity based on his mathematical analysis of the simple pendulum, which formed the basis for the design of his pendulum clock. In: Koyre, A. (1953) An Experiment in Measurement. Proceedings of the American Philosophical Society, Vol. 97, No. 2, p 222–237, p 234. As for Newton's law of universal gravity, the universal gravitational constant necessary to use the law to calculate absolute values of gravitational attraction between bodies was not determined experimentally until 1798 when Henry Cavendish used a torsion balance to find the force of gravitational attraction between two lead spheres.

[86] Gases and liquids can only be compressed, not stretched. Compressibility is the equivalent property to the elasticity of solids.

[87] Newton used the northern colonnade at Nevile's Court, Trinity College, where, as Lucasian Professor, he had rooms. "Professor Val Gibson recreates Newton's famous speed of sound experiment in Nevile's Court" https://www.youtube.com/watch?v=Gy7HqToiBvo.

[88] The air within the pump warms up because there is not enough time for the heat generated *during* the compression stroke to escape to the surroundings before the next compression. The effect is used in a *fire piston*, a device that ignites tinder inside a small bore cylinder by rapidly compressing the air within with a piston.

to dissipate. This alters the compressibility of the gas, and compressibility, as we see from Newton's formula, is a key property that affects the speed of sound through a gas. The oversight was eventually spotted and corrected in 1816 by the most ardent Newtonian of the time, Pierre-Simon Laplace, the leading French mathematician and astronomer of his day.[89]

The Entwined Histories of Light and Sound

Newton's theoretical treatment of sound, particularly the speed at which it propagates, was infinitely more successful than that of light. He stead-fastly refused to accept that light could be a wave of some sort from the moment he first came across the idea in Robert Hooke's *Micrographia* in 1664, when he was an undergraduate in Cambridge, until his dying day, by which time he should have known better. Newton was particularly intrigued by Hooke's account of his experiments on colours that can be seen in thin transparent films such as soap bubbles and flakes of mica. Hooke attributed the phenomenon to light's "vibrative motion". Newton immediately spotted a problem: if light is a vibration, he wrote in his notebook, "Why then may not light deflect from streight lines as well as sounds?"[90] He formulated an alternative theory that held sway in the science of optics until the 1820's, and which put the more promising wave theory on hold for the entire 18th century. Light, he insisted, is a stream of "corpuscles", i.e. minute particles. Moreover, they differ in size, the largest of which were responsible for the sensation of redness and the smallest for that of blueness.

Newton was a brilliant experimentalist, one of the most accomplished in the entire history of science, and the scope, ingenuity and precision of his optical experiments and discoveries are simply astonishing. All the more so when one considers that his equipment consisted of little more than imperfect prisms and lenses made from poor quality glass, sheets of card, and that his only source of light was sunlight over which he had very little control.[91] Yet, as became increasingly evident during the early decades 19th century, Newton

---

[89] Finn, B.S. (1994) Laplace And The Speed Of Sound. ISIS, Vol. 55, No.179, p 7–19.

[90] Newton, I., Additional Ms. 3958(3).1, Cambridge University Library, Cambridge, UK.

[91] The narrow beam of sunlight on which he relied for his experiments on colour would move steadily and unstoppably across the floor of his darkened chamber as the sun crossed the sky, forcing him to have to constantly adjust the position of his apparatus, all the while making careful observations and measurements of the spectrum that was the object of his investigation. Moreover, these experiments could only be performed on clear days, so his research would have been frequently interrupted by cloudy English skies. Finally, to obtain a long shaft of sunlight, experiments would have to be confined to times when the sun is close to the horizon, e.g. in Winter or in early morning or late afternoon in Summer.

inadvertently undermined his optical discoveries by insisting on explaining all the properties of light in terms of corpuscles. He claimed that light refracts when it passes between air and glass or water because corpuscles are attracted to the surface of a transparent body and so alter the direction in which they are travelling, that the colours in a rainbow are due to the different sizes of the corpuscles present in sunlight, and that Hooke's colours are the result of an ad hoc process he termed "fits of easy transmission or reflection" whenever corpuscles encounter a thin transparent film. Yet he failed to establish a satisfactory link between his corpuscles and their putative properties principally because, unlike his account of sound, he didn't provide his corpuscular theory with the necessary mathematical underpinning.

In fact, for more than a decade following the publication of his letter to the Royal Society in 1672 in which he first made his discoveries and ideas about light and colour public, he found himself embroiled in several disputes about the nature of light.[92] Exasperated and annoyed by what he considered to be petty, ill-informed criticism, he more or less abandoned natural philosophy—i.e. physics—until the astronomer Edmund Halley paid him a visit in Cambridge in the summer of 1684. Halley succeeded in reawakening Newton's interest in problems he had long since abandoned, especially that of planetary orbits. In an unparalleled act of creation, over the next two years Newton wrote what is the most important and influential book in the entire history of science, *Philosophiæ Naturalis Principia Mathematica*.[93] In this work he not only explained the orbital motion of planets and comets in terms of a purely mathematical account of gravitational attraction, he also gave a definitive mathematical account of the dynamics of moving bodies and much else besides, including the nature and velocity of sound waves.

The success of *Principia* meant that by the time that Newton got around to publishing a comprehensive account of his researches into light in his *Opticks* of 1704, his reputation was such that reservations about his corpuscular theory had been all but set aside. Rival accounts that postulated that light involves the movement of the medium through which it propagates were largely ignored by the majority of natural philosophers for another century. In any case, the experimental techniques Newton had employed in his optical

---

[92] Newton, I. (1671/2) A Letter of Mr. Isaac Newton, Professor of the Mathematicks in the University of Cambridge; containing his New Theory about Light and Colors: sent by the Author to the Publisher from Cambridge, Febr. 6. 71/72. Philosophical Transactions of the Royal Society, 19 Feb. No. 80, p 3075: http://www.newtonproject.ox.ac.uk/view/texts/diplomatic/NATP00006 (accessed 21/07/21).

[93] Newton, I. (1687) Philosophiæ Naturalis Principia Mathematica. London.

researches were quite rightly held up as a template of how go about investigating natural phenomena, and which bolstered acceptance of his corpuscular theory of light.

Several years after having formulated his ideas on the nature of light, however, Newton learned that under some circumstances light does indeed "deflect from streight lines". This is the phenomenon of diffraction, which had been discovered in 1665 by a Jesuit natural philosopher, Francesco Maria Grimaldi. Shortly before his death, Grimaldi had noticed that the disc of light formed on a wall within a darkened room when a beam of sunlight enters it through a very small circular hole is larger than expected. In other words, after passing though a tiny aperture, light spreads beyond the geometric edge of the beam. Recognising that this was a new phenomenon, one that cannot be explained in terms of refraction, Grimaldi coined the term from *diffractio*, chosen because it suggests the spreading or breaking up of light.

Newton repeated the experiment a few years later, having witnessed a demonstration of the phenomenon by Hooke in 1675 at an evening meeting of the Royal Society. But the experimental evidence for diffraction did not change his mind about nature of light, and he attributed the phenomenon to combination of attraction and repulsion between the corpuscles and the edge of the aperture through which they travelled. He even renamed the effect *inflection*.

But, when it came to sound, Newton was fully aware that its properties are a consequence of its wave nature, and included a helpful diagram in *Principia* to illustrate the diffraction of water waves as they pass through a narrow aperture.[94] An almost identical diagram is used to illustrate the diffraction of light in modern textbooks of elementary optics. Moreover, where Mersenne, Galileo, Hooke and Huygens all spoke of sound as a series of pulses, Newton was the first natural philosopher to prove that sound is a series of linked vibrations of a medium, i.e. as a wave. There's none so blind as will not see (Fig. 2.4).

It is frequently claimed that Huygens was the author of the modern wave theory of light. In 1677 he developed a mathematical theory of the propagation of a pulse through a medium to explain the reflection and refraction of light. The method he devised to calculate the path taken by a pulse is still employed in optics and is known as Huygens' Principle. But he assumed that these pulses are uncoordinated, each travelling through a medium independently of the others, so they lack two of the most important properties of a wave: wavelength and frequency. And without these properties his theory was

---

[94] Newton, I. (1687) Philosophiæ Naturalis Principia Mathematica. Book II, Section VII, Proposition 41, Theorem 32 & Proposition 42, Theorem 33.

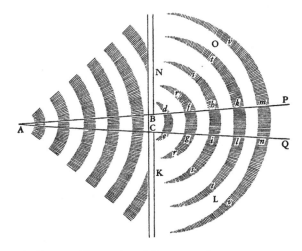

Fig. 2.4 Newton's diagram illustrating diffraction of waves in water after they pass through a narrow opening. A is the source of the waves and BC is the opening through which they pass[95]

unable to account for colours, as Huygens readily admitted. Newton knew of Huygen's theory—Huygens had sent him a signed copy of his *Traité de la Lumière*—but he refused to accept either its premise or its conclusions. And that was reason enough for the majority of European natural philosophers to ignore all alternatives to his corpuscular theory of light until well into the 19th century.

The idea that light is a wave was eventually accepted by the scientific establishment during the 1820's following the combined though uncoordinated research of Thomas Young and a French engineer, Augustin Fresnel. In addition to being an accomplished mathematician, Fresnel proved to be an exceptionally gifted experimentalist, which added weight to his ideas about the nature of light. Young's explanation of light in terms of a wave arose piecemeal out of his study of the nature of sound, but it failed to convince the legions of Newtonians either in Britain or France, despite the compelling experimental evidence with which he supported his hypothesis.

What first led Young to question Newton's corpuscular theory was his belief that some of the phenomena that Newton explained in terms of corpuscles, such as the colours seen in thin films of soap, can be explained more convincingly in terms of the interactions between waves. In fact, it was his

[95] Newton, I. (1687) Principa Mathematica, Prop 41, Theorem 32.

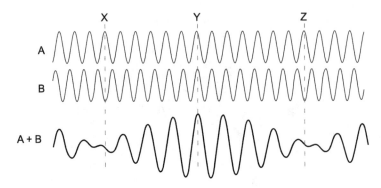

Fig. 2.5 Beating. The difference in frequency between waves A and B is 2 cycles. When they are in step as at Y, the resultant combination is maximum, i.e. sound is loudest. When they are out of step at X and Y the resultant sound is minimum

analysis of the interactions between waves that led Young to devise an experiment which he believed provided unassailable evidence that light is wave rather than a particle.

The idea for the experiment arose from his research into an acoustic phenomenon that was widely known though not fully understood in his day. This is the regular rise and fall in loudness that is heard when two pure tones that are very close in frequency occur together, and which is known as beating. Musicians had been using the phenomenon since the early 16th century to tune organs by ear, because tones that have the same frequency do not *beat*. But no one before Young had fully understood what causes beating. He realised that the rise and fall in loudness was the result of two tones that differ slightly in pitch strongly reinforcing one another when their vibrations coincide and partially cancelling one another when they are out of step. The result is a periodic rise and fall in loudness at a rate equal to the difference in pitch between the tones (Fig. 2.5).

He reasoned that if light is a wave it should exhibit a similar effect. In fact, if two waves are completely out of step they will completely cancel one another out. In this experiment, which must be carried out in a darkened room, a narrow beam of light is split in two by passing it through a pair of very narrow and closely spaced apertures. The recombination of the individual beams creates a pattern of alternating light and dark patches on a screen placed several metres from the apertures.[96] As Young pointed out, this

---

[96] In fact, to produce alternating light and dark bands, the source of illumination must be monochromatic (e.g. light from a laser). Sunlight is composed of a broad range of frequencies (which we perceive as distinct colours), so the interference between two beams results in multi-coloured bands. You sometimes see these bands when half closing your eyes so that the light from a bright source must pass through your eyelashes.

pattern is more readily explained in terms of waves than in terms of particles: patches of light occur where the waves from adjacent slits are in step and darkness where they are out of step. Young called the phenomenon the "principle of interference".[97] The principle applies equally to sound, as we shall see in Chap. 5, and to waves travelling across the surface of water, as we shall see in Chap. 3

What finally swung scientific opinion in favour of the wave theory of light was Fresnel's mathematical account of the diffraction of light, which employed Huygens' Principle to make precise calculations of the effects of diffraction when a wave encounters an edge and thus provide experimentally testable predictions. Ironically, the success of the wave theory led to a complete rejection of any form of a particle theory of light. The two aspects, wave and particle, were eventually united in the early years of the following century in Einstein's 1905 explanation of the photoelectric effect.[98]

Air is not the Only Medium

Early measurements of the speed of sound were all carried out in the open, and thus applied only to air. But during the 18th century chemists discovered several previously unknown gases including hydrogen, oxygen, nitrogen and carbon dioxide, and established that air is itself a mix of different gases.[99] The availability of these new gases prompted Ernst Chladni, to determine the velocity of sound in these and other gases.[100] His method was to compare the pitch of an organ pipe in air with that of an identical organ pipe filled with one of the newly discovered gases.[101] He knew from Newton's mathematical account of sound that the density of a gas affects the speed of sound and that consequently the pitch of an organ pipe filled with hydrogen will be higher than that of an identical organ pipe filled with air, whereas the pitch of one

[97] Steffens, H.J. (1977) The Development of Newtonian Optics in England. Science History Publications, New York, p 107–36.

[98] Einstein, A. (1905) On A Heuristic Point Of View Concerning The Production And Transformation Of Light. In: Stachel, J. (ed.) (1989) Einstein's Miraculous Year, Five Papers That Changed the Face of Physics. Princeton University Press, p 177–98.

[99] Oxygen was discovered independently by Priestly, Scheele and Lavoisier between 1772 and 1774. Priestly called the gas "dephlogisticated air" Lavoisier named it "oxygen" Hydrogen was discovered by Henry Cavendish in 1766 (he called it "inflammable air"), carbon dioxide ("fixed air") by Joseph Black 1755 and nitrogen ("phlogisticated air") by David Rutherford in 1772.

[100] Chladni, E.F.F. (2015) Treatise on Acoustics: The First Comprehensive English Translation of E.F.F.Chladni's 1809 Traité d'Acoustique (trans: Beyer, R.T.). Springer. For Chladni's account of the speed of sound in gases see Sect. 196, p 145–46.

[101] In addition to oxygen, nitrogen, carbon dioxide and hydrogen, Chladni also used nitrous oxide.

filled with carbon dioxide will be lower. And knowing the length of the organ pipe and the pitch of the note it produced, Chladni could calculate the speed of sound in the gas in the organ pipe. He made these discoveries known in *Die Akustik*, the first book devoted entirely to acoustics.

While performing these measurements, Chladni noticed to his surprise that the sound produced when the organ pipe was filled with hydrogen was far less loud than when filled with denser gases.[102] In fact, this had already been observed a couple of decades earlier. Joseph Priestly, the English natural philosopher and chemist who discovered oxygen, had inserted a clockwork bell in a glass vessel, removed the air within by means of a vacuum pump, filled it with various gases and found that "In inflammable air (i.e. hydrogen) the sound of the bell was hardly to be distinguished from the same in a pretty good vacuum…"[103] And around the time that Chladni was making his measurements, others had discovered that hydrogen has a pronounced effect on the timbre of the human voice. Given that a mixture of hydrogen and air is highly explosive, these days helium is used in place of hydrogen to demonstrate the effect. We'll come back the question of how and why the density of a gas affects the loudness of sounds and alters the voice in Chap. 6.

Chladni had developed the technique of using an organ pipe filled with air as an acoustic yardstick a year or two before his experiments with gases, when he had measured the speed of sound in various metals including tin, silver, copper, glass, iron, and several species of wood. His method was to take a long, narrow, uniform rod of the material, clamp it firmly at it centre and stroke one of its free ends with a resin-coated cloth, causing it to vibrate and emit a loud, high-pitched tone.[104] The pitch of the resulting sound was compared to that of an organ pipe of the same length, open at both ends and filled with air.[105] From the pitch of the sounding rod and its length he was able to calculate the velocity of sound in the material of rod and published his results in 1797. In every case, he found that the velocity of sound in a solid is always several times greater than it is in air.

A century after Chladni's discovery that the pitch of an organ pipe depends among other things on the gas within it, the phenomenon was exploited to detect firedamp. Firedamp is a highly explosive mixture of methane and air

---

[102] Chadni, E.F.F. (2015) Treatise on Acoustics: The First Comprehensive English Translation of E.F.F.Chladni's 1809 Traité d'Acoustique (trans: Beyer, R.T.). Springer, Sect. 198, p 147–48.

[103] Priestly, J. (1781) Experiments and Observations Relating to Various Branches of Natural Philosophy With a Continuation of Observations on Air, Volume II. J. Johnson, London, p 295–99.

[104] The vibration of the rod is due to stick and slip friction between the resin on the cloth and the rod. See Chapter 6 for more on this.

[105] The air in an organ pipe open at both ends vibrates in the same manner as a rod fixed at its centre, i.e. maximum movement (an antinode) at each end and no movement (node) at the centre.

and was a huge problem in coal mines in the days when the only source of illumination was a naked flame.[106]

The presence of firedamp could be detected using two whistles that have the same pitch in pure air. Inside a mine, one whistle is supplied with pure air and the other with the air from the mine. If the latter contains methane, its pitch will rise because the speed of sound in methane is some 30% greater than it is in pure air. The difference in pitch between the whistles will be heard as beats. The number of beats is directly related to the amount of methane within the mines atmosphere. The whistle was variously known as a firedamp whistle or firedamp indicator.[107]

Chladni's interest in the vibration of rods and columns filled with gases arose from his study of patterns created on the surface of a thin, square sheet of glass by stroking its edge with a violin bow. By adjusting the motion of the bow, it is possible to make the glass emit loud tones that correspond to the natural frequencies of the sheet. To reveal the pattern of vibration responsible for the sounds he sprinkled the surface of the glass with very fine, dry sand on. Vibrations of the glass sheet cause the sand to accumulate along nodal lines, i.e. where the surface of the glass is stationary. These patterns are so varied and eye catching that, despite his pioneering work on the speed of sound, Chladni's name is now invariably associated with these patterns, which have come to be known as *Chladni figures*. The figures are certainly striking and seemingly endlessly varied, but the purpose of the experiments was scientific, i.e. to reveal the complex vibrations of a thin, stiff sheet of glass or metal when it produces sound.[108]

Chladni earned his living by travelling around Europe demonstrating his collection of acoustic apparatuses and performing acoustic experiments, and became particularly renowned for his eponymous figures. He a spent a couple of years in Paris between 1808 and 1810, during which he had an audience with Napoleon, who was so taken by Chladni's figures that he awarded him a gratuity of 6000 Francs on condition that he translated his *Die Akustik* into French. While working on the translation, Chladni asked for help to find French equivalents for German words for acoustic phenomena and was informed that "Notre diablesse de langue ne veut pas se prêter á l'expression de toutes les idées possibles. Il faut même quelquefois sacrifier une idée aux caprices de la langue."[109] ("Our devil of a language does not lend itself to the

---

[106] Humphrey Davy's 1815 invention of the safety lamp went a long way to preventing explosions.

[107] Jeans, J. (1923) Science and Music. Cambridge, p 120.

[108] Waller, M.D. (1961) Chladni Figures, A Study in Symmetry. G. Bell and Sons.

[109] Stöckmann, H.-J. (2007) Chladni meets Napoleon. The European Physics Journal, Special Topics, 145, p 15–23.

expression of all possible ideas. Sometimes it is even necessary to sacrifice an idea to the caprices of the language.").[110]

This offhand remark gives a misleading impression of the state of science in France at the time, for no other nation then matched the French in the application of mathematics and Newtonian physics to scientific issues. During the early decades of the nineteenth century they were in a class of their own and the natural philosophers of other nations, in particular those in Britain, only caught up with them in the 1830's, when a new generation of British mathematicians abandoned Newton's outdated mathematical methods and adopted wholesale those of the French mathematicians. Undeterred by the lack of suitable French words and unable to match the mathematical rigour of French science, Chladni pressed on with the translation and published it in 1809 as *Traité d'Acoustique*.

French mathematical rigour was soon brought to bear on Chladni's figures by Sophie Germain, a gifted, self-taught French mathematician. Her early interest in mathematics had dismayed her parents, who did what they could to thwart her interest in the subject. And as a woman she was not eligible for entry to any institution where Frenchmen were taught mathematics and science. Fortunately, her talents came to the attention of several eminent mathematicians who were prepared to offer her help and guidance. Inevitably, however, her piecemeal mathematical education meant she was unable to reach her full potential as a mathematician.

Shortly after the publication of the *Traité d'Acoustique*, the *Institut de France* offered a prize of "a gold medal valued at 3000 francs" to anyone who could come up "the development of a mathematical theory of the vibration of elastic surfaces, and a comparison of this theory with experiments."[111] In other words, what was sought was a comprehensive mathematical account of Chladni figures.

The *Institut* was the powerhouse of French science in the Napoleonic era. It was the republican successor to the royalist *Académie Royale des Science*, which had been abolished in 1793 during the phase of the French Revolution that came to be known as the Reign of Terror because it was considered to be an elitist institution and its incumbents hostile to the revolution. The *Académie* was, in the memorable phrase of one of the Institut's founders, "gangrenées d'une incurable aristocratie".[112]

---

[110] Chladni sought French equivalents for the German terms *Schall*, *Klang* and *Ton*, but there is only a single word for these in French, namely *son*, i.e. *sound*.

[111] The full text of the announcement of the prize is included in the appendix to Chladni, E.F.F. (2015) Treatise on Acoustics, p 187–90.

[112] The claim was made by Abbé Henri Grégoire (1750–1831), one of the authors of the *Declaration of the Rights of Man*.

The competition attracted only one entry, that from Sophie Germain. She was not awarded the prize because the judging committee of the *Institut* found her mathematical analysis of Chladni figures was flawed and incomplete. However, the contest was left open and she submitted two further solutions, neither of which provided a complete answer of the issue though they were improvements on that offered in her original essay. She was grudgingly awarded the prize for the last of these solutions, though, apparently, she did not deign to collect it. In fact, no mathematician of the time could have improved on her efforts because the mathematical methods available to them were not up to providing the answers sought by the prize setters of the *Institut*. Chladni figures continue to fascinate mathematical physicists and delight laypeople to the present day.[113]

One of the leading lights of the *Institut* during Sophie Germain's lifetime was the applied mathematician, Jean-Baptiste BiotIn 1808 he took advantage of the installation of a network of cast iron water pipes in Paris to compare the speed of sound in cast iron to that in air. The section of pipe he used was 951 m long. A bell at one end of the pipe was struck with a hammer and two sounds were heard at the far end: the first travelling rapidly through the cast iron and the other much more slowly through the air within the pipe. Given that both the length of the pipe and the speed of sound in air were known, it was only necessary to measure the elapsed time between the two sounds to obtain the information required to calculate the speed of sound in cast iron.[114]

Allowing for the unquantifiable effect of the lead solder used to join sections of the pipe, Biot calculated that the ratio of the speed of sound in air to that in cast iron is 1:10.5. The result agreed well with Chladni's value for the speed of sound in iron obtained with his vibrating rods ten years earlier. During the experiment, Biot held a whispered conversation through the pipe with his assistant at the far end and although their exchanges were delayed by the five and a half seconds it took the sound wave to travel the length of the pipe and back, they found no difficulty in hearing one another because the sound of their voices was confined within the narrow tube. At the same time, their voices echoed back and forth within the pipe up to six times.[115]

---

[113] Waller, M. D. (1961) Chladni Figures, A Study In Symmetry. G. Bell and Sons, p xvii–xx.

[114] L = length of pipe, c_1 = speed of sound in air, c_2 = speed of sound in cast iron, t_1 = time taken for sound to travel though air, t_2 = time taken for sound to travel though cast iron. Then $c_1 = L/t_1$ and $c_2 = L/t_2$. Whence $t_1 - t_2 = L/c_1 - L/c_2$ from which c_2 can be calculated because both L and c_1 are known and t_1-t_2 can be measured with a stopwatch.

[115] Biot, M. (1809) Expériences sur la propagation du son à travers les corps solides et à travers l'air, dans des tuyaux très-alongés". Mémoires de Physique et de Chimie de la Société D'arcueil, p 405–23.

Sound in Water

In principle, Newton could have calculated the speed of sound in water along with that of the speed of sound in air, but the elasticity of water was not known in his time. Water, of course, can't be stretched, so its elasticity is measured by squeezing it, which yields a value for its compressibility. The first successful experiment to measure the compressibility of water took place some forty years after Newton's death in 1727.[116] And it wasn't until 1808 that Thomas Young became the first person to publish a calculation of the speed of sound in water based on those measurements. This gave a speed of sound of "4900 feet in a second" (1493 m/s).[117]

Young did not carry out any experiments to confirm this result. In any case, direct measurement of the speed of sound in a liquid is much more challenging than that in air because the speed of sound in a liquid is always many times greater than in a gas. This makes it much more difficult to measure accurately the time taken for sound to travel through water unless the distances involved are correspondingly greater than in air.

The first systematic measurements of the speed of sound in water were made in Lake Geneva in November 1826, by Daniel Colladon a Swiss physicist, with the help of his father. As we shall see in Chap. 4, the purpose of this experiment was not directly concerned with sound, it was to confirm the results of measurements of the compressibility of water that Colladon and his friend and colleague, the mathematician J. K. F. Sturm, had made in Paris the year before.

After a number of preparatory trials, none of which produced reliable results, Colladon came up with a workable arrangement. Colladon senior was stationed in a small rowing boat 200 m from the Swiss shore of the lake while his son took up position in another boat 200 m from the French shore. The distance between the boats was 13.5 km. The sound source was a small bell suspended a metre and a half beneath the surface of the lake from the boat on the Swiss side. The experiment had to take place at night because the only method to coordinate their actions was for Colladon to fire a rocket to signal his father to strike the bell.

The experiment was almost abandoned at the outset because Colladon had to take these rockets across the border between Switzerland and France. To

[116] Canton, J. (1764) Experiments and observations on the compressibility of water and some other fluids. Philosophical Transactions, 54, p 261–62.

[117] Young, T. (1845) Lecture XXXI, On the Propagation of Sound. In: Young, T. (1845) A Course of Lectures on Natural Philosophy and the Mechanical Arts, Vol 1. Taylor and Walton, London, p 295.

his dismay, the French customs officer initially refused to allow them through on the grounds that the importation of gunpowder was forbidden by French law.

> He repeated over and over again, 'The powder cannot pass.' And he showed me at the same time his orders. "You say these rockets are made with powder; consequently I am not able to let you pass under any consideration without encountering the risk of losing my job." I then understood his objection and said 'Yes, they are made with powder, but the work of constructing them destroyed the powder and now one can no longer use them as ordinary powder.' The man was evidently satisfied by this explanation, and said: 'Ah, if that is so, I can then let you pass with them,' and he gave me back my rockets.[118]

Striking the bell automatically ignited a small amount of gunpowder that created a flash of light bright enough to be visible from the French side. The sound of the bell was picked up via trumpet-shaped metal horn immersed in the water. By noting the interval between the flash and the arrival of the sound at the horn, Colladon came up with an average value for the speed of sound in water of 1435 m/s at 8.1 °C, the average temperature of the lake on the three days on which the measurements were made.[119] The results agreed well with the values obtained by calculations of the speed of sound based on the measurements of the compressibility of water made in Paris.

Since these pioneering measurements of the speed of sound in gases, liquids and solids, far more accurate methods have been developed and applied to a vast range of substances. But these refined techniques are of interest only to specialists and so we shall not pursue the history of the measurement of the speed of sound further (Fig. 2.6).

Sound and the Vacuum

The idea that nature abhors a vacuum, which came to be known as *horror vacui*, was taken for granted by almost all natural philosophers until the middle of the 17th century. The case against the possibility of a vacuum rested principally on objections first raised by Aristotle some two thousand

---

[118] Colladon, J-D (1893) Souvenirs et Memoires—Autobiographie de Jean-Daniel Colladon. In: Lindsay, R. B. (1973) Acoustics: Historical and Philosophical Development Benchmark Papers in Acoustics, Vol 1. Dowden, Hutchinson & Ross, p 197–201.

[119] Colladon, J-D, Sturm, J.F.K. (1827) Mémoire sur la compression des liquides et la vitesse du son dans l'eau. Annales de Chimie et de Physique, 36, p 113.

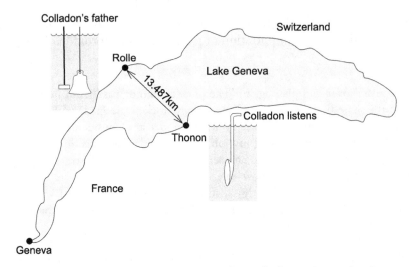

Fig. 2.6 J.D. Colladon's 1826 determination of speed of sound on Lake Geneva. His father struck the underwater bell with a hammer at Rolle and Colladon listened for its sound through a large horn lowered into the water at Thonon

years earlier. He argued that the very idea of a vacuum is incoherent, for in what sense can nothing be said to exist? Moreover, in an empty void there can be no directions or dimensions, i.e. no up or down nor larger or smaller. But the objection that carried most weight with natural philosophers stemmed from Aristotle's account of motion. He assumed that the natural state of a body is rest and that without a force to drive it forward, a body will either remain stationary or quickly come to a halt.

Numerous everyday experiences appear to confirm this: you have to make an effort to overcome the weight of a suitcase if you want to lift it off the ground, take your foot off the accelerator and your car will soon come to a stop, stop peddling and your bicycle slows down, etc. But, by definition, in a vacuum there is no medium to oppose a driving force, which implies that in a void the slightest force would cause a body to move with infinite speed. A stone would fall to the ground in an instant, whatever height from which it was dropped. A further difficulty with a vacuum, according to Aristotle, is that light requires a medium if it is to propagate through space. A vacuum would thus be opaque and we would be unable to see the stars or the sun. Since these things are never observed, logic dictates that all space must be filled with a medium and therefore a vacuum cannot exist.[120]

---

[120] Aristotle, 350 BC, Physics, Book IV, 6, 7, 8, 9. http://classics.mit.edu/Aristotle/physics.4.iv.html (accessed 23/03/2020).

The lengths to which Aristotle went to prove that a vacuum is impossible were driven principally by his desire to refute atomism, a doctrine to which he was opposed. A hundred years earlier, two Greek natural philosophers, Leucippus and Democritus, had made a vacuum an essential feature of their account of nature. According to this, the material world consists of atoms, particles too small to be visible to the naked eye, which are in constant motion. Hence the need for a vacuum if they are to move about. The idea that the world consists of atoms and the void is set out most memorably by Lucretius in *De Rerum Natura*, a work that was unknown in post-Roman Western Europe until 1417, when a copy was discovered in a monastery library.

But thanks to Aristotle's stature and influence, atomism never caught on in the ancient world, and his account of nature prevailed for the next 2000 years. Even then, René Descartes, the leading advocate of the mechanical philosophy that underpinned the scientific revolution of the 17th century, and a fervent opponent of the prevailing Aristotelian world-view, insisted that a vacuum is impossible. And, as we shall see, if you are willing to give him the benefit of the doubt, it could be said that he wasn't entirely wrong.

Not for the first time, however, a seemingly inconsequential practical problem led to a major discovery that had a bearing on an apparently unrelated issue. Towards the end of the 16th century the fashion for gravity-fed fountains and the need to pump water from deep wells and mines, threw up a puzzling fact: no mechanical pump could (or indeed can) raise water more than 9 or 10 m above the level of the source. Galileo was asked for his opinion on the matter but his reply was somewhat tentative and showed that he had not freed himself entirely from the idea of *horror vacui*.[121] He answered that there is a limit to nature's abhorrence of a vacuum, which is why above a certain height a column of water will break under its own weight.[122]

Galileo touched on the issue briefly in his *Dialogues Concerning Two New Sciences* published in 1638.[123] A year or two later, possibly as a result of reading the *Dialogues*, Gasparo Berti, a mathematician and astronomer then residing in Rome, devised an ingenious experiment to investigate the phenomenon. His apparatus consisted of a narrow, vertical tube approximately 12 m long, made of lead and fastened to the side of his house with ropes. The bottom of the tube was closed off by a brass stopcock and rested

[121] Letter from Giovanni Batista Baliani to Galileo, dated 27 July, 1630.

[122] Galileo Galilei (1914) Dialogues Concerning Two New Sciences, translated from the Italian and Latin by Henry Crew and Alfonso de Salvio. The Macmillan Company, p 16–17.

[123] Galileo Galilei (1914) Dialogues Concerning Two New Sciences, translated from the Italian and Latin by Henry Crew and Alfonso de Salvio. The Macmillan Company, p 16.

within a large wooden cask filled with water. The upper end, which was accessible from a window, was capped by a large glass flask cemented to the lead tube to create an air-tight seal. The top of flask had a narrow opening that could be sealed with a close-fitting stopper. The device was, in effect, a water barometer, though its purpose was not to measure air pressure but to find answers to questions about the vacuum.

To perform the experiment, the brass stopcock at the base was closed, the stopper removed from the glass flask and water poured in until it completely filled the apparatus. The stopper was replaced, and when the brass stopcock was opened some water drained out of the tube into the cask leaving the flask empty. The question was, what now occupied the flask? Could it really be devoid of any kind of matter and be a vacuum,. One of those present was Athanasius Kircher, a Jesuit scholar and professor of mathematics at the Jesuit College in Rome. He accepted that air is necessary for the propagation of sound—an assumption made by Aristotle two thousand years earlier—and suggested it should be possible to discover whether there was a vacuum in the empty flask by suspending a small bell inside it (Fig. 2.7).The bell could be made to ring by manipulating its clapper remotely with a magnet held against the outer surface of the glass flask. According to Emmanuel Maignan, a Minim friar and, like Mersenne, an enthusiastic natural philosopher, who wrote an account of the experiment, Kircher said that if "… the sound of this blow could be heard, it would be all over with the vacuum, since sound cannot be produced in one."[124]

When the experiment was eventually performed (by Kircher, not Berti), Kircher heard the clapper strike the bell and concluded that the apparently empty space within the flask was not devoid of matter and so was not a vacuum. He surmised that because the magnetic influence of the loadstone could pass though the glass of the flask it was equally possible that air could get inside the flask, say through tiny pores in the glass. Kircher concluded "even if a vacuum were possible in the nature of things, nevertheless sound could not occur in it. For since sound is an affection of the air—in fact air is the material cause of sound—, when that is lacking, sound also must be lacking; and on the other hand, we have clearly shown from the proposed experiment that a vacuum cannot be admitted in the nature of things."[125] In fact, the reason that the sound of the bell could be heard was that it was in contact with the side of the flask; sound travels through solids with far less loss than it does through air.

---

[124] Middleton, W.E. Knowles (1964) The History Of The Barometer. The Johns Hopkins University Press, p 14.

[125] Kircher, A. (1650) Musurgia Universalis sive Ars Magna Consoni et Dissoni, Vol I. p 11–13.

Fig. 2.7 Gasparo Berti's water barometer experiment. The barometer is on the right of the building. The length of the tube was approximately 12 m. Note the small bell M and tiny hammer N in the upper chamber that was used to discover whether the propagation of sound requires air[126]

---

[126] Image in Public Domain, Wikimedia Commons. In: Schott, G. (1664) Technica curiosa, sive, Mirabilia artis, Würzburg.

Some five years later, in 1643, at the bidding of Evangelista Torricelli, Vincenzo Viviani his erstwhile pupil, constructed what was in effect the very first mercury barometer in order to further investigate the issue. This type of barometer, which was in general use until well into the 20th century, consists of a narrow glass tube approximately one metre long, open at one end and closed at the other and containing mercury. To perform Torricelli's experiment, which was a standard school laboratory demonstration until concerns about the toxicity of mercury caused it to be abandoned, the tube is held with the open end uppermost and completely filled with mercury. The open end is closed off with a finger, the tube is carefully inverted and the open end submerged in a glass trough filled with mercury. The finger is removed and the level of mercury in the tube drops some 25 cm, leaving an empty space at the top of the tube.[127]

Torricelli claimed that this space does not contain air because none is seen to enter the tube as it is inverted. He therefore inferred that the space above the mercury must be a vacuum and that the weight of the mercury remaining in the tube after inversion is supported by the weight of the atmosphere. Furthermore, Torricelli noticed that the height of the mercury column varied slightly from day to day, which meant that the volume of the space was not constant: "Nature would not, as a flirtatious girl, have a different *horror vacui* on different days", he wrote, and correctly attributed the rise and fall of the mercury to changes in the pressure of the atmosphere (Fig. 2.8).[128]

Further compelling evidence that the mercury in a barometer is supported solely by atmospheric pressure, and that the space above the mercury is completely devoid of matter was provided in an experiment devised by Blaise Pascal. Pascal was one of the leading mathematicians of the 17th century, the inventor of the first mechanical calculator and the author of *Pensées*, one of the most sublime and thought-provoking works of Christian theology. He was one of Mersenne's circle of savants, and it was from Mersenne that he learned of Torricelli's experiments in 1646. A couple of years later, Pascal was able to confirm Torricelli's claims about the effect of atmospheric pressure when he persuaded his brother-in-law, Florin Périer, to take a barometer to the summit of Puy-de-Dome, an extinct volcano near the French city of Clermont-Ferrand. When the height of the mercury column in the barometer on the summit was compared to that at the foot of the mountain it was

---

[127] Toricelli, E. (1644) Letter to Michelangelo Ricci in Rome. In: Maggie W.M.A (ed) (1963) Sourcebook in Physics. Harvard University Press, p 70–73.
[128] Toricelli, E. (1644) ibid.

Fig. 2.8 This barometer was made by Vincenzo Viviani and used by Evangellista Torricelli. Being filled with mercury it is much shorter than Berti's water barometer (1 m compared to 12 m, reflecting the difference in density of the substances used)[129]

found to be approximately 8 cm less, which implied that it is the weight of the atmosphere that supports the mercury in the tube.[130]

Unaware of these experiments, possibly because he wasn't in Mersenne's address book, in 1654 Otto von Guericke, burgomeister of Magdeburg, a German city on the river Elbe, set about constructing an air pump, a device capable of removing air from a sealed container. Guericke had been intrigued since his student days by the many metaphysical puzzles thrown up by Copernicus' heliocentric theory. Among these was the question of whether or not the universe beyond the Earth's atmosphere is a void, an issue that had come to a head in 1577 when the Danish astronomer, Tycho Brahe, proved by carefully measuring the motion of a bright comet seen that year as it moved against the background of stars that it had passed through several of the concentric crystalline spheres within which it was supposed that planets are embedded and carried about the sun. The obvious conclusion to be drawn from his discovery was that these crystalline spheres are a fiction.

After several failures, von Guericke eventually perfected his pump, and in 1657 he demonstrated the efficacy of his invention in spectacular fashion

[129] Image in public domain. In: Opere di Evangelista Torricelli (1919) eds Loria G, Vassura G. Faenza, Italy: G. Montanari, vol. III, p 186.

[130] Pascal, B. (1648) Experiments with the Barometer. In: Maggie W.M.A (ed) (1963) Sourcebook in Physics. Harvard University Press, p 73–5.

in the Bavarian town of Regensburg before an audience that included the Holy Roman emperor of the time and his entire court. Two identical hollow hemispheres each approximately 50 cm in diameter and made of brass were pressed against one another using an air-tight seal made of leather and the air enclosed within them removed with his air pump. This been done, the only thing that held them together was atmospheric pressure. Each hemisphere was harnessed to a team of horses, which were then driven in opposite directions in order to pull the hemispheres apart. Even though his air pump would not have been capable of removing all the air, the weight of the atmosphere pressing on the hemispheres was sufficiently large that the teams were unable to separate them.[131]

Earlier, possibly soon after he constructed his first air pump in 1654, von Guericke performed the bell-in-a-vacuum experiment to discover the effect of an absence of air on the transmission of sound. He used a simple alarm clock as a source and found that the sound of its bell diminished gradually as the air was pumped out of the vessel within which it had been sealed. However when he placed his ear to the vessel, he perceived "a duller sound or noise … but by no means a ringing",[132] which he correctly attributed to direct contact between the clock and inner wall of the vessel.

When news of this experiment reached the Anglo-Irish natural philosopher Robert Boyle in Oxford, England, in 1657, he lost no time repeating it with a much-improved version of von Guericke's air pump designed and built by his young laboratory assistant, the 19 year old Robert Hooke.[133] Realising that sound from the vacuum within the bell jar might be audible unless care was taken to isolate its source from its surroundings, they suspended a small alarm clock by a thread within a glass bell jar and removed as much air as they could with Hooke's pump. Removing air from the vessel was a laborious affair given the design of the pump, but eventually a point was reached when they could no longer hear anything, though they could see that the alarm was ringing. The ringing became audible when air was allowed back into the vessel.[134]

---

[131] Given that the diameter of each hemisphere was approximately 50 cm, the force exerted on each hemisphere by the pressure of the atmosphere was approximately 20,000 N (equivalent to weight of 2 tonnes). It has been conjectured that the pump enabled Guericke to remove some 90% of the air from within the sphere.

[132] Guericke, O. von (1672) Experimenta Nova (ut vocantur) Magdeburgica de vacuo spatio. Johannes Jansson Waesberge, Amsterdam, p 91.

[133] Guericke didn't publish an account of this experiment until 1672, so Boyle probably learned about his experiment from reading the account in Caspar Schott (1657) Mechanica hydraulico-pneumatica, p 442–488.

[134] Boyle, R. (1682) New Experiments Physico-Mechanical, Touching the Spring of the Air and its Effects, Made, for the most part, in a New Pnuematical Engine. London. See Experiment 27:

In fact, the effect on the loudness of sound as air is pumped out of a vessel is not the cut and dried affair that these early experiments suggest, for sound within the vessel becomes inaudible long before it is fully evacuated. Why sound from within partially evacuated vessel can't be heard outside the vessel is something that we will consider in the next chapter in connection with a phenomenon known as acoustic impedance.

If the experiment that involved dropping cannon balls supposedly performed by Galileo at Pisa undermined Aristotle's account of motion, those of Torricelli, Pascal, von Guericke and Boyle were just as consequential because creating a vacuum was the best, indeed the only, hard evidence that the natural philosophers of the 17th century could point to in support of atomism, a doctrine that underpinned their mechanical philosophy.[135] Atomism, you will recall, requires a vacuum. These experiments, however, did nothing to sway the Aristotelian professors, of which there were huge numbers in European Universities, all of whom vehemently opposed the ideas of the mechanical philosophers. They expended gallons of ink in attempts to discredit the experiments by means of logic alone, but succeeded only in convincing one another. Chalk and talk was no match for the combination of innovative experiments and mathematical analysis of the iconoclastic mechanical philosophers.

As for the nature of a vacuum, one of the central tenets of Descartes' version of the mechanical philosophy was that all physical events must involve direct contact between bodies. This, in turn, requires that space is a plenum, i.e. that it is entirely filled with matter, something that makes a vacuum impossible. But according to Descartes the only necessary characteristic of matter is simply that it occupies space, not its solidity or ponderability. The apparently empty void between the sun and Earth is filled with this intangible Cartesian matter and sunlight is actually a centrifugal pressure exerted on the plenum by a spinning sun. Moreover, the implication of making extension the sole defining property of matter is that Cartesian matter must be absolutely rigid. The further implication is that this centrifugal pressure acts instantaneously across the vast distance between the sun and the Earth, where it presses on our eyes and results in sensations of light. How this pressure is

---

Touching the Propagation of Sound, p 103–108. For an annotated commentary see "Robert Boyle's Experiments In Pneumatics. In: Conant, J.B., Nash, L.K (eds) (1957) Harvard Case Histories In Experimental Science Vol 1. Harvard University Press, p 30–38.

[135] There is no evidence in any of Galileo"s writings that he ever dropped any cannon balls from the Leaning Tower, or indeed any other tower, though other natural philosophers (e.g. Simon Stevin in 1586) did so. Galileo proved mathematically in his "Dialogues Concerning Two New Sciences, Third Day", that, allowing for air friction and experimental errors, all bodies fall at the same rate regardless of weight.

perceived by the mind as brightness and colour posed a intractable problem for Descartes and his followers, arguably one that has yet to be resolved, namely the issue of how incorporeal minds interact with inert matter.

Descartes' conception of matter was not accepted by all natural philosophers of the 17th century. English natural philosophers were particularly resistant to it. Newton made a point in his *Principia* of demolishing Descartes' theory that the planets are carried around the sun in their orbits within the vortex of imponderable matter that swirls around it. In its place he substituted the almost equally absurd notion of universal gravitation whereby every particle of matter attracts every other particle without the need for a medium to transmit the force. Unlike Descartes, however, Newton was able to give mathematical form to his idea, even though, when pressed, he always wisely refused to be drawn on the nature of gravity. Indeed, he was well aware of the problem posed by action at distance and said so when pressed.

> That Gravity should be innate, inherent and essential to Matter, so that one body may act upon another at a distance thro' a Vacuum, without the Mediation of any thing else, by and through which their Action and Force may be conveyed from one to another, is to me so great an Absurdity that I believe no Man who has in philosophical Matters a competent Faculty of thinking can ever fall into it.[136]

A Newtonian universe consists of atoms within a vast, fixed void, arranged in its present form and constantly maintained in good order by God. The Cartesian universe didn't require God's intervention, at least not on a daily basis as Newton's did, because the laws of nature were assumed to have been laid down by God at the moment of creation, and thereafter nature had adhered to them rigidly.

And, of course, Newton's claim that light is composed of a stream of material particles not only provided an alternative to Descartes' hypothesis that light is pressure, it was a rival to the pulse theories of light of Hooke and Huygens, theories which also require a medium within which propagation can take place. Newton's ideas about light prevailed, principally because of his success in accounting for planetary motion in terms of his mathematical theory of gravity rather than the merits, if any, of his corpuscular theory. As we saw above, the idea that light is a wave took a back seat until it was revived in the early 19th century following the work of Thomas Young and Augustin Fresnel. But because they both conceived light as a material

---

[136] Isaac Newton to Richard Bentley, Feb 25, 1692/3. In: Bernard Cohen, I., Westfall, R.S. (1995) Newton, Texts, Backgrounds, Commentaries. W.W. Norton & Co. p 336–39.

wave, like sound but with enormously higher frequencies, their wave theory required a medium, which came to be known as the luminiferous aether. Moreover, as Young realised to his dismay, if light is a mechanical wave, whatever its form, its enormous velocity would entail that this aether has to be absolutely rigid. How planets could move through such a medium remained an open question until the following century.

Even when light came to be regarded as an electromagnetic wave towards the end of the 19th century following the discoveries of Michael Faraday, James Clerk Maxwell and Heinrich Hertz, the aether was still considered necessary for its propagation, though its nature was never fully spelled out. Several attempts to detect it by the American physicist Albert Michelson were inconclusive, and the need for a medium to explain the propagation of light was quietly dropped following the publication by Albert Einstein of his Special Theory of Relativity in 1905, in which he declared an ether to be unnecessary for the transmission of light though space.

However, although the Newtonian void is a space in which there are no atoms of ordinary matter, i.e. matter made of electrons and protons, and the luminiferous aether is an anachronism, the cosmologists of our day believe that the universe is *entirely* filled with two mysterious forms of invisible matter, which they have named *dark matter* and *dark energy*. Moreover, measurements of the motion of galaxies suggest that dark matter and dark energy together make up 95% of the universe. Descartes would be surely delighted to learn that physics has not entirely abandoned the idea of a plenum. But since science has yet to discover the nature of dark matter or dark energy, perhaps this is to put Descartes before the facts.

3

Sound Science

Abstract An account of the science of sound and the hearing system. It covers the physical properties of sound, the purpose and evolution of the hearing system, the workings of the mammalian hearing system, sound location, the cocktail party effect and biosonar in bats and whales.

Mind Over Matter

Does a falling tree make a sound when there is no one to hear it? It's a question philosophers are supposed to ask themselves, the sort of question that can give rise to the uncharitable thought that they have too much time on their hands. Let's be clear: if we are to avoid confusion and get to the heart of the matter, the answer should be an emphatic "no!" Enter that particular philosophical mire and you may never find your way out. What's worse, seduced by the sedentary pleasures of philosophical rumination, you may not want to.

To be fair to philosophers, none ever posed this question in quite that form.[1] But for a very long time it was taken for granted—principally on the authority of Aristotle—that qualities perceived by our senses are inherent in the objects and events in which they are observed. In other words, colours,

---

[1] The question of whether the physical world can exist independently of perception was famously posed by George Berkely, the Anglo-Irish philosopher and Bishop of Cloyne in his 1734 *A Treatise Concerning the Principles of Human Knowledge*.

J. Naylor, *Now Hear This*,
https://doi.org/10.1007/978-3-030-89877-9_3

smells, sounds and the like all exist independently of perception. The Aristotelian assumption that sound is somehow embodied in the pulse of air created when bodies collide with one another, with the implicit corollary that a falling tree necessarily makes a sound even when no one is present to hear it, was taken for granted by everyone until the seventeenth century.

Aristotle's failure to draw a clear distinction between a physical event and the resulting sensation in the case of sound is understandable because when we hear a sound the experience is not of vibrations within air but of something altogether different: an auditory sensation. Moreover, unless one is listening through headphones or, God forbid, hallucinating, sounds don't appear to be located within one's head, they invariably seem to come from somewhere in one's surroundings. As we shall see, the reason for this is that the evolution of hearing was driven in no small measure by the survival value of being able to locate the direction of the source of a sound relative to the ears that hear it. This, in turn, depends on the brain creating the illusion that sounds are events outside the body despite the fact that they are sensations within the brain. Hence the seemingly reasonable inference that a sound is independent of its perception.

The scientific revolution of the seventeenth century, which resulted in a wholesale rejection of Aristotle's ideas about the natural world, was made possible, in part, when natural philosophers such as Galileo Galilei and René Descartes divorced sensation from matter and focused their attention exclusively on the latter.[2] For this new breed of self-styled mechanical philosopher, sensations are the result of interactions between matter and the various senses. Consequently, they reasoned, the material world itself must itself be colourless, odourless and silent, unknowably unlike the world we see, smell and hear. How an immaterial mind is able to experience a material world, the so-called mind–body problem, was left to the philosophers and, more recently, to psychologists and neuroscientists, all of who continue to wrestle with the issue.[3] And even though neuroscientists are now able to correlate particular sensations with activity in specific areas of the brain, the precise relationship between brain activity and consciousness, i.e. between matter and perception, continues to elude everyone.

---

[2] Natural philosophers were not referred to as scientists until the second half of the nineteenth century. Michael Faraday, whose discoveries and ideas transformed physics and chemistry, always insisted that he be known as a natural philosopher. The term "scientist" was coined in 1833 by the English polymath and Cambridge don, William Whewell at the request of Samuel Taylor Coleridge who demanded a more suitable and inclusive term in place of "natural philosopher". But the term did not come into general use in the English-speaking world until the 1860's.

[3] Descartes, having drawn a distinction between mind and matter, floundered when he came to explain how an immaterial mind is able to experience the material world. He claimed that the bridge between them located in the pineal gland.

Although the mind–body problem remains unresolved, since the seventeenth century it has been accepted that sound is a sensation brought about when vibrations that lie within a particular range of frequencies impinge upon the ear.[4] A falling tree generates vibrations in the air and in the ground. But without a means to intercept and convert these tremors into the electrochemical signals that are the medium of exchange of the nervous system, and a cortex to process those signals into perceptions, the world would be utterly silent. In any case, most of the vibrations generated when a tree strikes the ground are inaudible to humans, being either too sluggish or too brisk for human ears.

It's just the same with light. Photons emitted by a source of light, or reflected by the surfaces they encounter, are pictured as minute bundles of energy devoid of colour or brightness, which are the sentient hallmarks of light. And just as the human ear is insensitive to the majority of vibrations that it encounters, so too the human eye is blind to all but a tiny fraction of the photons of differing energies that reach us from objects near and far. We cannot see very high energy or very low energy photons, which science classifies as ultraviolet or infrared respectively. Indeed, as we shall see, compared to the range of vibrations to which the ear responds, the eye is sensitive to a minute fraction of the total range of energies of the photons that span the electromagnetic spectrum.[5]

The acoustic equivalent of infrared is known as infrasound and that of ultraviolet as ultrasound. And just as we can sense infrared as warmth, in some circumstances we can sense infrasound as a silent and sometimes unpleasant sensation within our body. There is also a parallel between ultraviolet and ultrasound: although we neither see nor feel ultraviolet, nevertheless it does affect us because it is the cause of snow blindness and sunburn. And although we can't hear ultrasound, these high frequency vibrations are differentially absorbed or reflected by flesh and bone, and are used in ultrasonic imaging of our innards, to break up kidney stones and even to gently warm internal organs. Furthermore, as we shall see, we are an exception among mammals because these very high frequencies are audible to almost all mammals.

---

[4] Although almost all the sounds we hear are airborne, we hear our own voices through the bone of our skull as well as through air, which is why a recording of one's voice does not sound the same as the one we are familiar with. The recorded voice invariably seems less sonorous.

[5] Taking the entire electromagnetic spectrum, from very low frequency radio waves to the highest frequency gamma radiation, the fraction occupied by the visible spectrum is approximately 1 in 10,000,000.

But despite the mechanical philosophers of the seventeenth century concluding that there is a world of difference between sound as a sensation and the physical vibrations that are its cause, as far as most of us are concerned, Aristotle is alive and well. Sounds are invariably experienced as events located within one's environment, not within one's head with the corollary that sounds are due to a special category of aerial vibrations. This is misleading because *all* aerial vibrations are in principle audible to anyone with heathy ears as long as they satisfy two conditions. First that they are sufficiently energetic to stimulate the ear and second that their frequency lies between 20 Hz and 20 kHz. Mersenne, as we saw in chapter two, had already realised as much sometime in the early decades of the seventeenth century.

The Science of Vibrations

Do you recall the baleful tagline that was used to promote the film *Alien*[6]: "In space nobody can hear you scream"? It's quite true, though it might have been helpful to have inserted "the vacuum of" between the first and second words. It's the absence of matter in a vacuum that mutes your scream because there is nothing for your vocal chords to set into vibration as they can within Earth's atmosphere. As we saw in the previous chapter, although all natural philosophers, beginning with Aristotle, accepted that sound requires a medium in which to propagate, this wasn't confirmed until the seventeenth century, when the celebrated bell-in-a-vacuum experiment was independently performed by several of the leading natural philosophers of the time.

A vibrating body, be it solid, liquid or gas, communicates its motion to the surrounding medium by alternately compressing and dilating it. Compression forces the molecules of the medium together more closely than before, which leads to a local increase in pressure. The opposite occurs when the medium is dilated: its molecules move further apart and the local pressure is reduced. As a result, the sustained back and forth motion of the source creates recurring rise and fall in pressure in the surrounding medium that spreads out in every direction like a three-dimensional ripple.

The result is known as a longitudinal wave because the vibration travels through the medium in the same plane as the back and forth motion of its molecules. And of all such waves, the most common example is the one we experience as sound. Should the particles of a medium move up and down,

---

[6] *Alien* was directed by Ridley Scott and released in 1979.

as they do in the case of waves travelling across the surface of water or along a vibrating string, the wave is said to be transverse.

Vibrations can travel though solids either as longitudinal waves or as transverse waves because, unlike molecules in gases and liquids, those in a solid are bound to one another by strong electric forces so that they act as if linked to their immediate neighbours by tiny springs. Moreover, a transverse wave in a solid will set up a longitudinal wave in the surrounding air, which is why we can hear sounds from a solid body even when it is vibrating transversely.[7] In fact, as we shall see in chapter six, transverse waves travelling through a solid are the source of an unusual sound that can be heard from slinky springs, railway lines and frozen lakes. Light, when considered as an electromagnetic wave as opposed to a particle, is also classed as a transverse wave because the fluctuations of the electromagnetic field occur perpendicularly to its direction of travel. Unlike sound, however, light does not require a medium in which to propagate. This is why you can see the sun but you can't hear it (Fig. 3.1).

Fortunately, all waves, whether transverse or longitudinal, material or immaterial, can be usefully described in terms of just three attributes: wave speed, frequency and wavelength. The speed of a wave is the rate at which it travels through a medium and is measured in metres per second. The frequency of a wave is the rate at which the medium vibrates as the wave travels through it, i.e. of compression followed by dilation and back to compression in the case of longitudinal waves, and is measured in cycles per second. The scientific unit for frequency is the Hertz, and one cycle per second is one Hertz. Wavelength is the distance the wave travels during one cycle, and is measured in metres.

The relationship between these attributes was formulated by Newton and published in 1687 in his magnum opus, *Principia Mathematica*.[8]

$$\text{Wave speed} = \text{frequency} \times \text{wavelength}.$$

The speed of a wave, however, is determined by the properties of the medium through which it travels, not by its frequency or wavelength, which as we saw in the previous chapter is another of Newton's discoveries. This means that in a given medium, the frequency and wavelength of a wave are inversely related: high frequency waves have small wavelengths and vice versa.[9]

---

[7] To hear the difference in the sound made when a solid object vibrates either transversely or longitudinally, search online for videos of "ringing aluminum rod".

[8] Newton, I. (1687) Philosophiæ Naturalis Principia Mathematica. The formula is given in proposition 50 in Book II, Sect. 8.

[9] Wave speed is designated by the symbol *c*, which stands for *celerity* (derived from Latin for *swift*).

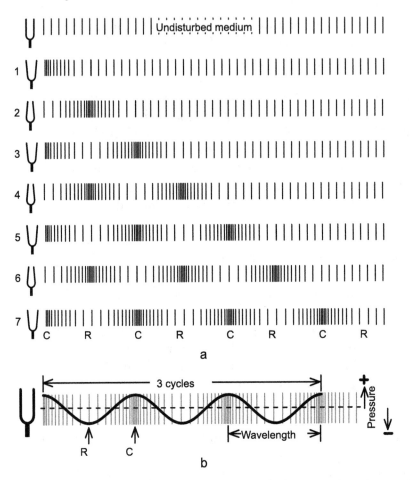

Fig. 3.1 **a** The progress of a longitudinal wave through the atmosphere (or any gas) occurs as the result of a series of compressions followed by rarefactions. In these diagrams C stands for compression (pressure momentarily above local atmospheric pressure) and R stands for rarefaction (pressure momentarily below local atmospheric pressure). **b** The variations in pressure are related to frequency and wavelength of the wave.

And in the case of a gas, the speed at which longitudinal vibrations can propagate depends only on its temperature and the mass of the individual molecules. Molecules in a gas move independently of one another so a change in pressure can be transmitted only through collisions between its molecules. This limits the speed at which a pressure wave, i.e. a vibration, propagates within a given gas to the average speed at which its molecules are travelling, a speed that is determined by the temperature of the gas. In fact, the speed at which vibrations propagate through in a gas is less than the average speed of its molecules because molecules travel in random directions, not just the direction in which the vibrations are propagating.

In a gas composed of identical molecules, all of which have the same mass, this speed increases when the temperature of the gas increases and vice versa. In the case of air, which is a mixture of nitrogen and oxygen, and which is the medium in which sounds are usually heard, the speed at which vibrations propagate—which we usually refer to as the speed of sound—is 344 m/s at 20 °C.[10] But for lighter gases it is much greater: in hydrogen at 20 °C the speed is 1350 m/s. And for denser gases it is less: in carbon dioxide—the major constituent of the atmospheres of Mars and Venus—at 20 °C it is 267 m/s.[11] And because we experience these aerial vibrations as sound we refer to the speed at which they propagate through a gas as the speed of sound.

You might expect that the speed of sound in a gas would also be affected by pressure, i.e. greater pressure, greater speed. But the speed of molecules in a gas is unaffected by pressure, so it is not necessary to specify altitude when giving the speed of sound in air.[12] The statement that "the speed of sound at *sea level* is 340 m/s" is misleading because it implies that far above the ground its speed will be different due to the reduction in atmospheric pressure that occurs with increasing altitude. In fact, the speed of sound does vary with altitude, but that's because up to an altitude of approximately 10 kms, where the troposphere gives way to the stratosphere, i.e. at the tropopause, the atmosphere is usually warmest at or near ground level and becomes steadily cooler with altitude, reaching a low of approximately -60 °C at the boundary between the two layers. And, of course, at sea level the speed of sound on a warm day will be *slightly* greater than it is on a cold day, as G. L. Bianconi discovered in 1740.

But at the top of the stratosphere, some 50 kms above the Earth's surface, the speed of sound is almost the same as it is at sea level. Although atmospheric pressure at that altitude is approximately 1% of what it is at sea level, the temperature of the air is only slightly less than that at sea level due to the absorption of a portion of the sun's ultraviolet radiation by the ozone in the

---

[10] The ratio of molecular mass of nitrogen to oxygen is 28 to 32. Speed of sound at 20 °C in nitrogen is 354 m/s and in oxygen it is 326 m/s, reflecting that their molecular masses are not the same.

[11] The speed of sound near the surface of Mars lies between 220 m/s and 240 m/s because its surface temperature averages -50 °C. Martian surface atmospheric pressure is 700 Pa (less than 1% of Earth's surface pressure). At the surface of Venus, where the pressure of its atmosphere is 90 times that of Earth and its temperature is 450 °C, the speed of sound is 420 m/s.

[12] Speed of sound in a gas depends on the ratio of pressure to density. But a change in pressure brings about a corresponding change in density, so the ratio of pressure to density remains constant.

stratosphere.[13] As we'll see in the next chapter, the variation in the speed of sound due to temperature differences within the atmosphere plays an important role in determining the greatest distance at which it is possible to hear very loud sounds.[14]

Even allowing for the effect of temperature, the speed of sound in air is almost a million times less than the speed of light.[15] This huge disparity is one of the factors that tend to reinforce our bias towards the visual world. We can usually assume that an object, even if it's moving, is where we see it because it takes an insignificant amount of time for its light to reach us.[16] But sound takes approximately 3 s to travel 1 km through air so there will always be a noticeable time lag before a sound reaches a listener from a source, even one a few metres away. This delay is the principle cause of many delightful and intriguing acoustic effects of which distant thunder is perhaps the most arresting and certainly the best-known example, and of which the echo is the most varied and fascinating.

Given that the speed of light is 300,000 km/s, to all intents and purposes we see a stroke of lightning at the instant it occurs, however far away it is. But the thunder may take several seconds to reach us, and may persist for several seconds more as sound from the more distant parts of the lightning channel reaches us. We shall return to the subject of thunder at the end of chapter six. As for echoes, where to begin? The subject deserves, and gets, an entire chapter to itself.

The speed of sound in liquids and solids is in every case much greater than it is in gases. In a liquid the speed depends on its compressibility and density, and in a solid on its stiffness and density.[17] The extraordinarily high speed of sound in diamond—approximately 12 km/s, 35 times greater than it is in air—is due to a combination of its exceptional stiffness and low density. We'll explore some unusual and intriguing consequences of the variation in the speed of sound in a liquid in both Chaps. 4 and 6.

---

[13] The ultraviolet absorbed within the stratosphere is known as UV-C and UV-B, both of which are harmful to living organisms. The ultraviolet that reaches the ground is known as UV-A.

[14] Rule of thumb for calculating the speed of sound in air at a given temperature: *Speed of sound in air = 331.4 + 0.6T_C*, where T_C is temperature of air in degrees Centigrade.

[15] Speed of light is ~3 × 10^8 m/s, speed of sound at 20 °C is ~340 m/s. Speed of light: speed of sound = ~882,353.

[16] This is not true of mirages. In a *superior mirage*, objects that are beyond the horizon appear above it, albeit often so compressed, elongated and inverted as to be all but unrecognisable. The pools of water characteristic of an *inferior mirage* are actually portions of blue skylight refracted in the direction of our eyes by the layer of warm air in contact with the ground.

[17] The degree to which a solid can be compressed or stretched is a measure of its stiffness. A liquid can only be compressed, it can't be stretched.

But when all is said and done, the medium in which we hear sounds is invariably air. Vibrations in air, as in all other media, can range in frequency from a mere fraction of a cycle per second to millions of cycles per second. However, as we have already noted, only a particular range of these frequencies stimulates human hearing and are classed as sound. Nominally this extends from 20 Hz (perceived, if at all, as a deep, visceral rumble) to 20,000 Hz (a faint reedy, pitchless whistle, assuming you are able to hear it). In practice, few adults can hear sounds much below 40 Hz or above 14,000 Hz, unless the source is extremely loud and their ears are in tip-top condition.[18]

Moreover, as we all discover sooner or later, as we age our ability to hear high frequencies diminishes, which eventually renders one deaf to high-pitched sounds. And, arguably, the most significant consequence of this loss is that it affects our ability to understand speech clearly. Not being able to hear high frequencies can also make it difficult to determine where a sound is coming from. In any case, as we shall shortly see when we come to look at the workings of the ear, all ears, however youthful, are far less sensitive to low and high frequencies than they are to intermediate ones, so sounds of different frequencies that have the same energy are not equally audible, even to a healthy ear.

Vibrations below 20 Hz are classed as infrasound. Although infrasound is inaudible to human beings, it can sometimes be felt as pressure on one's eardrum, chest and abdomen and has been known to cause varying degrees of discomfort, including nausea. Nevertheless, some animals communicate at infrasonic frequencies, most notably elephants. Man-made sources of infrasound include the very lowest frequency bass pipes of musical organs, heavy traffic and air conditioning units. There are also innumerable natural sources of infrasonic vibrations, among which are earthquakes, volcanic eruptions, large waterfalls, tornadoes, not to mention the impact of the philosopher's tree as it strikes the ground.

Vibrations above 20,000 Hz are classed as ultrasound. Although humans are unable to hear ultrasonic frequencies, almost all mammals not only hear ultrasonic frequencies, some have evolved to make active use of it to navigate and to hunt for prey. Dogs and cats can hear the lower end of the ultrasonic spectrum (which can be produced by a dog whistle) and dolphins

---

[18] Without access to a device capable of generating sounds of specific frequencies it is not easy to have a sense of what different frequencies sound like or whether you can hear them. Apps that can generate pure tones of any audio frequency as well as white noise are available for both Android and Apple phones and tablets. Try "apps:*Tone Generator Pro*" by Performance Audio for iPods and iPads. Hook your device up to a portable Bluetooth speaker (or a *vibration speaker*, a device that will turn just about any surface into a loudspeaker) and you have the beginnings of an acoustics laboratory.

are able to produce and hear frequencies between 20,000 and 150,000 Hz, which they employ to locate and identify fish by listening for reflections of the high frequency chirps that they emit as the hunt. We'll return to the subject of echolocation at the end of this chapter. Nor are animals the only source of ultrasound. Audible sibilant sounds such as *sss* and *sh* include ultrasonic frequencies. Whispered speech is particularly sibilant, which may be the reason why it is recommended that one should speak in undertones rather than in whispers if one wants to avoid distressing small mammals.[19] Metallic sounds such as jingling keys, squealing brakes and scraping a plate with a knife are all sources of ultrasound.

We saw that the wavelength of a sound in a given medium depends on a combination of its frequency and the speed at which it travels through the medium. Taking the average velocity of sound in air to be 340 m/s, the wavelength of airborne sounds range from approximately 17 m at 20 Hz to 2 cm at 20,000 Hz. In water, in which sound travels almost four and a half times faster than it does in air, all wavelengths are correspondingly greater: 76 m at 20 Hz and 9 cm at 20,000 Hz. This increase in wavelength is one of the factors that make echolocation under water far more difficult than it is in air. Moreover, as we shall see in this and later chapters, wavelength also plays an major role in our ability to locate sounds in air, in the propagation of sound around corners and the acoustic qualities of many types of echoes (Fig. 3.2).

Sounds and Waves

We almost never hear sounds in their original state, unaffected by the surroundings in which they are heard or the medium through which they travel. Thoreau ruminated that "All sound heard at the greatest possible distance produces one and the same effect, a vibration of the universal lyre, just as the intervening atmosphere makes a distant ridge of earth interesting to our eyes by the azure tint it imparts to it. There came to me in this case a melody which the air had strained, and which had conversed with every leaf and needle of the wood, that portion of the sound which the elements had taken up and modulated and echoed from vale to vale."[20]

It is, indeed, very difficult to engineer a situation in which sounds reach us in a pristine state. As we shall see in chapter five, the only environment in which it you can be assured of hearing a sound in its raw state is within an anechoic chamber, a specially constructed room that has walls that

---

[19] Sales, G. and Pye, D. (1974) Ultrasonic Communication by Animals. Chapman and Hall, p 8.

[20] Thoreau, H.W. (1893) Walden or a Life in the Woods. Houghton, p 198.

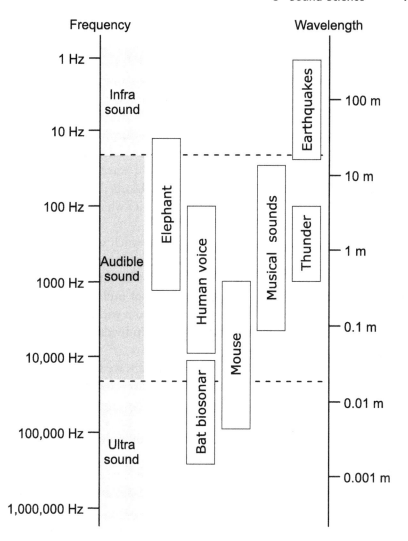

Fig. 3.2 The audio spectrum of some sounds and their sources in air

do not reflect sound. But anechoic chambers are research laboratories and are not usually open to the public.[21] However, an isolated mountain peak or a large open field, preferably covered in an acoustically absorbent layer of freshly fallen snow, make passable alternatives because of an absence of vertical surfaces to reflect sounds. You may, of course, prefer to wait until you can visit an anechoic chamber on an open day rather than trudge to the top

---

[21] Some research establishments allow occasional public access, particularly on open days, so it is always worth inquiring if it is possible to visit the anechoic chamber of a university or research centre near you.

of a mountain just to listen to the sounds in the absence of reflections. But do take a companion with you to compare notes should you wish to gauge the how a lack of reflections affects the quality of sounds in general and voices in particular.

Either way, it is worth seeking out an environment in which to experience sounds in their pristine state because it provides a benchmark against which to notice and understand the effects of the surroundings in which we usually hear them. And as noted in the opening chapter, the reason we don't hear sounds in their original state is that, being a wave, they can be modified by being reflected, refracted, diffracted or absorbed by the objects they encounter. Each of these processes affects the quality what we hear to a greater or lesser degree.

In the first place, in almost every situation in which we hear a sound, it is accompanied by one or more reflections. Reflection, occurs whenever a sound meets a surface capable of sending it off in another direction, much like a ball bouncing from a floor or a wall. The effect of multiple reflections on a sound is particularly apparent in a large, empty room within which reflections succeed one another too rapidly to be perceived individually so that they are heard as a prolonged reverberation.

Another way in which the direction in which a wave travels can be altered is through refraction, which occurs when a wave changes its speed. The circumstance in which this happens varies from one type of wave to another. The speed of wind-driven waves travelling across the surface of open water depends on depth, being less in shallow water than in deep water; light slows when it passes from a rarefied transparent medium such as air to an optically denser one such as water or glass; and, as we have already noted, sound travels marginally faster in warm air than in cold air, and a great deal faster in both liquids and solids than in gases. In the case of air-borne sound, its speed is also affected by wind. However, the effect of temperature and wind on sound is slight and so, unlike reflection, the refraction of sound in the open is usually noticeable only over distances of several hundreds of meters or more. The effect of refraction on the propagation of sound is the subject of the Chap. 4.

The law of refraction, which enables the change in direction of a refracted wave of whatever type to be precisely calculated based solely on its change in speed, was originally discovered in connection with light in the early years of the seventeenth century, first by the Dutch mathematician Willebrord Snell, and independently a few years later by René Descartes. Descartes was accused of plagiarism by some of his contemporaries, not least because his explanation

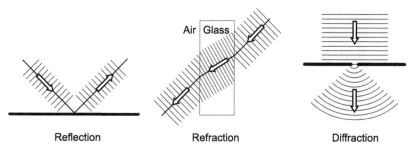

Fig. 3.3 Reflection, refraction and diffraction of waves. Refraction occurs when wave speed changes. Diffraction occurs when a wave passes through a narrow opening or encounters an edge

of why light refracts is, frankly, unconvincing. But it has since been established that he did arrive at the law independently of Snell. In any case, Snell never published his findings, and the version of the law that we use today is that of Descartes. It made its debut in his successful explanation of the shape and size of the rainbow, which was published in 1637.[22] But it was another two hundred years before anyone got around to applying the law to sound.

Diffraction is the last of these properties of waves to be discovered and proved to be the most difficult to understand and explain. It occurs when a wave meets an obstacle that has dimensions that are less than its wavelength or when a wave passes through an opening that is similar in size to its wavelength. In these situations the wave wraps around an object or spreads out after passing through an opening. As we saw in the previous chapter, diffraction was originally discovered as an optical phenomenon by Francesco Grimaldi in 1665. But neither he, nor his far better known co-discoverer, Newton, understood the significance of the phenomenon, which is that it can be adequately explained only by assuming that light is a wave. As we shall see in the next chapter, we encounter diffracted sounds several times a day without giving the matter a second thought (Fig. 3.3).

Sounds Combined

When two travelling waves meet they can produce some striking effects. Perhaps the clearest example of what can happen is the pattern created when two sets of expanding ripples, spreading out across the surface of an otherwise still body of water, cross paths. The resulting mesh-like pattern, which

---

[22] For Descartes' explanation of the rainbow see René Descartes (1637/2001) Discourse on Method, Optics, Geometry and Meteorology (trans: Olscamp, P.J.). Hacket Publishing Company, p 332–45.

persists only while the two sets of ripplescross one another, is due to the up and down motion of one ripple combining with that of the other. Where these are in step, so that the trough or crest due to one ripple coincides with the trough or crest of the other, the result is a more pronounced trough or crest; the ripples are said to have interfered constructively with one another. But if both ripples have the same height, where they are completely out of step, so that a trough of one coincides with a crest of the other, the result is a momentary patch of undisturbed water. Here the ripples have cancelled one another out, an effect known in physics as "destructive interference". But don't take my word for it: toss a couple of small, similar sized pebbles simultaneously into a still pond so that they strike the water a metre or so apart and observe what happens (and take a photograph or make a slow motion video of the pattern formed as they intersect in case you might want to mull over the event at leisure).

The patterns produced by intersecting ripples was one of the phenomena that led Thomas Young to devise the experiment that provides some of the most compelling evidence for the wave theory of light because he realised that the pattern of alternating bright and dark stripes that is observed where two identical beams of monochromatic light meet would be impossible if light consisted of Newtonian corpuscles. Young later invented a device that enabled him to demonstrate the patterns produced by ripples in a controlled manner. The device is still in use to teach students of physics about the properties of waves and is known as a ripple tank. It consists of a shallow rectangular tray filled with water that is agitated by one or more vibrating prongs to create trains of ripples that travel across the water's surface (Fig. 3.4).[23]

Destructive interference of sound is more difficult to notice than that of light, principally because sounds are more efficiently reflected than lights.

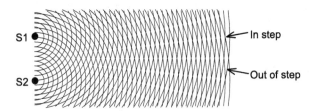

Fig. 3.4 Interference patterns are produced when two sets of identical waves intersect. Where they are in step they reinforce one another, when they are out of step the cancel one another out

[23] Paul Falstad's "Ripple" is a digital version of Young's ripple tank: http://www.falstad.com/ripplejs/ (accessed 26/07/2021).

Unwanted reflections will usually mask the silences due to destructive interference. The only place where it is possible to supress all reflected sounds is within an anechoic chamber. Yet, as we shall see in the chapter five, there are situations in which it is possible to hear the effects of interference quite clearly as long as one knows what to listen for.[24]

Destructive interference of sound is employed in noise-cancelling earphones. A microphone in the headset picks up ambient sounds and its built-in electronics inverts and feeds them to the earpiece. The inverted sound emitted by the loudspeaker in the earpiece interferes destructively with the ambient sound. In theory this should make any ambient sound inaudible. However, it is very difficult to cancel out all frequencies to the same degree, so in practice in very noisy surroundings it is still possible to hear some external sounds. High frequencies present a particular problem because the electronic circuitry in the earphones may not be able to react fast enough to deal with them. Noise-cancelling earphones have another advantage: their use reduces potential damage to the ear because one doesn't have to turn up the volume of whatever one is listening to in order to counter ambient noise.

Sounds Impeded

Whether a wave is reflected or transmitted when it encounters a new medium is not a cut and dried issue because in almost all situations it is partially reflected and partially transmitted, i.e. some of the wave's vibrational energy passes into the new medium, undergoing a slight change in speed and direction as it does so, while the remainder is reflected. This is very obvious in the case of light: whenever you see your reflection in a pane of clear glass you can also be seen by someone on the far side because while some light is reflected at surface of the glass the remainder travels through it and emerges on the far side. Partial reflection of light was something that Newton was unable to explain satisfactorily in terms of his corpuscular theory of light, despite devising an ad hoc process to explain the colours in soap films which he referred to as "fits of easy transmission or reflection".[25] The implausible assumptions of this hypothesis were widely accepted until 1800, when Young

---

[24] The acoustic version of Young's double-slit experiment makes use two speakers fed with the same audio frequency. Walking past the speakers one hears a rise and fall in loudness. Ideally the demonstration should be carried out in an anechoic chamber or outdoors far from surfaces such as walls that will reflect sounds from the speaker and make the effects of interference less obvious.

[25] Newton, I. (1675) An Hypothesis explaining the Properties of Light. In: Birch, T. (1757) The History of the Royal Society, vol. 3. London, p 249 (also online at The Newton Project: http://www.newtonproject.ox.ac.uk).

pointed out that partial reflection can be explained far more satisfactorily by assuming that light is a wave.

Where sound is concerned, the degree to which it is reflected or transmitted depends on the effect that pressure has on the motion of the molecules of the medium through which it is travelling, a property known as acoustic impedance. In a gas, a very small change in pressure will cause considerable molecular motion because individual molecules are free to move independently of their neighbours. In a liquid, molecules are packed together much more tightly than they are in a gas, so a far larger pressure is necessary to have the same effect. And in a solid, in addition to greater density, there is also the stiffness of the material to overcome. Hence the acoustic impedance of a gas is several thousand times less than that of a liquid or a solid. This difference affects the degree to which sound is reflected or transmitted when it encounters a new medium, say when a sound in air meets water or glass. The greater the difference in impedance between any two media, the greater is the proportion of sound that is reflected at the boundary between them.

Another way to visualise impedance is in terms of a collision between two bodies, one of which is far more massive than the other. When the lighter body collides with a stationary heavier one, the lighter body bounces off and the heavier one hardly moves. Very little of the energy of the lighter body is transferred to the heavier one. In this example the lighter body represents a sound wave travelling through air and the heavier one a denser medium such as water or glass.

The reason why it difficult to hear airborne sounds clearly when one's head is immersed in water unless the original sound is very loud is because that the acoustic impedance of water is approximately 3500 times greater than air. As a result only a tiny fraction of an airborne sound enters the water, almost all of it having been reflected by the surface.[26] An even greater proportion of airborne sound is reflected by glass because the acoustic impedance of glass is some ten times greater than that of water. And where a sheet of glass is concerned, due to reflections of sound within the glass, high frequency sounds are attenuated more than low frequency ones, something that you will notice when you close a window. Not only is the loudness of the external sound reduced, it also changes its timbre because the balance between frequencies is altered in favour of lower frequencies. Hence closing a window reduces traffic noise to a rumble because the hiss of tyres doesn't make it through the glass.

---

[26] The percentage of sound reflected when sound in air meets water can be calculated from the acoustic impedances of air and water, and is equal to 99.89% of the incident sound. This corresponds to a reduction of almost 30 dB as sound travels from air to water.

The situation is almost the reverse with light. The surface of a body of clear water typically reflects approximately 2% of the incident light, the remaining 98% being transmitted. In the case of glass the figures are 5% and 95% respectively. Here is the reason why closing a window is far more effective as a barrier to unwanted sound than it is to unwanted light.[27]

A difference in impedance is also the reason why sound from a source within a bell jar ceases to be audible long before a vacuum is fully established. The residual air continues to transmit sound, but the impedance of low-pressure air is more than a hundred times less than that of air at normal atmospheric pressure. The increased difference in impedance between the residual air and the glass of the bell jar means that almost no acoustic energy is transferred from the air to the glass and as far as one is able to judge, the bell becomes inaudible outside the jar.

Sounds on Other Worlds

Conditions on Mars are not dissimilar to those within an evacuated bell jar because the pressure of the Martian atmosphere at ground level is less than 1% of that of the Earth's atmosphere at sea level.[28] Accordingly, the acoustic impedance of the Martian atmosphere is approximately 1% that of the Earth's atmosphere with the implication that Martian sounds, were it possible to hear them directly, would be very faint compared to the same sounds on Earth. At the same time, the absorption of sound by the thin Martian atmosphere is at least a hundred times greater than that of the Earth's atmosphere, so that a sound such as a stroke of lightning powerful enough to be heard at a distance of several kilometres on Earth would be inaudible at more than a few hundred metres on Mars.[29]

A specially designed microphone was carried on the ill-fated Polar Lander that was launched in 1999 and which crash-landed on arrival, destroying the probe. The purpose of the microphone was not simply to listen to the Martian soundscape, delightful and intriguing though this might be. The

---

[27] The ability of a sheet of glass to cut out unwanted sounds can be improved by using so-called "acoustic glass". This consists of a sandwich of two sheets of glass separated by a thin layer of transparent viscoelastic resin (polyvinyl butyral or *pvb*). Sound is reflected at each of the glass surfaces and is absorbed as it travels through the resin.

[28] This is the same as atmospheric pressure approximately 120 km above sea level on Earth. A school vacuum pump in good working order should easily attain such low pressures.

[29] Leighton, T.G. and Petculescu, A. (2009) The Sound Of Music And Voices In Space Part 1: Theory. Acoustics Today, July, p 17–26.

quality of sound recorded by a microphone can reveal a great deal of information about the composition and properties of a planet's atmosphere as well as detecting winds blowing around the lander itself. There were plans to use the same design of microphone on a later 2007 European Mars mission that was cancelled due to funding difficulties.[30] And a microphone on the Phoenix Lander that successfully reached the Martian surface in 2008 was never switched on because there were fears that it might lead to problems with the probe's electronics. However, NASA's engineers have not given up: a microphone was installed aboard the Mars Perseverance Rover that landed on Mars on 18th February, 2021. The first sound it picked up was a brief, faint rumble as Martian "air" flowed past the microphone.[31] You don't have to travel to Mars to record wind noise, but at least this showed the microphone was in working order. Several weeks later the microphone picked up sound from the rotor blades of Ingenuity, the mission's helicopter, as it flew around Perseverance. It also picked up the noise of the Martian soundscape, wind blowing across the rim of nearby crater.[32]

Microphones were successfully deployed on Venus. Two of the Soviet Union's Venera probes, numbers 13 and 14, of the 16 that landed on Venus between 1961 and 1984 carried microphones that briefly recorded the sound of Venusian thunder and wind.[33] Sounds on Venus are inherently louder than those on Earth because the impedance of the Venusian atmosphere is almost 70 times greater than that of the Earth.[34] The clearest sound recordings from another world were obtained from a microphone on the Huygens probe that landed on Titan, the largest of Saturn's moons, in 2005 and which recorded the sound of its descent through its atmosphere (i.e. wind by any other name).[35]

Recently it has been possible to reproduce how organ pipes and the human voice should sound on other planets based on our knowledge of the properties of their atmospheres. The resulting sound files have been posted on the internet and are well worth listening to.[36]

---

[30] This was the *Netlander* mission.

[31] Martian "air" is, of course, carbon dioxide.

[32] Video of the event is available here: https://www.youtube.com/watch?v=y5niGi4k9vQ (accessed 21/07/2021).

[33] Horowitz, S.S. (2012) The Universal Sense. Bloomsbury, p 264.

[34] Like the atmosphere of Mars, the atmosphere of Venus is almost entirely composed of carbon dioxide, but Venusian atmospheric pressure at ground level is some 90 times that of the Earth's atmosphere and 9000 times that of the Martian atmosphere.

[35] "Sounds of Titan": https://www.youtube.com/watch?v=ohMJsFAay2M (accessed 19/07/2021).

[36] Leighton, T.G., Petculescu, A., (2009) The Sound of Music and Voices in Space Part 2: Modeling and Simulation. Acoustics Today, July, p27-29. For the sounds themselves: https://acousticstoday.org/off-world-sounds-sound-files/ (accessed 19/07/2021).

The Evolution of the Ear

Disputes about evolution often centre on the eye. "How can mere random mutations lead to such a complex organ?", a sceptical Creationist will demand. In *On The Origin of Species* Charles Darwin himself admitted the difficulty: "To suppose that the eye...could have been formed by natural selection, seems, I freely confess, absurd in the highest degree.",[37] but then went on to sketch out how the eye might have evolved though a succession of random mutations reinforced by prevailing conditions. Opponents of evolution don't appear to have read beyond Darwin's opening sentence on the subject of the eye, even supposing they have opened the book in the first place, and stubbornly maintain that nature alone cannot fashion a complex organ as well suited to its purpose as the eye without the intervention of a supernatural agency.

But the reality is that the eye is far from being optically perfect. Herman von Helmholtz, the foremost authority on the optics of the human eye during the latter half of the nineteenth century, asserted that "Now it is not too much to say that if an optician wanted to sell me an instrument which had all these defects, I should think myself quite justified in blaming his carelessness in the strongest terms, and giving him back his instrument."[38] But he went on to point out that although "...the eye in itself is not by any means so complete an optical instrument as it first appears: its extraordinary value depends on the way in which we use it: its perfection is practical, not absolute, consisting not in the avoidance of every error, but the fact that all its defects do not prevent its rendering the most important and varied services...".[39]

It seems as if Helmholtz's reservations have been overlooked, not least by people who insist that the creation of the world is the result of a once and for all event that took place a few thousand years ago at the behest of a supernatural being. Almost a century and a half later, Steve Jones, a British geneticist, felt it necessary to emphasise that not only is the human eye one of many types of eye that have come about through gradual evolution but also that all of them are only as good as they need to be to satisfy the demands made of them by the host's life style and environment. However marvellous it appears,

---

[37] Darwin, C. (1859) On the Origin of Species by Means of Natural Selection, or the Preservation of Favoured Races in the Struggle for Life. John Murray, London See: "Organs of extreme perfection and complication" in chapter VI: Difficulties on Theory.

[38] Helmholtz, H. (1885) Popular Lectures on Scientific Subjects. (trans: Atkinson, E). D Appleton, New York, p 219.

[39] Helmholtz, H. (1885) Popular Lectures on Scientific Subjects. (trans: Atkinson, E). D Appleton, New York, p 227.

every type of eye is a compromise between optics and biology.[40] In any case, the formation of an image on the retina by the cornea and lens is only the first of many steps that result in the mental event that we experience as sight. Almost all the processing that converts the nerve signals from the eye into a form that enables us to perceive the world visually is done in the visual cortex located at the back of the brain rather than in the eye itself. And it is that processing that overcomes the eye's optical imperfections.

But if the eye is a marvel, albeit flawed in many respects when considered exclusively as an optical instrument, the ear is its equal as a sensory organ and arguably surpasses it in terms of complexity and acuity. As for its flaws, we might compare the ear to the microphone, not always to the latter's advantage as we noted in connection with the so-called *cocktail party* phenomenon in chapter one. Indeed, it can be said of the ear that, like the eye, it is only as good as the needs of an organism dictate.[41] Moreover, evolution operates blindly, exploiting errors that arise when organisms reproduce, errors that occasionally prove to be of advantage in the struggle for survival. This make-do process, which applies in all evolutionary developments, means that, as in the case of eyes, biology has played a much more important role than physics in the evolution of ears. Eyes and ears, along with every other body part, are not deliberately designed to fulfil a particular purpose as, for example, are cameras and microphones.[42]

Darwin, incidentally, said almost nothing about the ear, possibly because far less was known about hearing than about vision during the years when he was writing *On the Origin of Species*. Here again Helmholtz blazed a trail due to his ground-breaking research into the psychophysics of the ear, for not only was he the leading authority of the day on the eye, he was also the leading authority on the perception of sound and its relationship to music.[43] And in 1857 he was the first person to propose a feasible mechanism whereby the ear is able to detect and separate sounds of different frequencies,[44] having already partially solved the parallel problem of how the eye perceives colours.[45]

---

[40] Jones, S. (1999) Almost Like A Whale: The Origin Of Species Updated. Black Swan, p 139.

[41] This is probably true of all senses: they are all only as good as they need to be given the organism's needs and environment.

[42] Manley, G.A., Lukashkin, A.N., Simões, P., Burwood, G.W.S., Russell, I.J. (2018) The Mammalian Ear: Physics and the Principles of Evolution. Acoustics Today, vol 14, issue 1, p 8–16.

[43] Helmholtz, H. (1895) On The Sensations Of Tone As A Physiological Basis For Theory Of Music (trans: Ellis A.J.). Longman Green & Co.

[44] Helmholtz, H. (1873) The Mechanism Of The Ossicles Of The Ear And Membrana Tympani (trans Buck, A.H., Smith, N.). William Wood & Co., New York.

[45] The outlines of the correct mechanism of colour perception— the so-called trichromatic theory of colour vision — was first sketched out by Thomas Young in a lecture to the Royal Society in 1802. But it was several decades before Young's ideas on colour perception were taken up and developed

We can get some idea of the extraordinary complexity of the ear by comparing it to the eye. The purpose of the eye is to convert electromagnetic energy into electrochemical nerve pulses; that of the ear is to do the same thing with vibrations in the form of minute though rapid periodic changes in the atmospheric pressure acting on the eardrum. The eye, of which the mammalian eye is but one of several types found in the animal kingdom, evolved over many hundreds of million of years from patches of light-sensitive cells capable only of distinguishing brightness from darkness into a complex organ that can form an image on an array of receptors composed of thousands of individual cells that are sensitive to both to the energy and the frequency of narrow range of electromagnetic waves.[46] But to see this image requires a brain to process the electrochemical signals generated by these cells and so the eye and those parts of the brain devoted to vision have evolved in tandem.

It's the same with the ear. Whether a creature lives in water or on land, if it is to hear, it must possess a means of converting vibrations within the surrounding medium into electrochemical signals and have a brain capable of turning these into auditory sensations. In land dwelling vertebates the conversion occurs within the inner ear, an organ that differs in design from one class of creature to another.

We noted in chapter one that, without exception, all vertebrates, i.e. fish, amphibians, reptiles, birds and mammals, can hear, although not all equally well. Nor do they all hear the same range of frequencies. Hearing—the ability to experience a succession of minute, periodic variations in ambient pressure as an auditory sensation—probably developed first in jawless fish that inhabited ancient seas some 400 million years ago, and continued to evolve in their terrestrial descendants, who first emerged from the sea around 360 million years ago.[47]

But the fossil record suggests that knowing which way is up was initially more important than being able to detect external vibrations, because hearing in fish evolved from a more ancient sense, known as the vestibular system. In all vertebrates this is located within the bone of the skull and enables them to keep track of the movements of the head relative to their surroundings. The

---

further, first by James Clerk Maxwell in 1855, and later, independently, by Helmholtz in 1860. But in 1892, Ewald Hering, a German physiologist, pointed out that both their accounts miss out an important step: the output of the retinal cells are combined in such a way that they supress one another. Hering's idea, which has since been confirmed, is known as "opponency" and is the reason why the human eye is not able to perceive red and green as a single colour.

[46] The differential sensitivity of the three types of cone cells to the range of frequencies of the visible spectrum is the basis of the eye's ability to perceive colour by pooling the relative response of each type of cone to the light falling on the retina.

[47] Very low frequency vibrations are also experienced at the surface (i.e. skin) of a body as a tactile sensation.

rotation of the head is detected by three narrow, fluid-filled tubes, known as semicircular canals, each of which is aligned in one of the three cardinal spatial axes: vertical, horizontal and side to side. Rotating one's head causes the fluid to displace specialised microscopic hairs, known as stereocilia, within chambers at the base of each canal, which in turn trigger electrochemical signals that travel along nerve fibres to the brain.[48] Two further organs—the saccule and the utricle —sense whether one's head as a whole is moving up or down or forward or backwards. Each consists of tiny fragments of calcium carbonate known as otoliths that rest on a patch of stereocillia.[49] When the body begins to move, the otoliths, which are denser than their surroundings, lag behind as the array of stereocilia brush against them resulting in an electrochemical signal that enables the brain to sense whether one is moving up or down (sensed by the saccule) or forward or backwards (sensed by the utricle).

The vestibular system is a sixth sense, of which we are by and large unaware until it goes wrong. You probably assume that the principle reason you don't fall over when you are standing still or walking or running is entirely due to vision: when your eyes are open you see your surroundings and this enables you to maintain your balance in most situations. But whenever you move your head, your eye movements are under the control of your vestibular system. When the semicircular canals sense that the head is rotating in one direction, they automatically signal the eye muscles to move the eyes in the opposite direction, thus keeping the image of whatever you are looking at on the same bit of the retina. If you want proof of how effective this reflex is, continue reading this page while shaking your head up and down or from side to side. These movements hardly interfere with your ability to keep your eyes "on target" because your vestibular system is supplementing the voluntary movements of your eyes as they scan across the page. And to reinforce the role that the vestibular system plays in stabilising your eyes, try reading when you move the book up and down rather than your head. The reflex also explains why you are largely unaware that your head is bobbing up and down when you are walking, whereas a hand held video made while you are walking reveals just how jerkily you move.

The semicircular canals were discovered in the mid sixteenth century by an Italian anatomist, Gabriello Fallopio. But their purpose and significance wasn't recognised until the middle of the nineteenth century when a French anatomist, Pierre Flourens, severed the semicircular canals in several live pigeons. He had assumed that the canals were part of the hearing system and

---

[48] Strictly speaking, given the structure of these hair cells, *stereocilia* should be known as *stereovilli*.
[49] "oto" = ear, "lith" = stone.

was surprised that following their excision, pigeons could still hear but they couldn't hold their heads up or walk straight. And 1861, a French doctor, Prosper Ménière, found that congenital disorders of the vestibular system in humans adversely affects not only balance but also hearing, a condition that now bears his name. Julius Caesar almost certainly suffered from Ménière's disease rather than epilepsy, as did Jonathan Swift.[50]

The vestibular system can also be impaired by an inflammation of the inner ear, incapacitating the sufferer, who may be unable to stand up, raise his head or keep his eyes steady without feeling dizzy and nauseous until the infection is treated.[51] Having myself once suffered for several weeks from this unpleasant condition, I have first hand experience of just how distressing and disabling this is. The slightest motion of my head sent the room spinning and made me nauseous. Nor could I stand without holding onto someone, let alone walk unaided. Although the worst effects abated within a couple of days, it was several months before I completely recovered. For several weeks I was wary of turning my head while walking because it made me slightly dizzy when I did so. Alcohol in excess has the same effect because it diffuses from the bloodstream into the fluid within the canals, creating convection currents within the fluid that stimulate the stereocilia to respond as if one's head is spinning.

Over eons, the displacement of stereocilia due to bodily movements was adapted to detect external vibrations. In fish this takes place within the saccule at the base of the semicircular canals. Unlike land animals, however, fish have no opening through which vibrations can reach their hearing organ. Nor do they need one because being composed mainly of water the acoustic impedance of their flesh is almost identical to that of water, which makes it transparent to vibrations within the medium in which they live. Were it not for the otolith's, these vibrations would pass through their saccule without trace. As it is, although the vibrations travel unimpeded through the fluid within the saccule, the otolith, which is much denser than its surroundings, remains stationary as the array of stereocilia brush against it. This, in turn, produces minute electrochemical pulses that travel along the auditory pathways of the brain to its auditory centres. In some bony fish hearing is enhanced by a small, air-filled pouch, known as the swim bladder, which is used to control buoyancy. The swim bladder lies next to the saccule and, being air-filled, its volume fluctuates in response to external vibrations. This

---

[50] Cawthorne, T (1958) Julius Caesar And The Falling Sickness. Proceedings of the Royal Society of Medicine 51 (1), p 27–30.

[51] McCredie, S. (2007) Balance: In Search of the Lost Sense. Little Brown Book Group.

magnifies the movement of the saccule and extends the range of frequencies that can be heard by fish such as goldfish and herring, among others.[52]

Having evolved in a watery environment, the organ that enabled primeval fish to hear had next to adapt to an altogether less suitable sonic medium: air. The problem is that due to the low acoustic impedance of air compared to water, vibrations in air are too feeble to pass easily into flesh, let alone enter a liquid-filled inner ear. Instead, aerial vibrations are almost entirely reflected at the surface of a creature's body, which implies that the first land vertebrates that evolved from fish had very limited hearing in air. The only way that sound could reach the inner ear of these ancestral creatures was through the jaw, which rested on the ground because, despite having limbs, they were unable to lift their body clear of the ground. This would, of course, have limited them to hearing low frequency sounds propagated through the ground, such as the shuffling movements of prey or predator.

But hearing has proved to be such a vital sense for survival that tens of millions of years after leaving the sea, modifications of features present in the anatomy of these early land vertebrates were adapted to overcome the problems posed to their primitive hearing organ by being immersed in a medium a thousand times less dense than water.

If feeble aerial vibrations are to make it into the inner ear it is necessary not only that it be connected to the outer surface by an air-filled passage, i.e. the ear canal, but also that the minute variations in air pressure due to those vibrations be considerably amplified. And the way in which this has been achieved is one of the marvels of evolution. In a process that began some 250 million years ago, bones that were once part of the jaw hinge have been refashioned and adapted in stages by trial and error over millions of years to transfer vibrations from a stiff membrane, the so-called eardrum, to the inner ear, which itself underwent considerable development from its fishy ancestry.[53]

These auditory bones, known as ossicles, are the smallest bones in the body—human ossicles are about the size of a grain of rice. Not only do ossicles transfer vibrations from the eardrum, they also increase the pressure exerted on the inner ear because the cross sectional area of the stapes, the ossicle that is attached to the membrane that covers the entrance to the inner

---

[52] In those fish with the most sensitive hearing, the vibrations of the swim bladder are transmitted to the inner ear by tiny bones, known as Weberian ossicles.

[53] The eardrum almost certainly began as a membrane on the surface of the head, and only later migrated to the base of the ear canal.

ear, is a fraction of that of the eardrum. Hence the tiny variations in air pressure exerted on the eardrum result in much larger pressures acting on the inner ear, which in land vertebrates differs greatly from that of fish.

Remarkably, ears able to cope with airborne sounds did not evolve until some fifty million years *after* the ancestors of modern amphibians, reptiles and birds had diverged from the ancestors of modern mammals.[54] Equally surprisingly, the ancestors of amphibians, reptiles and birds were themselves already separate groups when they began to acquire the ability to hear faint airborne sounds—they could probably already hear loud sounds. One result of this is that the majority of land-dwelling vertebrates—i.e. amphibians, reptiles and birds—have only one ossicle, known as a columella. Mammals are unusual in that they have three ossicles hinged together to form a series of levers that enable them to hear sounds that have much higher frequencies than is possible with a single ossicle. The combination of reduced area and the increased leverage due to three hinged ossicles increases the pressure exerted on the inner ear some thirty times compared to that exerted on the eardrum.[55] Even this three-boned arrangement can't make up for the loss due to the acoustic mismatch between air and water, though it gets more than half way there. It also enables mammals to hear much higher frequencies than any other vertebrate creature. Ancestral mammals, it is conjectured, required more acute hearing than other land dwelling vertebrates because they were primarily nocturnal, and so relied on hearing to a much greater degree than their diurnal reptilian counterparts both to search out prey and to avoid been preyed upon.[56]

Life on land wasn't for every mammal, however. Beginning some 30 million years ago, one group of mammals returned to the sea in stages and eventually evolved into whales and dolphins.[57] Immersed as they were in dark and murky waters, hearing became far more important to their ancestors than vision. They have retained their mammalian inner ear, but because the ear canal was long ago closed off, external vibrations reach the inner ear through their lower jaw. The jaw is an excellent conductor of sound, as we shall see when we take a look at biosonar in the final section of this chapter.

---

[54] Independent evolution of organs and body parts in different organisms that have the same function is very common and is known as *convergent evolution*. For example, insects, birds and bats all evolved wings even though they do not share a common ancestor.

[55] In the human ear the area of the eardrum is $80mm^2$ and that of the stapes is $3mm^2$, so there is a 27 fold increase in pressure between eardrum and the opening to the inner ear.

[56] Shubin, N. (2008) Your Inner Fish: A Journey Into The 3.5 Billion-Year History Of The Human Body. Allen Lane, p 159–164.

[57] The fossilized remains of the terrestrial ancestor of whales and dolphins was originally discovered in 1970's in the Himalayan foothills of Northern Pakistan. These foothills were once on the coast of the Asian landmass before the Indian subcontinent collided with Asia some 50 million years ago.

As we have already noted, the inner ear of terrestrial vertebrates differs from that of fish. In addition to having to cope with feeble vibrations, it also has to be capable both of being sensitive to a wider range of frequencies and of greater discrimination between frequencies if it is to improve their chances of survival. This form of inner ear is known as a cochlea and is a hollow, fluid-filled tube within the bone of the skull just behind the ear and adjacent to the semicircular canals of the vestibular system. The cochlea contains the Organ of Corti, a sensory organ that converts vibrations into sensations, and which was discovered in 1851 by the Italian anatomist Alfonso Giacomo Gaspare Corti. Although it is found in every terrestrial vertebrate as well as in aquatic mammals, the shape of the cochlea differs from one class of vertebrate to another. In birds the cochlea is more or less straight and in mammals it is a shell-like spiral tube, similar in shape to a snail shell, from which it takes its name. The coiled cochlea in mammals is an evolutionary adaptation that enables a longer Organ of Corti to be accommodated within a tiny space in the skull. This, of course, is one of the reasons why mammals can hear a much greater range of frequencies than other vertebrates.[58]

The combination of eardrum and ossicles that transfers variations in air pressure acting on the outer surface of the eardrum via the ossicles to the cochlea is only the first step in a series of physiological and neurological processes that results in auditory perception. To begin with, the complex variations in air pressure that reach the eardrum has to be separated into its constituent parts, i.e. into individual frequencies. At the same time the amount of pressure exerted on the eardrum must somehow be gauged, enabling us to judge loudness. Both these things are achieved mechanically, i.e. through the movements of the relevant parts of the inner ear, movements that have then to be converted into electrochemical impulses that are sent to the auditory centres of the brain where they result in auditory perceptions, i.e. in the sensation we experience as sound (Fig. 3.5).

Sounds Untangled

When we listen to an orchestra in full flow most of us can usually make out a few individual instruments such as violins and trumpets. But the pressure wave that arrives at the eardrum is a composite of all their vibrations. It's the same for almost every sound we hear: they are all a combination of multiple frequencies, each of which is likely to differ in intensity and duration

---

[58] Manley, G. (2015) Aural History. The Scientist Magazine, Sept, p 36–42.

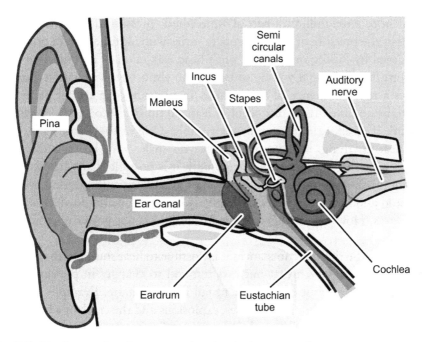

Fig. 3.5 The human hearing system showing its important features

from all the others. Moreover, in many situations several distinct, unrelated sounds reach the ear at the same time, which is at the root of the troublesome *cocktail party* effect. Hearing a sound that has a single frequency is extremely rare and will invariably be man-made, such as a simple whistle or a vibrating tuning fork. The hearing system did not evolve to cope with monotone sounds. Music, which together with animal vocalisations such as birdsong and human speech, is the most deliberately fashioned of all sounds, not to say the most complex, typically consists of several harmonically related frequencies and variations in intensity. Almost all other sounds— from a clap of thunder to the babble of a turbulent stream — are a mixed bag of unrelated frequencies.[59]

The change in pressure exerted on the eardrum by sound is in most cases almost vanishingly small. To give an idea of just how small, the variation in air pressure air acting on the eardrum of someone listening to conversation in a normal voice at a distance of one metre is in the order of 0.02 Pa.[60] This is

---

[59] You can learn a great deal about the composition of sounds with an audio spectrograph such as *Spectrum View*, an app for iOS devices by Oxford Scientific Research, that display the frequency spectrum of sounds (rather like a spectrometer reveals the different colours present in white light).

[60] 1 pascal (Pa) is the pressure due to a force of 1 N acting over an area of 1 square meter. 1 N is the approximate weight of a medium sized apple.

equivalent to a ten millionth part of atmospheric pressure at sea level, which, given that the pressure of the atmosphere varies with altitude, can in principle be achieved by raising or lowering one's head a fraction of a millimetre. But you won't hear a thing if you do so because in almost every situation hearing requires a rapid sequence of fluctuations in pressure over a brief period of time; to be audible these must occur at least 20 times a second, which would be heard, if at all, as a low frequency rumble. One of the reasons why you won't hear anything, however vigorously you shake your head, is that you can't do so at the required rate. Intriguingly, it is possible that when walking or running, one's ears are subjected to very low frequency infrasound because one's head and hence one's ears are regularly bobbing up and down by several centimetres. However, in this case the infrasound is not powerful enough to be felt.

In fact, in the right circumstances it is possible to hear sound in the absence of any vibrations. Given that our ears respond to changes in pressure, all it takes is a very brief but rapid rise and fall in air pressure. Examples of such sounds include sonic booms, thunder, explosions and the crack of a bullwhip. All of them are all caused by a brief spike in air pressure known as a shock wave and which is the subject of the final section of Chap. 6.

Of course, we are able to hear sounds that are far fainter than those of a normal conversation. A whisper involves fluctuations in pressure one hundred times smaller than that involved in conversation. In fact, in terms of energy, the sensitivity of the ear closely matches that of the eye, which is remarkable considering that the means by which the ear detects these minute variations in pressure is entirely mechanical.[61] This degree of sensitivity depends on the extremely delicate structures within the inner ear that are easily damaged by loud sounds. As we noted in chapter one, hearing loss, which occurs naturally with age, is hugely accelerated by exposure to loud sounds for prolonged periods. The sound of a jet engine at full blast at a distance of 50 m involves pressure variations some ten thousand times greater than that of normal conversation, i.e. 200 Pa. Nor will your unprotected ears long survive exposure to such a sound, even through the maximum pressure it exerts on the eardrum is 500 times *less* than standard atmospheric pressure.

Decomposing the complex pressure wave pressing on the eardrum into its constituent frequencies depends on a thin, narrow, tapering membrane,

---

[61] To compare sensitivities the minimum amount of energy per second that will produce a sensation is measured. In both organs this is approximately 10^{-16} J/s. Incidentally, the human eye cannot "see" individual photons; at least ten photons are required to produce a sensation of light. The eye is a chemical device. Photons are absorbed by molecules in retinal cells and temporarily alter their chemistry. It is this change that is the source of the signals that are sent to the visual cortex.

approximately 1 mm wide and 3.5 cm long, known as the basilar membrane, which lies within the fluid-filled space of the cochlea. This membrane runs the length of the cochlea and is much stiffer and narrower at the end nearest the entrance to the cochlea where the stapes is attached than it is at its far end. Variations in pressure within the fluid in the cochlea due to the back and forth motion of the stapes cause a wave to travel along the basilar membrane rather like the pulse that moves along the length of a bull whip when it is cracked. The position along the membrane at which the pulse reaches its maximum amplitude depends on the frequency of incoming vibrations. The stiffer part of the membrane responds to high frequency pressure waves and the floppier part, furthest from the stapes, to low frequency pressure waves.[62] The movement of the membrane is converted into electrical impulses by tiny bundles of miniscule rods known as stereocilia arranged in an array of some three and a half thousand cells, known as *inner* hair cells. These are distributed along the entire length of the Organ of Corti, which is attached to the surface of the basilar membrane.[63] The motion of the basilar membrane causes the stereocilia to brush against a rigid plate that lies parallel to the basilar membrane, and known as the tectorial membrane, and as they move to and fro they act like switches that trigger minute electrochemical impulses that are conveyed along auditory nerves that connect each individual hair cell directly to the brain.[64]

If a sound consists of two distinct frequencies, one low and the other high, then the stiffer end of the basilar membrane vibrates in step with the high frequency and the floppier end does so in step with the lower frequencies. Groups of inner hair cells at either end of the Organ of Corti are activated by the motion of the basilar membrane and the brain receives signals from each set.

The cochlea's ability to discriminate between neighbouring frequencies is enhanced by a second array of hair cells on the basilar membrane. These are known as outer hair cells. There are three times as many of these as there are inner hair cells, and they are arranged in groups of three on the basilar membrane in an array that runs parallel to that of inner hair cells. Their role

---

[62] How the ear untangles a complex sound into a series of individual vibrations was discovered by Georg von Békésy in 1928, for which he received the Nobel Prize in Physiology in 1961. For details see Békésy's Nobel Lecture "Concerning The Pleasures Of Observing, And The Mechanics Of The Inner Ear" https://www.nobelprize.org/uploads/2018/06/bekesy-lecture.pdf (accessed 23/08/2021).

[63] Alfonso Corti, an Italian physiologist, described the organ named after him in a paper published in 1851.

[64] Each eye has some 120 million rods and six to seven million cones but the optic nerve that connects them to the brain has only one million fibres, whereas each hair cell within the cochlea is individually connected to the brain.

is to amplify the movement of the stereocilia of the inner hair cells in order to limit the area of the basilar membrane that resonates in response to the pressure wave travelling through the fluid within the cochlea.

The Organ of Corti is, in effect, an auditory prism, which in some circumstances enables the ear to achieve something that the eye cannot do without help. We cannot separate out individual colours in lights and pigments without the aid of a prism or similar device, though one can make an educated guess—something that painters used to preparing and mixing pigments become rather good at. The aural equivalent to a rainbow, in which the spectrum of sunlight is separated, albeit imperfectly, into its constituent colours, occurs within the cochlea. Moreover, our aural palette is far more extensive than our visual one: the ear is sensitive to almost ten octaves, whereas the eye barely manages one octave.[65]

But exactly how the array of inner hair cells in the Organ of Corti encode and signal the vibrations of the basilar membrane is still something of an open question, as we shall now see in connection with pitch.

The Paradoxes of Pitch

Pitch is arguably one of the most misunderstood of the qualities that we perceive in sounds. It is often said to be the aural equivalent of colour because in both cases the perception appears to be determined by the physical frequency of the stimulus. But the relationship between stimulus and sensation is more complicated than this implies. In the case of light the high frequency end of the spectrum of visible light is sensed as violet and the low frequency end as red. We can, however, perceive colours that are not present in a light source and as a consequence colours that appear identical to the eye may have different spectral compositions. Such colours are known as metamers. A well known example of a metameric colour is yellow, which exists in its own right as a pure colour in the spectrum of sunlight, but can also be reproduced in RGB television screens and computer monitors, neither of which have yellow LEDs, by mixing light from red and green LEDs in suitable proportions.[66]

---

[65] Visible light wavelengths lie between 400 nm (perceived as violet) and 700 nm (perceived as red).
[66] Individual LEDs in a TV are too small to be resolved separately by the eye. So although the light from red and green LEDs do not overlap, eye movement in effect fuses them together. This type of colour mixing is known as partitive mixing and is also the basis of printing coloured images (CYMK) and was employed by Pointillist painters such as George Seurat in the belief that this would yield brighter canvases because by not mixing colours together individual colours remain pure, i.e. saturated.

A similar phenomenon occurs with sound because two or more sounds that are composed of different groups of frequencies can have the same pitch. This is why different musical instruments, each of which produces a different range of harmonically related frequencies that depend on their design, can play the same notes. More surprisingly, it is possible to hear a pitch that is not physically present in a sound that is composed of two or more frequencies.

Part of the explanation for these puzzling facts is that pitch is all too often conflated with frequency, especially by physicists, who are, by and large, uninterested in psychoacoustics. They gloss over the fact that pitch is a sensation whereas frequency is a measure of a physical process. That is not to say that pitch and frequency are unrelated: high frequencies are perceived as high-pitched sounds and vice versa. Indeed, the pitch of a pure tone, i.e. one that is due to a source emitting a single, steady frequency, is taken to be the same as that frequency. But, as we have already noted, pure tones are very much the exception and almost every sound that has an identifiable pitch is a combination of several steady frequencies that are whole number multiples of one another, which collectively are known as *harmonic partials*.

Surprisingly, where musical tones are concerned, however many harmonic partial tones are present, the pitch one hears is always that of the lowest frequency, known as the *fundamental*; the higher frequencies are known as *overtones*. For example, the fundamental tone of a sound consisting of a series of frequencies such as, say, 100, 200, 300 and 400 Hz is 100 Hz. That is the pitch which would be heard in this example. Overtones add colour and depth to the perceived pitch, i.e. they contribute to the timbre of the sound.

The commonest sources of such sounds are musical instruments, which, of course, have been designed to create sounds that have specific pitches. However, although the pitch of the note produced by, say, a flute and a clarinet may be the same, they differ in timbre. The timbre of a note played on a flute differs from that of the same note played on clarinet because, among other things, the same combination of frequencies is not present in the note produced by each of these instruments due to their different design. A flute is effectively a tube open at both ends while a clarinet is tube closed at one end (at the mouthpiece) and open at the other. As a result, all the possible overtones of the fundamental are present in a note played by a flute, whereas with a clarinet the even-numbered overtones are missing. The difference is obvious in the timbre of each instrument.

All this points to the fact that pitch is not a property of physical vibrations. It is derived in the brain from the harmonic relationships between the frequencies that are physically present. This enables us to hear the fundamental whether or not it is physically present in the sound, i.e. the

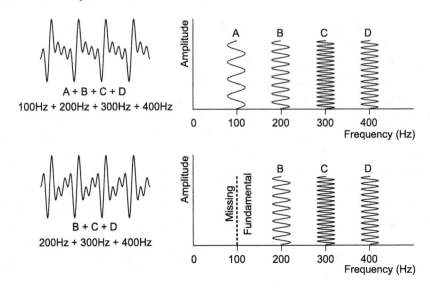

Fig. 3.6 Missing Fundamental. These two diagrams show that the absence of the fundamental has little effect on the overall envelope of sound. The combination of A + B + C + D has a very similar waveform to B + C + D and the difference in timbre between them is hardly noticeable

corresponding frequency is not physically present in the source. Hence the pitch of a sound composed of 200, 300 and 400 Hz is perceived to be 100 Hz, though its timbre will be perceptibly different from the sound that physically includes the fundamental (Fig. 3.6).[67]

The ability of our auditory system to recreate the missing fundamental or *residue pitch*, as it is also known, i.e. the frequency that is not physically present or is very weak in the source, can be exploited to produce low frequencies.[68] A small loudspeaker, such as those in telephones and earphones, that have dimensions that are much less than those of the wavelengths of the sounds it is required to produce, reproduce low frequencies only faintly. The fact that we nevertheless hear low frequencies distinctly when using such devices is due to the phenomenon of residue pitch. The inability to create low frequencies also affects some musical instruments such as the viola and cello, the strings of which resonate weakly in the lowest part of their range. In these instruments, the higher harmonics are much louder than the fundamental, which, nevertheless is heard quite distinctly.

---

[67] Listen to a missing fundamental: https://www.youtube.com/watch?v=AZ8qZCGg4Bk (accessed 19/07/2021).

[68] "Residue pitch" was coined in 1940 as an alternative to "missing fundamental" by a Dutch physicist, J.F.Schouten.

The distribution of frequencies is not the only thing that determines the timbre of a sound. Just as important is the initial rate at which the sound builds up to its maximum loudness, known as its attack, as well as the rate at which it decays once the attack phase has ceased and how long it persists, known as its sustain. Together these define the so-called envelope of a sound. Sounds with a fast attack, i.e. that reach maximum loudness rapidly, include explosions, hand claps and plucked string instruments. Those that increase slowly include distant thunderclaps, the human voice and a note played on a French Horn.

Even though it is usually very brief, the attack phase has been shown to play an important role in our ability to identify musical instruments from their sounds alone. Experiments in which the attack phase of a selection of musical instruments was removed have shown that doing so results in confusion: a tuning fork can be mistaken for a flute, a trumpet for a cornet and a clarinet for an oboe.[69] And given how important hearing is to survival, with the consequent need to identify a sound as rapidly as possible, there is every reason to suppose that the attack phase is equally important in identifying everyday sounds, though little research appears to have been done on this issue.

Alfonso Corti's investigation of the anatomy of the cochlea made it possible to begin to understand how the ear perceives pitch. In 1863, a few years after the publication of Corti's discoveries in 1851, Helmholtz proposed that pitch is determined by which part of the basilar membrane vibrates most in response to a sound. This, in turn, stimulates only those hair cells in the immediate vicinity of the vibration. Hence the perceived pitch is determined by the position of the group of stimulated hair cells along the Organ of Corti.[70] This has come to be known as the *place theory of hearing*. An alternative theory, originally suggested by August Seebeck, who in 1841 discovered the phenomenon of residue pitch using sirens, proposes that the pattern of stimulation is equally important, especially in the perception of high frequencies. This is known as the *temporal theory of hearing*. Physiologists now accept that pitch perception depends on both mechanisms, though the exact details are still being worked out.[71]

Something else that demonstrates that the relationship between stimulus and sensation is not straightforward is that it is not possible to assign a

---

[69] Winckel, F. (1967) Music Sound and Sensation, A Modern Exposition. Dover Publications, p 34.

[70] Helmholtz, H., (1895) On The Sensations Of Tone As A Physiological Basis For Theory Of Music (trans: Ellis A.J.). Longman Green & Co.

[71] Note that a similar situation occurs in the eye. Two mechanisms are responsible for colour perception: the trichromatic response of the cones & the opponent pooling of the signals from the cones before they arrive at the visual cortex.

distinct pitch to all audible frequencies, something that you are sure to have noticed. Pure tones that have clearly identifiable pitches lie between 40 and 4000 Hz. Sounds with a frequency below 40 Hz are heard as low pitched hums, always assuming that they are loud enough, given that the ear is relatively insensitive to low frequencies. And the reason why it is not possible for the ear to discriminate between very high frequencies is that only about a quarter of the basilar membranethe portion closest to the entrance to the cochlea—responds to frequencies above 4000 Hz. Hence very high frequencies all sound similar, namely a piercing, high pitched whistle, because only 20% of the basilar membrane is devoted to 80% of the audible spectrum (i.e. from 4000 Hz to 20,000 Hz.)

The antithesis of pitched sound, however, is not silence, it is a chaotic jumble of unrelated, irregular frequencies, otherwise known as noise. As we noted in chapter one, with the exception of the biophony, pure tones and harmonic sounds are extremely rare in the geophony of the natural world and in the soundscape due to human activity, i.e. the anthropophony. Consequently, almost every sound we hear is discordant to a greater or lesser degree.

The reason why these discordant noises don't all sound the same is that each is composed of unique combination of frequencies, some of which will be more energetic (i.e. louder) than others. Sounds that consist mainly of inharmonic low frequencies will be heard as a boom (e.g. distant thunder) and those that are composed mainly of inharmonic high frequencies will be heard as a squeak (e.g. a hinge in need of oil).

Many natural sounds, however, consist of a wide range of frequencies, all of similar energy, spanning much of the audible spectrum. They are heard as a pitchless hiss and are often referred to as *white noise*.[72] But though such sibilant sounds are common—as we noted in chapter one you can create a hiss by clenching your teeth and steadily expelling air through them—there are no true examples of white noise in nature because none is composed of an infinite number of audible frequencies, all having the exactly same amount of energy, which is the technical definition of white noise. Real-world sounds that are taken to be white noise include those due to wind in trees, heavy rainfall, waterfalls, surf, effervescent medicines and the flame of a blowtorch. Although these sounds are all composed of a broad range of frequencies, the distribution of frequencies and their relative loudness varies from one source to the next, as you will easily notice if you are able to listen to two or more of these supposedly white noises in close succession.

---

[72] White noise is the acoustic analogue of white light. White light is the simultaneous perception of a combination of all the frequencies present in visible light.

In practice, the distribution of frequencies in many of these sibilant sounds is closer to that of so-called *pink noise*. Pink noise, like white noise, is an idealised version of a full spectrum of sound frequencies. But with pink noise the energy is not uniformly distributed across the spectrum. Instead, the amount of energy in each octave is the same, so that the low frequency end is louder than the high frequency end of the spectrum. Listen to examples of both white and pink noise through headphones from any of the many samples available online and hear the difference for yourself. To my aging ears, heavy rainfall sounds much more like pink noise than white noise.[73] And as we shall see in chapter five, when one of these pitchless broadband sounds is combined with its reflection it is possible to alter the distribution of frequencies through destructive interference and consequently to alter its timbre noticeably.

How Big is that Sound?

Loudness is a sensation determined by the size of the vibrations of the basilar membrane, which in turn depend on the changes in pressure acting on the eardrum. As with pitch, the link between loudness and the physical stimulus is anything but straightforward. In the first place, the ear is not equally sensitive to every frequency, however healthy or youthful it may be. Sounds at either end of the audible spectrum must be far louder if they are to be equally as audible as those in the mid range. This is particularly true for very low frequency sounds, though unless you have had the opportunity to compare pure tones of equal energy across the entire audible spectrum, you may not be aware of this.[74]

The ear is least sensitive to frequencies between 20 to 200 Hz, and most sensitive to those between 3000 and 4000 Hz. In other words, a pure tone of 100 Hz is far less audible than one of 3000 Hz even when they should be equally loud by any objective measure such as the variations in air pressure acting on the eardrum.[75] One reason for our greater sensitivity to

---

[73] There are several types of "coloured" noised: in addition to white and pink, there is brown and black. Tone generator apps can also create "coloured" noise. Many of these apps are sold as aids to sleep.

[74] A similar situation occurs with the eye: the retinal cone cells are not equally sensitive to every colour in the spectrum. They are least sensitive to the violet end and most sensitive to the greeny-yellow section in the middle of the spectrum.

[75] You can confirm this with a tone generator app: gradually increase the frequency without altering the volume of the sound and the apparent loudness increases, reaching a maximum around 3000 Hz, above which its loudness decreases.

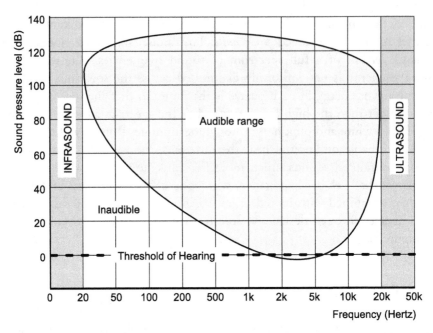

Fig. 3.7 The sensitivity of the Human ear. The range of sounds audible to the healthy human ear extends from 20 Hz to 20,000 Hz. The human ear is most sensitive to frequencies between 1.5 and 3 kHz. Sounds below 1.5 kHz must be much louder to be equally audible. A 50 Hz sound must be some 50 dB louder than one of 500 Hz if it is to be perceived as equally loud

these intermediate frequencies is that the adult ear canal, or meatus, though which sound reaches the eardrum acts as a resonant chamber with a resonant frequency that lies between 3000 and 4000 Hz, making those frequencies noticeably louder than they would otherwise be.[76] Another reason is that one of the primary functions of the ear is to detect the direction from which a sound reaches the ear, which it does by exploiting high frequencies present in sounds composed of a wide range of frequencies (Fig. 3.7).

To complicate matters, the ear's response to loudness is not linear. That is to say that a sound that we judge to be twice as loud as another is in fact much more than twice as large measured in terms of its energy or the pressure it exerts on the eardrum. Being a sensation, however, it is not possible to measure loudness objectively, it can only be estimated. Fortunately there is

---

[76] Assuming the average ear canal to be a narrow tube 3 cm long, open at one end and closed at the other, the corresponding fundamental wave for such a tube is ¼ λ (λ = wavelength). So ¼λ = 3 × 10^{-2}, and λ = 12 × 10^{-2} m. The corresponding frequency, f, can be calculated from 340 m/s = 12 × 10^{-2} m × f from which f = 340/12 × 10^{-2} = 2,800 Hz (close enough to 3000 Hz). The longer the mear canal, the lower its resonant frequency.

a rule of thumb that appears to apply to all sensations that was discovered in 1860 by Gustav Fechner, a contemporary of Helmholtz, and like him a physicist and physiologist. Known as Fechner's law, the rule states that to double the perceived strength of a sensation it is necessary to increase the strength of the stimulus tenfold.[77]

In the case of the ear, this means that it has to be sensitive to a huge range of sound pressures. Indeed, the smallest change in air pressure acting on the eardrum that will create a barely audible sound in a healthy ear is 1,000,000,000 (one billion) times less than that of a sound loud enough to cause pain in one's ears.

Obviously, any scale used to measure such a huge range of pressures cannot be linear if it is to be practicable, which is why the one used in connection with sound is logarithmic. The basic unit of this scale is the bel, which for convenience is subdivided into 10 decibels—hence the scale is known as the *decibel* scale rather than the *bel* scale. Each successive step in this scale represents a tenfold increase in intensity: 2 bels (i.e. 20 decibels) is 10 times greater than 1 bel (10 decibels), 3 bels is $10 \times 10 = 100$ times greater than 1 bel, and so on. The zero of the scale (0 bels) was established by testing large numbers of people to determine the faintest sound that is just audible by the average healthy ear.

Zero on the decibel scale does not imply total silence, i.e. a complete absence of vibrations, because these could in principle be heard by ears more sensitive than ours. It represents a threshold below which the average healthy human ear cannot hear anything. However, someone with particularly acute hearing can hear sounds of slightly less than zero decibels at frequencies where the ear is most sensitive (1000–3000 Hz). And in an anechoic room, very sensitive microphones can pick up "sounds" of –10 dB. The upper limit of this scale is 194 dB, which is due to a pressure difference of one atmosphere between the inner and outer surfaces of the eardrum. Such a pressure difference is large enough to kill a person, let alone destroy their hearing. Larger pressure differences are produced by shock fronts due to, among other things, explosions.

There is nothing uniquely auditory about this scale, however, because it is simply a scale of magnitudes. Such scales are useful when the range of quantities that are to be represented is huge. Both the electromagnetic spectrum and the Richter scale of earthquakes employ a logarithmic scale. And a logarithmic scale is a very convenient way to illustrate the dimensions and developmental history of the universe.

---

[77] An alternative version of Fechner's law is that the perceived strength of perception is proportional to the logarithm of the stimulus.

Not only does the decibel scale make it possible for scientists and engineers to deal with the huge range of pressures exerted by audible vibrations on the ear, it also provides them with a way to estimate their subjective loudness. The reason is that an increase or decrease of 1 bel corresponds to a tenfold change in intensity, which according to Fechner's law is experienced as a doubling or halving in perceived loudness. Furthermore, since a sound that causes pain is 1,000,000,000,000 times more intense than the faintest audible sound, there is only a 12-fold change in sensation over this enormous range of pressures. A sound that causes pain is 12 orders of magnitude greater than the faintest audible sound, i.e. 12 bels or 120 decibels. Although a difference of 1 bel is easily noticeable, the ear is surprising poor at discriminating fractional differences in loudness. The minimum difference in loudness that the ear can detect in ideal conditions depends on the state of one's hearing, but it is usually taken to be between 0.3 and 0.6 dB.

The decibel scale also provides an explanation for something you may have noticed when you hear sounds in the open, far from large bodies such as buildings, hills or cliffs that reflect sound. As vibrations travel away from a source in all directions their energy spreads over an ever-increasing area, causing their intensity to diminish. Doubling the distance from a source results in a fourfold decrease in intensity, tripling the distance results in a further diminution to 1/9th the original intensity, quadrupling the distance reduces intensity to 1/16th, and so on.[78] But the perceived loudness of the sound diminishes with distance at lesser rate in line with Fechner's law. Hence although the physical intensity of a sound diminishes rapidly, there isn't an expected corresponding decrease in its perceived loudness. This can make it frustratingly difficult to get away from sounds one doesn't want to hear because the further you are from a source, the greater is the distance you must put between you and it to reduce its loudness by the same factor. Loudness is thus subject to the law of diminishing returns: you have to increase the distance between you and the source by an ever increasing amount in order to bring about the same reduction in loudness. And as we shall see below, this affects our ability to locate the source of a distant sound.

As we have seen, loud sounds can permanently impair hearing because they can damage the stereocilia in the Organ of Corti. There are only three and a half thousand of these in each cochlea, and once damaged they cease to generate signals, which leads to selective deafness, i.e. one ceases to hear particular ranges of frequency. In fact, the cochlea has the least number of

---

[78] Intensity, I, of a source varies with distance as $I \propto 1/r^2$, where r is the distance from the source.

receptors of any of the sense organs, so one can ill afford to lose even a few of them; the eye has 10,000 times more receptors than the cochlea.

One likely reason why we lose high frequency hearing sooner than low frequency hearing is that the section of the basilar membrane that respond to high frequencies is next to the opening to the cochlea, hence is exposed to the most energetic vibrations from the eardrum.

Hearing Lost and Found

We saw in chapter one that in some circumstances sight can be partially restored to people who have either been born blind or who lost their sight in early childhood. But restoration is possible only if the retina, the array of light sensitive cells that covers the back of the eye ball, is intact. If the retinal cells are damaged or missing, removing cataracts from the eye cannot restore sight. And as things stand, with one exception, sensation can't be restored where the relevant sensory cells are no longer in working order or have been destroyed. That exception is hearing, and the means of restoration is an electronic device known as a cochlear implant that is inserted as deep as possible into the cochlea. The process of insertion is not without risk because it can destroy any remaining functioning hair cells, so the procedure is used only in cases where the auditory equivalent of the retina, the array of hair cells that line the basilar membrane, is not in working order because of malformation or disease.

The implant itself is a bundle of wires that feed electrical signals to a couple of dozen electrodes. The electrodes are arranged so that each one is as close as possible to a group of auditory fibres, nerves that in a healthy ear transmit the electrochemical signals produced by hair cells when they are stimulated by sounds. The electrical current fed to the electrodes is produced by a speech processor lodged behind the ear on the surface of the skull. The processor receives sounds from a microphone and converts them into electrical signals that are transmitted to the implant.

Each cochlea in a healthy ear has approximately 3000 hair cells responsible for the electrochemical signals that are processed by the brain's auditory system to be experienced as sounds, so replacing these with a couple of dozen electrodes cannot recreate detailed hearing. Moreover, the implant can't be inserted all the way to the apex of the cochlea, so it doesn't reach the nerves at the low frequency end of the basilar membrane.[79] To overcome this, low

---

[79] The average length of the spiral cavity within the human cochlea is 35 mm. Depth of insertion of the implant is between 18 and 26 mm.

frequency sounds picked up by the microphone are fed to electrodes at the far end of the implant — well short of the far end of the basilar membrane— and so are experienced as having a higher pitch than they would by a healthy cochlea.

Cochlear implants are optimised for speech sounds, which means that apart from rhythm, most music comes across as incoherent noise. When the implant is first turned on, usually a month or more after surgery, patients report that speech sounds squeaky and high pitched, like that of a cartoon character. This, of course, is a consequence of the inability of the electrodes to reach and stimulate the auditory nerves at the far end of the basilar membrane. But the brain soon adapts to sounds which to a healthy ear can be both unpleasant and, initially, difficult to decipher.[80] A successful cochlea implant can enable someone who is profoundly deaf to understand up to 80% or 90% of a conversation.

It is infants who are born deaf who benefit most from a cochlear implant. People born deaf or who lose their hearing in early infancy are classed as being prelingually deaf. In other words, they are deaf at the stage at which children learn to speak—a process that requires that their hearing is unimpaired. Hence it is vital that deafness is diagnosed as soon as possible following birth so that a child who is found to be deaf is given a cochlear implant to put it on an equal footing with hearing infants during the critical period in which spoken language is acquired. Providing prelingually deaf adults with cochlear implants, however, does not lead to the same degree of improvement in speech recognition as it does in infants. Like the restoration of sight in those born blind, if the necessary auditory pathways within the brain have not been stimulated during the critical period of development, these either atrophy or are taken over by other senses, particularly by vision. They cannot be reactivated or remade at a later stage.[81]

Hearing, of course, is not confined to speech. It is an essential source of information about events within one's environment, which cochlear implants provide to everyone who has them, including prelingually deaf adults. And given that locating sounds requires two ears, it is becoming common practice to implant both ears at the same time.

---

[80] Listen to simulations of speech and music as heard through a cochlear implant on the *Auditory Neuroscience* website.

[81] Denworth, L. (2014) I Can Hear You Whisper: An Intimate Journey Through The Science Of Sound And Language. Dutton Books.

Locating Sounds

Human beings are exceptionally good at pinpointing the direction from which a sound reaches the ear, far better than almost every other creature whose hearing has been studied. This comes as a pleasant surprise given how poorly our other senses seem to compare with those of many vertebrates. Our sense of smell cannot compete with that of a dog and our sight is not as acute or colourful as that of birds, especially that of birds of prey.[82] But where hearing is concerned it appears that we outperform most of the animal kingdom, at least as far as determining the direction of the source of a sound is concerned.[83]

The ability to determine sound direction depends on having ears on opposite sides of the head so that what is heard by one ear differs slightly from what is heard by the other, the aural equivalent of binocular vision. The difference is greatest when the source of a sound is directly to one side or other of the head and least when it is directly in front, behind or above the head.[84]

When the source is to the side of the head, sound reaches the ear nearest the source a fraction of a second before it reaches the other one. The delay is due to the time it takes sound to travel the distance between the ears and is known as the "interaural time difference" or ITD. Taking the width of the average adult human head to be 20 cm and the speed of sound as 340 m/s, the maximum delay is 0.0006 s, i.e. approximately half a millisecond — an interval so brief that it puts the proverbial "split second" to shame.

At the same time, the listener's head can prevent sound reaching the ear on the side furthest from the source because, being a wave, only low frequency sounds diffract around the head to reach that ear. The cut-off is approximately 2000 Hz because the wavelength of a sound with a lower frequency than that is greater than the width of a human head and will wrap around it so that both ears receive sounds of the same intensity—with, of course that all-important interaural time delay. Above 2000 Hz, however, diffraction

---

[82] A major reason why these creatures outperform us is because their respective senses possess far more sensory cells than ours. Dogs have up to a billion olfactory sensors (and require an elongated snout to accommodate them) whereas we have some five million. Polar bears have an even more acute sense of smell than dogs. A bird of prey can have up to seven times as many colour sensitive cells in it retina as does the human eye. Nevertheless, senses are only as good as they need to be and where our sense of smell is concerned, we are the only creatures capable of *retronasal olfaction*, which enables us to detect flavours. See: Shaw, B. (2018) The Smell Of Fresh Rain, The Unexpected Pleasures Of Or Most Elusive Sense. Icon Books, p 123.

[83] Heffner, H.E., Heffner, R.S. (2016) The Evolution of Mammalian Sound Localization. Acoustics Today, Spring, Vol 12, issue 1, p 20–27.

[84] These facts were first established and explained by Rayleigh in 1876. See: Strutt, J.W. (1877) Perception Of The Direction Of A Source Of Sound. In Strutt, J.W. (1899) Scientific Papers, Vol 1, 1869–1881. CUP, 1899, p 314.

becomes progressively less pronounced and the ear furthest from the source falls increasingly within the head's acoustic. When a sound consists of a broad range of frequencies, as do most sounds, the absence of these higher frequencies results in a small but perceptible difference in its intensity. As with the time-delay, the difference in intensity, which the auditory system perceives as loudness, is used to determine the direction of the source of a sound. The difference in loudness in this situation is known as the "interaural level difference", or ILD. Unlike the time-delay, however, loudness differences are noticeable only if a source is not too distant because, as we saw in connection with Fechner's law, the perceived difference in loudness between one ear and the other becomes ever smaller the further one is from the source of a sound. In the absence of a noticeable ILD, the hearing system has to depend entirely on ITDs. At the same time high frequencies are absorbed and scattered to far greater degree by air and by objects than are low frequency sounds, which provides a clue as to how far the source is from a listener (Fig. 3.8).

Determining sound direction based on time-delay works best for sounds that diffract around the head to reach the ear farthest from the source, i.e. for sounds with frequencies below 2000 Hz. Differences in loudness only come into their own above 4000 Hz because above this frequency diffraction around the head is negligible. This is why we find it particularly difficult to locate sounds that are composed of frequencies that lie between 2000

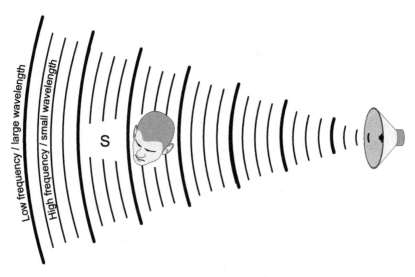

Fig. 3.8 Locating sound direction. The sound source here produces of a broad range of frequencies. S is the sound due to the listener's head that blocks high frequencies present in the sound

and 4000 Hz, a region of the audible spectrum where the auditory mechanisms that detect both time delays and loudness differences are least effective. Burglar and car alarms are difficult to locate because they typically emit sounds within this range of frequencies. The difficulty is further increased because alarms usually emit pure tones.[85]

As you would expect, differences in timing and loudness are always greatest when the source is directly to one side or the other of the head. If the source is directly in front, behind or above the head, these differences disappear and it becomes more difficult to locate the source. To overcome this, one unconsciously cocks one's head in order to increase the difference in what is heard by each ear. The ear-flap, or pina, plays an important role here because it partially shields the ear canal from sounds that come from behind or above the head, while its curved ridges selectively funnel high frequency sounds into the ear canal. As proof, if the grooves and whorls within the flap of pina are covered, say by pressing your fingers firmly against each ear so that only the ear canal is left open, it is very difficult to accurately locate the direction a brief sound when its source is directly in front of one, either above or below one's head, with one's eyes closed. Incidentally, your sensitivity to ITDs and ILDs may not be the same depending on the state of your hearing, though this is something you will become aware of only under laboratory conditions.[86]

Loudness is encoded by the brain as the rate at which the neurons in the first stage of the auditory system fire in response to electrochemical signals from the cochlea. The louder the sound, the greater the rate at which neurons fire. ILDs are established by comparing the firing rate of neurons that receive signals from the left hand cochlea with that of neurons receiving signals from the right hand cochlea. The difference in firing rates indicates on which side of the head the sound is loudest. ITDs, on the other hand, depend on the time that elapses between left and right neurons firing.

The ability to determine sound direction declines with age because we loose the capacity to hear higher frequencies. Unsurprisingly, people who are hard of hearing or deaf in one ear, find it difficult to locate sounds reliably because they can't perceive the necessary timing and loudness differences. You can confirm this by sticking your finger in one ear when listening to broadband noise such as traffic while trying to determine where the sound is

---

[85] Hear this for yourself. "Edward Boas what direction is that sound coming from?" http://www.edboas.com/science/sound_direction/index.html (accessed 21/04/2121).

[86] You can test your spatial hearing on https://auditoryneuroscience.com/spatial-hearing (accessed 15/1/2021).

coming from: it should convince you that two ears are better than one if you want to successfully locate a sound.

But given that these facts are true of all land-dwelling vertebrates, why should humans be particularly good at locating sounds? Part of the explanation is that we are predators and have eyes that face forward, giving us a relatively narrow field of view. The ability to determine sound direction accurately compensates because it enables us to accurately direct our narrowly focused gaze at the source of a sound. Creatures that are prey have eyes on either side of their head giving them a much wider visual field than predators have. Prey also have better peripheral vision than predators do. Together, these advantages in vision compensate for their comparatively poor ability to locate sound sources, reducing the need to evolve a hearing system that is as good at sound location as that of a predator.

Sound location is also affected by the medium in which sounds are heard. As divers will attest, locating sounds underwater is much more difficult than in air. In large part this is because the speed of sound in water is some four and a half times greater than in air, which means that the interval between sound reaching one ear and reaching the other when we are immersed is far too small to be discernible by our brain.[87] This makes underwater sound location all but impossible for humans, though, as we shall see, whales and dolphins have evolved ways of coping with this. Remarkably, the accuracy with which a dolphin can locate a sound source in water is twice as good as that of humans.

Locating the direction from which a sound reaches the listener also depends on being able to identify the "same" sound in each ear. And this requires that the auditory system can identify that they are composed of the same frequencies having the same relative intensities. Recognition works best with transient sounds rather than those of a single frequency, such as a prolonged whistle. As we have noted, pure tones are rare in nature, which is why there was never any evolutionary pressure for the brain to cope with them. Natural sounds are almost all transient, i.e. bursts of complex, short-lived noise, such as a clap of thunder, the roar of a lion or the rustle of leaves.

Sound direction is a uniquely auditory sensation, which along with loudness and pitch, has been hard wired into us during the evolution of the hearing system so that perceiving the location of a source of sound usually requires little or no conscious effort on the part of the listener. The next time you are in an environment where there are several sound sources, such as in

---

[87] The corresponding delay in the case of a human skull is 0.00014 s, almost five times less than it is in air.

a field or woodland where there are birds or beside a busy road, close your eyes and listen attentively. You will have little difficulty in sensing the position of each source relative to you. Listen as a car drives past: even though you can't see it, you know more or less where it is relative to you moment by moment. But because we usually rely on both our eyes and ears when we are out and about, it isn't immediately obvious that sound location works completely independently of sight.

Perhaps the most extraordinary, and certainly the most fascinating aspect of sound location, is that without the hard-wired capacity to determine tiny differences in intensity, timing and pitch necessary to identify specific sounds with such precision in order to determine where they are coming from, vocal communication between animals would not be possible. Humans are so good at detecting these differences that they can identify individual voices and follow complex musical harmonies.[88] In fact, all creatures that communicate orally depend on being able create and recognise distinctive sounds of great complexity, as we shall shortly see in the case of frogs and penguins.

The difficulty of locating sources that emit sounds between two and four thousand hertz was believed to be the reason why the glass armonica, invented in 1761 by Benjamin Franklin fell out of favour. The instrument was based on the sound that is produced when the wetted lip of a wine glass is rubbed by a clean finger, a phenomenon described by Galileo, but known much earlier.[89] The armonica was made from a dozen or more glass bowls of different sizes and thicknesses threaded closely together with one within the other on a horizontal central shaft and partially immersed in a trough of water. The bowls were rotated with a treadle and kept wet by the water in the trough. The pitch produced by Franklin's instrument ranged from one and four thousand hertz, which apparently made it difficult listeners to locate its sounds. Hence the claim that it disorientated both musicians and audiences and even made some lose their minds. There is no substantive evidence for these claims and the instrument probably fell out of favour because it was difficult to play (it is essential that the player's fingers are free of grease, something you will have discovered if you have ever tried to get a wine glass to sing) and it was not loud enough to he heard clearly in concert halls.[90]

---

[88] Beament, J. (2005) How We Hear Music: The Relationship between Music and the Hearing Mechanism. Boydell Press. See Chapters 9 and 10 for a particularly clear exposition of how the ear determines direction and the role this ability plays in music.

[89] Galileo Galilei (1914) Dialogues Concerning Two New Sciences, translated from the Italian and Latin by Henry Crew and Alfonso de Salvio. The Macmillan Company, p 99.

[90] You can listen to the sound of a glass armonica online. See "glass armonica" https://www.youtube.com/watch?v=eEKlRUvk9zc (accessed 10/05/2021).

The Cocktail Party Phenomenon

Together with our ability to identify speech, the physiological and neurological mechanisms that enable us to determine sound direction offers a partial explanation for our ability to hear distinctly what someone is saying in a noisy environment, thus overcoming the troublesome *cocktail party effect*. The phrase was coined in 1953 by an electrical engineer, E. C.Cherry, who sought a solution to a problem encountered by air traffic controllers of the time.[91] In those days, the voices of pilots were all relayed to the control tower over the same loudspeaker, which made it very difficult for controllers to hear individual pilots distinctly. Subsequent experiments found, among other things, that speech that enters a person's ear from a particular direction, which relies on the hearing system's direction finding ability, will stand out from the overall hubbub of conversation, which is likely to be equally loud in both ears. The effect is known as binaural unmasking. People who are hard of hearing in one ear find it much more difficult to follow a conversation in these circumstances because their ability to locate sounds is poor.

Cocktail parties (though not cocktails) are perhaps rather *passé*, but the auditory challenge posed by a noisy environment has not gone away. In search of sustenance and conviviality we are prepared to endure boisterous bars and noisy restaurants. Spending time in such venues isn't compulsory, of course, nor is it essential to our survival; food and drink are always available in quieter surroundings, as is conviviality. But many creatures are forced to contend with a raucous environment for sake of the survival of their species. A female frog depends upon the loudness and quality of a male's croak to select a suitable mate, and must seek him out in the midst of the chorus of hundreds of other male frogs; somehow she is aware that failure to bag a mate in his prime may lead to a pond full of second-rate spawn.

But, arguably, it is King and Emperor penguins who must cope with the mother of all noisy cocktail parties. Neither of these creatures nest, so parents and their chicks don't have a permanent location where they can come together. Instead they rely entirely on their hearing to find one another within a colony that can often contain ten thousand and more of their fellow creatures, the majority of which will be calling out at any given time. The problem is mitigated to a degree because fledgling chicks wait in groups on the edge of the colony for the return of parents bearing the food their chick needs if it is to survive. Chick and parent find one another guided by the distinct calls that each emits, calls which differ slightly from the scores of

[91] Cherry, E. C. (1953) Some Experiments On The Recognition Of Speech, With One And Two Ears. Journal of the Acoustic Society of America, 25, p 975–79.

those in their immediate vicinity. So, not so much a noisy cocktail party as a chaotic restaurant in which customers and waiters wander around in the hope of bumping into one another. As with humans, these penguins rely on sound location and the ability to recognise the specific acoustic qualities of the calls of each individual.[92]

Understanding why we are able to identify a specific voice in a noisy environment may have a wider application. How is it that we can usually experience a jumble of individual frequencies as an identifiable sound? And how is it possible to perceive clearly more than one sound at a time? As I type these words I am aware of the click of the keyboard, the ticking of the clock on my desk and the intermittent whine of an electric drill being used by someone in a neighbouring garden. How does the brain segregate the sounds from each of these events given that they all arrive at the eardrum simultaneously? How does the brain "know" that a particular group of frequencies belong together? It's an issue known in neuroscience as *auditory scene analysis*, and one that is yet to be fully understood.

It is likely that the principles of Gestalt psychology provide a partial answer to the problem. These principles were originally developed by Viennese psychologists around the turn of the last century to explain how it is possible to identify individual objects given that the image of a complex scene formed on the retina is a just jumble of shapes and colours. According to Gestalt principles, elements within the image that have similar features (e.g. shape or colour) or are close to one another in space are perceived as belonging to the same object.[93] Analogously, it is argued, frequencies that arrive at the ear from a particular direction are grouped together, which is why our ability to determine where a sound is coming from plays a central role in hearing it as a distinct event. Other factors that enable us to identify individual sounds is the presence of frequencies that are harmonically related or that begin and end together. Useful though they have proved to be to psychophysicists, however, Gestalt principles are no more than rules of thumb. If the answer is to be found to how we are able to hear and identify distinct sounds, neuroscientists will have to discover the neurological processes that enable the auditory system to act in ways that conform to Gestalt principles, if indeed they do, and this research is still in its infancy.[94]

---

[92] Haven Wiley, R. (2015) Noise Matters: The Evolution of Animal Communication. Harvard University Press, p 92–94.

[93] There are several more Gestalt principles of perception than are mentioned here.

[94] Schnupp, J., Nelken, I., King, A. (2011) Auditory Neuroscience, Making Sense of Sound. MIT, p 267.

The Limits of Audible Sound

It requires considerable mental effort to convince yourself that sound is just a sensation even when you know that it is caused by noiseless vibrations. Like latter day Aristotelians, we unthinkingly conflate the sensation with its cause, which leads us to assume that vibrations that fall within a particular range of frequencies are uniquely sonorous, while vibrations outside this range are inherently inaudible.

The inability of natural philosophers to distinguish clearly between sound as a sensation and sound as a physical disturbance before the seventeenth century, the century during which physical nature was gradually but inexorably mechanised, meant that no one up to that time ever considered the possibility of sounds inaudible to the ear. The first person to do so seems to have been Marin Mersenne, who surmised that.

> ...all movements that occur in the air, in water, or elsewhere, can be called sounds inasmuch as they lack only a sufficiently delicate and subtle ear to hear them; and the same thing can be said of the noise of thunder and cannon with respect to a deaf person who doesn't perceive these great noises.[95]

But it wasn't until the end of the nineteenth century that there was any direct evidence that many other creatures hear vibrations that are inaudible to us—or, indeed, that in some cases they can see forms of light invisible to us.[96] But why don't all creatures hear the same range of frequencies? What sets the limits to our perception of physical vibrations as auditory sensations? The answer, in a word, is "survival": eat or be eaten.

We have seen that the ear's primary function is to warn of danger and prompt flight. This role is so central to survival that a sudden sound can trigger the most immediate and powerful instinctive response of all, the startle reflex. We've all experienced it: unexpected loud sounds such as gunfire can make one jump out of one's skin, particularly when the source is out of sight. Even when one is in control of a sonorous event, one finds oneself tensing in anticipation as, for example, one prepares to prick a balloon with a pin.

But simply hearing a sound is not enough. It is just as important to move away from a potential threat—or, indeed, towards a potential meal, so determining the direction of a sound source is just as important as hearing it. And

---

[95] Mersenne, M. (1636) Harmonie Universelle, Sebastien Cramoisy, Paris. Livre Premier De La Nature Et Des Proprietez Du Son, prop I, p 2.

[96] Many insects perceive ultraviolet and rattlesnakes can perceive infrared. So can humans, though not as well as rattle snakes. We perceive infrared as warmth in sunlight or from a heater, but we don't have an ability to form a clear image of the source using that sense of warmth.

this, as we have just seen, depends in part on the presence of high frequencies that are heard by one ear more distinctly than by the other one.

Of all creatures alive today, mammals can hear the widest range of vibrations, a legacy of their ancestry. For their (and our) diminutive, nocturnal ancestors that scuttled nervously around the gloomy forests of the Mesozoic era,[97] the era during which dinosaurs were the dominant creatures on land, hearing was arguably more important than sight in the struggle for survival. As a result, all mammals can hear sounds above 10,000 Hz, and the majority can hear sounds well beyond 20,000 Hz, which is the nominal upper limit of human hearing (Fig. 3.9).

And if there is one reason why the hearing range of most mammals extends beyond that of humans it is skull size. As we have seen, one of the most important cues to sound direction relies on high frequencies *not* reaching the ear furthest from the sound source because they can't diffract, i.e. wrap, around the head. And the smaller the skull, the higher the frequency must be if it is *not* to diffract around it. Hence mice have to hear sounds as high as 80,000 Hz to locate sounds as well as dogs, which, given their larger heads,

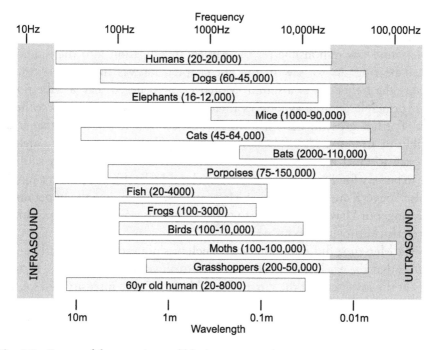

Fig. 3.9 Range of frequencies audible to some creatures

[97] Mesozoic era began 250 million years ago and ended 65 million years ago with the extinction of the dinosaurs. The earliest mammal fossils are some 210 million years old.

can hear sounds up to 45,000 Hz. Although the correlation between skull size and hearing range is not linear, it's close enough to show that the two are related. The upshot is that a creature hears the range of frequencies necessary to locate sounds based, in large part, on the size of its head. And for humans this ranges from 20 Hz to 20,000 Hz.

Here is the reason why human beings can't hear infrasound or ultrasound: these frequencies are not essential to our survival. We can locate the majority of sound sources that are important for our survival within the range of frequencies we identify as sound. As for speech and music, without which human society would not be possible, they don't require us to be able to hear the very lowest or the very highest frequencies of the range to which we are sensitive. But then, language and music came late in human evolution. In the case of language perhaps more than 100,000 years ago, based on the fact that we all share a common ancestry, that humans began migrating from their original homelands in Africa at least 60,000 years ago, and that every human society has both spoken language and music. Moreover, as we noted in the previous section, in order to develop and interpret the intricate sounds that make up spoken words and musical harmonies, our distant ancestors inadvertently exploited the pre-existing capacity for sound-location that relies on precise identification of time delays and loudness.

Mersenne was responsible for the earliest investigation into the limits of human hearing. As we saw in the last chapter, he was the first person to measure the frequency of a sound. Moreover, he believed that all vibrations are in principle audible as long as there is an ear sensitive enough to hear them. Further experiments with brass wires led him to conclude that it is possible to hear sounds of 16 Hz, though given his often rather careless approach to exact measurement, not to mention that low frequency sounds have to be very loud to be heard clearly, we should take this claim with a pinch of salt.[98] Several authors assert that Mersenne also determined the upper limit of hearing, but I have been unable to corroborate this claim despite consulting the references provided to his writings.[99]

What Mersenne did not know, and which was not established clearly until 1830, is that the ear is not equally sensitive to all audible frequencies. Particularly at the lower end of the range of audible sound, the intensity of a sound must be some 70 dB louder than the frequencies to which the ear is most sensitive, i.e. to those between 1000 and 4000 Hz, if they are to be heard as being equally loud.

---

[98] Mersenne, M. (1636) Harmonicorum libri, Liber Secundus de causis sonorum. Proposition xxxiii.
[99] Crombie, A.C. (1969) Mathematics, Music and Medical Science. Organon, 6, p 28.

Mersenne's estimate of the lower limit of audible sounds was not improved upon until Chladni conducted his own investigation a century and a half later. As a result of his experiments, he concluded that the range of sounds the ear can hear occupies some 9 octaves, with the lowest perceptible frequency set at 30 Hz and the highest between 8000 and 12,000 Hz.[100]

During the nineteenth century there were several futher attempts to determine the limits of audibility. Felix Savart, a French physicist, set the upper limit at 24,000 Hz based on experiments with toothed wheels of different diameters rotating against a stiff card to produce high frequency sounds. Using a different apparatus for low frequencies, he set the lowest audible frequency at 8 Hz. It was Savart who established that the ear is not equally sensitive to all audible frequencies when he found that low frequencies have to be very much louder than high frequencies to be audible. Estimates by Biot and Helmholtz placed the lower limit at 32 Hz.

A few years before Savartmade his measurements, William Hyde Wollaston, an English chemist, had already noticed that the sensitivity of the ear to high frequencies differs from person to person and often goes hand in hand with increasing age, a condition known as presbycusis. He found that that people whom one would not ordinarily judge to be deaf may nevertheless be unable to hear high frequency sounds audible to people with unimpaired hearing (i.e. those who can hear the full spectrum of audible sound.) When Wollaston was boy one of his relatives informed him that she "never could hear the chirping that commonly occurs in hedges during a summer's evening, which I believe to be that of the *Gryllus campestris* [a field cricket]." Another telling example that he came across was of a man who could not hear the chirp of the common house sparrow. It led Wollaston to surmise that since there is no limit on the rate at which air can vibrate, it is very likely that some creatures can both produce and hear sounds that are beyond our range of hearing.[101]

Wollaston's conjecture was confirmed some fifty years later by Charles Darwin's cousin, the long-lived polymath, Francis Galton. One of Galton's numerous interests was the limits of human hearing, and to investigate this he devised a whistle that can produce ultrasonic frequencies as high as 55,000 Hz. It was known as Galton's whistle and is now called a dog whistle. He used it to test the hearing of people and animals, usually without letting on that he was doing so, and discovered that:

---

[100] Chladni, E.F.F. (2015) Treatise on Acoustics: The First Comprehensive English Translation of E.F.F.Chladni's 1809 Traité d'Acoustique (trans: Beyer, R.T.). Springer, Sect. 238, p 179.

[101] Wollaston, W.H. (1820) "On Sounds inaudible to certain Ears". Philosophical Transactions part ii, p 306–14.

Of all creatures, I have found none superior to cats in the power of hearing shrill sounds; it is perfectly remarkable what a faculty they have in this way. Cats, of course, have to deal in the dark with mice, and to find them out by their squealing.[102]

As for infrasonic frequencies, they were first discovered when records of barographs around the world were studied following the eruption of Krakatoa in 1883. Further evidence of these low frequency waves was detected with much more sensitive barographs following the impact of the great Siberian meteor that fell to Earth near Tunguska in 1908.

Biosonar

Many blind people are adept at navigating their way around obstacles by listening for reflections of sounds that they make, either by tapping the ground with a cane or producing clicking sounds by flicking their tongue against their upper pallet. But the unrivalled masters of this technique, known as echolocation or biosonar, are insect-eating bats and toothed whales. The use of echolocation by these creatures, however, was not definitively established until 1940 in the case of bats, 1953 in the case of dolphins and later still for toothed whales such as sperm whales.

Bats are the second most common type of mammal on the planet after rodents, and are found on every continent except for Antarctica. Most bats feed on fruit, for which they forage using their eyes, and are known as megabats. But a third of all bat species eat insect and rely on their ears to hunt for their prey. The insect eaters, known as microbats, are nocturnal, yet they have no difficulty in finding and catching insects on the wing while avoiding collision with even the smallest obstacles in pitch darkness.[103]

The earliest attempt to solve the mystery of how nocturnal bats achieve these feats was made in 1794 by an Italian physiologist, Lazaro Spallanzani, who occupied the chair of Natural Philosophy at the University of Pavia. It's not known why Spallanzani became interested in this question, but the series of experiments he and his collaborators performed on them, cruel even by the standards of the time, eventually seemed to imply that microbats rely on hearing rather than vision.[104]

---

[102] Galton, F. (1883) Inquiries Into Human Faculty And Its Development. Macmillan, London, p 26–27.

[103] The correct name for microbats is *microchiroptera*.

[104] Several decades later, commenting on a later anatomical investigation to discover the mechanism of hearing, John Herschel wrote "…the details are too revolting to find a place in any but works of

To test the explanation of a bat's unusual power of perception that was current in his day, that they rely on touch, Spallanzani and his collaborators systematically mutilated bats by the dozen: they gouged out eyes, cut out tongues and destroyed noses. And to reduce their sense of touch, Spallanzani varnished their wings and covered their bodies in flour paste. But in every case but one, the bats that survived maiming were able to fly and hunt just as well after their mutilation as they had before. Only one procedure, the least cruel, rendered them helpless: plugging up their ears.

That experiment was performed by Louis Jurine, a Swiss physician and naturalist, who plugged the ears of bats with wax and found that this caused them to fly erratically and collide with objects. But if bats employed hearing to navigate, what were they listening to? One could accept that the beating wings of an insect in flight might be audible to a bat, but how were they able to avoid colliding with inanimate objects?

The possibility that bats might themselves be the source of the sounds to which they were listening would have been too much to expect from the naturalists and natural philosophers of that time. Spallanzani considered bats to be largely silent and never reported hearing bats squeak, even though they do so audibly when communicating with one another. Perhaps, he concluded tentatively, bats possess a sixth sense that somehow enables them to sense the presence of things without having to see or touch them.[105] Unfortunately, although Spallanzani changed his mind when he got around to studying Jurine's findings, he died before he was able to make public his agreement with Jurine's conclusions.

George Cuvier, the leading zoologist of the day, rejected Spallanzani's and Jurine's views out of hand and ignored their findings. Bats, Cuvier insisted dogmatically, don't have a sixth sense and don't employ their ears. Instead, he claimed, without a shred of evidence, they use a highly developed sense of touch located in their wings. This imagined sense of touch was supposed to be sensitive enough to detect tiny currents of air that might be set up as a bat flies very close to a solid object. Cuvier's conjecture more or less put paid to Jurine's discovery for more than a century. It wasn't just that one of the most respected and influential naturalists of the day had declared Jurine to be mistaken, there was also the difficulty of accounting for the source of the sounds that bats might be employing when foraging. Galton's discovery that

---

anatomy and physiology." In: Smedly, E. (ed) (1830) Encyclopaedia Metropolitana, Vol II. Baldwin and Cradock, London, p 810.

[105] In Spallanzani's day the existence, let alone the role of the vestibular system, our sixth sense, was unknown.

some mammals can hear frequencies inaudible to humans, with the implication that they are capable of producing equally high frequency sounds, was not made until the last decade of the nineteenth century.

Although some naturalists and physiologists during the early decades of the twentieth century were prepared to accept that microbats rely on hearing to navigate, hard evidence that they do so did not emerge until 1940, when two American graduate students, Robert Galambos and Donald Griffin, employed a newly invented device that both produced and detected ultrasonic frequencies to establish beyond doubt that microbats emit and hear the reflections of ultrasonic chirps, chirps that they produce when hunting.[106] Galambos coined a word for this process: "echolocation"; these days echolocation by animals is also known as "biosonar".

Thanks to Galambos and Griffin, and the countless researchers who have since followed up their discoveries, we now know that microbat chirps range from 20,000 to 100,000 Hz, which at the higher end can produce distinct echoes from objects with dimensions as small as 3 mm. Moreover, when hunting, they emit a series of closely spaced chirps that are so rapid and brief that they sound like a succession of clicks. The rate at which they chirp increases as they home in on a target, which enables them to get a precise fix on their prey. And as a bat closes in on its prey, the clicks are so closely spaced that they merge into a buzz. Some microbat species also employ the Doppler shift in the echo from a moving target such as a moth to gauge the direction in which it is moving.[107]

But, as we shall see in the next chapter, high frequency sound is strongly absorbed by air, which is why these bat chirps have to be exceptionally loud, on average 120 decibels, comparable to the loudness of a home fire alarm at a distance of 10 cm.[108] Unless a bat emits a very loud sound, the echo will be too feeble to be audible at distances of more than a few metres. Moreover, they can detect a distinct echo from an object—i.e. they can distinguish the reflection from their chirps—when they are within centimetres of it, which implies that they can process sound signals at rates that are several times greater than humans can.

Microbats are remarkably long-lived compared to mammals of comparable size and metabolic rate, which suggests that these loud sounds do not damage

---

[106] Galambos, R. (1942) The Avoidance of Obstacles by Flying Bats: Spallanzani's Ideas (1794) and Later Theories. ISIS, Vol 34, p 132–40.

[107] Denny, M. (2007) Blip, Ping & Buzz: Making Sense of Radar and Sonar. Johns Hopkins University Press, p 139–53.

[108] Ultrasound frequencies are strongly absorbed by fog, which may be the reason why microbats avoid flying in fog given that their ultrasonic chirps will not be returned. See Sales, G. and Pye, D. (1974) Ultrasonic Communication by Animals, Chapman and Hall, p 9.

the hairs in their cochlea; hearing loss would, of course, very quickly lead to starvation in a creature with the very high metabolic rate characteristic of a microbat.[109] One mechanism that protects their hearing is that when they chirp, a muscle within the middle ear contracts and prevents the stapes transmitting the full intensity of a vibration to the inner ear. The muscle is also present in the human ear and is known as the stapedius muscle. It comes into play in the presence of a loud continuous sound, for example when one is speaking. Unfortunately, the auditory reflex that causes the stapedius muscle's contraction is slow, so it does not protect the ear from sudden loud sounds. But how bats preserve their hearing over their unusually long lifetime it is not fully understood, though it would be desirable to do so because it might shed light on how to prevent hearing loss in humans.[110]

Bats' echolocation is so sensitive that it enables them not only to locate prey and avoid solid objects, it can also identify the nature of the reflecting surface. They have no difficulty in distinguishing a tiny insect from the leaf on which it may be resting. Of course, the echo is only the first step of the process. The real work is done in the auditory centres of their brain, a brain that in most microbats is about the size of a peanut. Yet this miniscule organ is able to recreate a precise and detailed auditory image of its immediate environment merely from echoes, something that is well beyond human capabilities.

Microbats don't have things all their own way, however. There is evidence that some moths have evolved ways to evade being caught. Recently, researchers in China have discovered that the scales on the wings of certain moths may absorb enough of the energy of a bat's chirp to significantly reduce the strength of the echo. This makes it more difficult for a bat to locate the moth and is, in effect, a form of stealth coating. Another way in which moths seem to be able to evade capture is to "jam" a bat's cries by emitting ultrasonic clicks when they find themselves targeted. These clicks confuse the echo reaching the bat making it more difficult to home in on the moth.[111]

This, of course, requires that these moths are able hear ultrasonic frequencies. But moths are by no means the only insects that feature on a microbat's menu. Crickets, another of their prey, use ultrasound to communicate with one another, i.e. they produce and hear ultrasound. But the ability to hear

---

[109] Microbats can devour several hundred insects per hour, which over the course of a single night can add up to 40% of their body weight.

[110] Horowitz, S.S. (2012) The Universal Sense. Bloomsbury. For information on bats see chapter 3: The High Frequency Club.

[111] Conner, W.E. (2013) An Acoustic Arms Race. American Scientist, Vol 101, No. 3, p 203–209.

these high frequencies is a comparatively recent development in the evolution of these insects, which have been around in one form or another since the beginning of the Permian period, some 300 million years ago. In 2012, Chinese palaeontologists working in Inner Mongolia found a 165 million year old fossil of a cricket from the Jurassic period. It was so well preserved that it was possible to reproduce an exact copy of the minute structures on its wings using a 3D printer, enabling scientists to recreate the sound it would have made when it rubbed one wing against the other. What they heard was a steady high pitched whistle with a frequency of 6.4kH, which is well within the spectrum of audible sound. This ancestral cricket was named *Archaboilus musicus* in recognition of the pure tone it produced. The advantage of audible sound over ultrasound for communication is that it travels further because low frequencies are not as readily absorbed as high frequencies as they travel through the atmosphere. But the advent of echolocating bats towards the end of the Cretaceous period 50 million years ago, seems to have been the trigger that led crickets to evolve the ability to produce and hear ultrasound in order to avoid being discovered.[112]

The other major group of creatures that employ echolocation include dolphins, porpoises, killer whales and sperm whales, which are known collectively as odontocetids or "toothed whales".[113] All these creatures are descended from that terrestrial mammalian ancestor that returned to the sea in stages some 50 million years ago. But the evolutionary factors that resulted in their capacity to echolocate remains the subject of conjecture though the absence of light in deep seas would have played a part.

Odontoceti echolocate using a series of brief ultrasonic chirps, just as micro-bats do. But unlike bats, *odontocetids* focus their chirps before they enter the surrounding water by passing them through a large fatty organ located in their forehead and known as a "melon". The melon acts as an acoustic convex lens that focuses the chirps into a narrow sonic beam. Being mammals, *odontocetids* have a cochlea, but there is no channel open to the outside through which sound can reach the inner ear; nor do they have a tympanic membrane. Instead, sound reaches their cochlea though the jaw.

Echolocation in water is, in principle, far more difficult than it is in air because the speed of sound in water is four and a half times greater than it is in air. Hence the wavelength of a given sound in water is correspondingly larger. Ultrasound with a frequency of 100,000 Hz has a wavelength

[112] Gu, J.J. & Montealegre-Z. F. et al. (2021) Wing stridulation in a Jurassic katydid (Insecta, Orthoptera) produced low-pitched musical calls to attract females. Proceedings of the National Academy of Sciences, USA, 109, p 3868–3873.

[113] Killer whales (orcas) are dolphins, the largest of that species.

of 3.4 mm in air and 1.2 cm in water. Consequently, any object that can be echolocated in water must be correspondingly larger than it would be in air. A further limitation is that fish are almost completely acoustically transparent because their flesh is largely composed of water, i.e. the acoustic impedance of fish flesh is almost the same as that of water. However, the air-filled swim bladder within bony fish is an excellent acoustic reflector because the impedance of air is so much less than that of water, so what dolphins and whales perceive is not the fleshy body of a fish but its swim bladder. Sperm whales feed on giant squid, which have no swim bladder, so the echoes they receive from squid come from the internal sheath of chitin, which serves to anchor the squid's muscles.

But despite the vast differences between *odontocetids* and microbats, such as their size and the medium in which they live, they employ identical techniques when hunting. Both hunt by emitting short-lived chirps, and increase the rate of emission as they close in on their prey. The major difference is that the intensity of sound emitted by a whale is several orders of magnitude greater than that of a microbat.

The auditory world of these creatures is largely beyond our imagination. Although our hearing works in much the same way as theirs, and is an irreplaceable source of information about our immediate surroundings, it is not optimised for echolocation. This makes it impossible for us to form a detailed three-dimensional picture of the physical world from sounds alone, as and microbats appear to do. Nevertheless, blind people can compensate to a remarkable degree for their lack of sight by exploiting their hearing, touch and smell to a far greater extent than do sighted people. This is especially true of people born blind.

But humans inhabit an acoustic world that is unquestionably more remarkable than that of these echolocators, for we make use of sound to communicate with one another though the medium of a highly structured language. Nor did our hearing evolve to enable this form of communication. Speech and language, which are so intimately linked that they must have evolved together, are probably too recent a development in the evolution of our species to have had a major influence on our auditory system. As we saw in chapter one, people born deaf are able to communicate with one another through sign language, which means that the language centres of the brain are independent of the auditory ones.

And this also applies to music, which probably made its debut even more recently than language, possibly no more than a few tens of thousands years ago. Instead, speech and music harnessed what was already there: a hearing system fine-tuned to particular features of natural sounds that enable their source to be located and identified, especially its capability to recognise and identify similarities and differences in transient sounds and determine the moment of their arrival at each ear very precisely.[114]

These capabilities are shared with all vocalising creatures, of course, not just with bats and toothed whales. What sets us apart from the rest of the animal kingdom is that the human cerebral cortex is uniquely able to identify these sensations as words and string them together meaningfully using a grammar that governs their use. And while some mammals, such as dogs or chimpanzees, may appear to understand a few spoken words, the only cerebral cortex able to deal with the complexities of language is the human one.[115]

But before the hearing system can get to work converting physical vibrations into sensations, those vibrations have to travel through whatever medium that lies between the ear and their source. And, as we shall now see, as they do so, they are absorbed by that medium to a degree that depends, among other things, on their frequency. This differential absorption affects the balance of frequencies that reaches the ear so that the timbre of what we hear is noticeably affected by how far we are from a source of sound.[116] Moreover, sounds do not always follow the most direct path, so that in some circumstances it is possible to see the source of a sound and not be able to hear it and vice versa.

---

[114] Beament, J. (2005) How We Hear Music: The Relationship between Music and the Hearing Mechanism. The Boydell Press, p 90–92.

[115] Pinker, S. (1994) The Language Instinct, How the Mind Creates Language. Penguin. See p 333–42 for a discussion on whether chimpanzees can learn and understand language.

[116] Compare this to the reddening of sunlight that has travelled through the lowest layers of the atmosphere when the sun is at the horizon. The mechanism responsible for the loss of high frequency light (i.e. blue/violet), however, is not the same as that which causes differential absorption of sound.

4

Sounds Abroad

Abstract The chapter deals with the passage of sound through air, water and earth over small and large distances and how the interaction between sound and its surroundings affects what we hear. Anomalous sound transmission that makes sounds audible hundreds of kilometres from the source is described and explained and a detailed account of the history and physics of the Doppler effect.

Noises from Afar

On the morning of 27th August, 1883, after months of fitful activity, the volcano on the Indonesian island of Krakatoa erupted explosively. Much of the island disappeared and, according to some estimates, up to 40,000 people perished in the aftermath. As usually happens during a explosive volcanic eruption, vast quantities of ash and noxious gas was ejected, much of which reached the stratosphere and where they remained for several years, leading to a slight global cooling, magnificent sunsets and the occasional spectacle of a "blue" sun and moon. These optical effects, although very pronounced, were by no means unusual; similar, if not always quite as spectacular, meteorological and optical effects have often been caused by volcanic eruptions, the most recent being that of Pinatubo in the Philippines, which erupted in 1991.[1] But Krakatoa is also renowned for something else: the vast expanse

---

[1] Meinel, A. & Meinel, M. (1983) Sunsets, Twilights, And Evening Skies. CUP.

© The Author(s), under exclusive license to Springer Nature
Switzerland AG 2021
J. Naylor, *Now Hear This*,
https://doi.org/10.1007/978-3-030-89877-9_4

of sea and land over which the sound of the eruption was heard. It was even heard at the far side of the Indian Ocean by the inhabitants of the island of Rodriguez, a distance of almost 4800 km from Krakatoa. Unsurprisingly, it is often claimed that the eruption produced the loudest sound ever created.

Yet, Krakatoa was not the world's most powerful eruption. In 1815, Tambora, a volcano on another island in the Indonesian archipelago, erupted with even greater force and was heard in Jakarta, 1,300 km away, where it was taken for gunfire. Indeed, the town's small garrison was mobilised as a precaution against a possible attack on the settlement.[2] The fact that Krakatoa, the lesser of the two eruptions, was heard at almost three times this distance suggests that it is not just the magnitude of an explosion that determines how far away it is heard, other factors are at work.

No man-made explosion, nuclear or chemical, has ever matched the power of Tambora or Krakatoa, but there have been numerous occasions when explosions and gunfire have been heard at distances of hundreds of kilometres. These are exceptions, however, because even the most ear-splitting sounds such as thunder and gunfire are usually inaudible at distances of more than 25 km.[3]

There are several reasons why sounds fade with distance. In the first place, sound usually spreads out from its source in all directions, particularly so in the open air, so that its energy is spread over an ever-increasing area as it travels away from the source. At the same time, the orderly vibrations that constitute sound are gradually converted into random vibrations of the molecules of the medium through which it is travelling: i.e. sound becomes heat. And the degree to which this occurs in air depends on the frequency of the sound and the humidity of the air. Dry air absorbs sounds of every frequency to a far greater degree than does humid air. The proverbial silence of a desert owes as much to the lack of atmospheric moisture as it does to the absence of life. But whatever the humidity of air, absorption is always greater for high frequency sounds than it is for low frequency ones.[4]

---

[2] Raffles, S. (1835) Memoir Of The Life And Public Services Of Sir Thomas Stamford Raffles, F.R.S., vol 1. John Murray, p 267.

[3] If you are standing at the seashore, the horizon is approximately 5 kms away. From a vantage point 50 m above sea level, the horizon is 25 km away, but may not be visible on a hazy day because haze can make it difficult to see where the ground meets the sky.

[4] The transfer of sound into heat occurs during the brief period of compression, during which molecules are pushed closer together leading to an increased number of collisions. Collisions between molecules transfer energy from kinetic energy, i.e. the energy of motion, to energy of vibration within the molecules as well as causing them to spin and tumble more rapidly. During the succeeding phase of rarefaction the space between molecules increases and so they collide less often, which allows their vibrational energy to become kinetic energy. The vibrations of low frequency sounds allow sufficient time for this to occur, which is why they attenuate very gradually, and indeed hardly at all at infrasonic frequencies. But at high frequencies there is not enough time to enable this transfer to

Hence at distances of more than 1000 m, thunder, which close to the stroke of lightning consists of frequencies that range from an inaudible infrasonic 4 Hz to an audible 2000 Hz, is reduced to an audible low frequency rumble (accompanied by unabsorbed though inaudible infrasound) because the higher frequencies have been converted into heat as they travel through the atmosphere. It is the same with gunfire and explosions: at a distance, only their lowest frequencies survive. In every case the sharp crack of an explosion fades to a muffled boom.

Nor is the state of the atmosphere the only thing that affects sounds. Sound is scattered and absorbed by the terrain across which it propagates, the obstacles it encounters together with atmospheric turbulence. And, as we shall now see, variations in air temperature and wind strength and direction also play an important role because they both affect the speed of sound

Silent Nights and Noisy days

You will often have noticed that sounds carry further at night than they do during daylight hours. It's such a marked phenomenon that it would have been known in prehistoric times. But the earliest attempt to explain it is to be found in an eclectic collection of questions and answers that was probably compiled during the third century BC. The work came to be known as the *Problemata* and was once attributed to Aristotle, though it was almost certainly assembled by his followers. "Why", the anonymous authors ask rhetorically, "are sounds heard better at night?" and answer that it's because nights are calmer than days: during the hours of daylight the air is stirred up by the sun's heat, scattering sound. After sunset, air cools down and becomes still, allowing sounds freer passage. A few centuries later, Plutarch, a Greek writer, came up with an alternative explanation: we hear sounds better at night because our hearing is keenest when we are deprived of our sense of sight by the absence of light.[5]

Neither of these views was questioned until well into the nineteenth century. William Derham who, as we saw in Chap. 2, spent several years systematically investigating the propagation of sound in the open air at the turn of the eighteenth century, and so might have been expected to come up with some fresh insights on the matter, concluded only that "sound seems

occur fully, which is why high frequency sounds attenuate far more rapidly than low frequency ones. Indeed the conversion of vibrations into heat is so pronounced at ultrasonic frequencies, that very high frequency ultrasound is used to warm subcutaneous tissue.

[5] Plutarch (1957) Moralia, Book VIII, Chap. 3, Sect. 3. (trans: Cherniss, H., Helmbold, W.). Loeb Classical Library, Harvard University Press.

stronger and somewhat acuter in the night, when there happens no noise, as is frequently the case in the day time."[6] His views on the subject were widely accepted; at any rate, no dissenters put pen to paper.

A century later, Alexander von Humbolt, the renowned German naturalist and explorer, revived the issue. Humbolt travelled extensively in northern Latin America and Mexico between 1799 and 1804, accompanied by his close friend, the French botanist Aimé Bonpland. During their exploration of the Orinoco river, Humbolt noticed that from a distance its cataracts sounded like waves breaking on a beach and was surprised to discover that "The noise [of the cataracts] is three times as loud by night as by day"[7] But at night the jungle through which the Orinoco flowed was anything but silent: "the air is constantly filled by an innumerable quantity of moschettoes, where the hum of insects is much louder by night than by day, and where the breeze, if ever it be felt, blows only after sunset."[8] He suggested that the improvement in audibility during the hours of darkness is not because nights are generally quieter, but because during the day the warmth of the ground heats the atmosphere and creates turbulent air currents that scatter sounds, making them less audible at a distance. Humbolt reminded his readers that a similar explanation had been offered in Aristotle's *Problemata.*[9]

Several decades later, Humbolt's conjecture appeared to be confirmed by research carried out by John Tyndall, the professor of physics at the Royal Institution in London who had succeeded Michael Faraday as its Director. Tyndall was an accomplished experimental scientist with wide interests and had been commissioned in 1872 by Trinity House to investigate the effectiveness of foghorns, sirens and guns as warning signals to help coastal shipping to avoid running aground in fog.[10] At the time it was assumed that fog, rain and falling snow all deaden sound because it was supposed that small drops

---

[6] Derham, W.I., (1708) Experiments and Observations on the Motion of Sound, &c. By the Rev. Mr. Derham, Rector of Upminster, and F. R.S. In: Hutton C., Shaw, G. Pearson, R. (eds.) The Philosophical Transactions Of The Royal Society Of London, From Their Commencements, In 1665, To The Year 1800; Abridged With Notes And Biographic Illustrations, Vol V from 1703 to 1712, London 1809, p 380–95, p 384, p 388.

[7] Humboldt, A., Bonpland, A. (1821) Personal Narrative Of Travels To The Equinoctial Regions Of The New Continent, During The Years 1799—1804, Vol V. Longman, London, p 69–72.

[8] Humboldt, A., Bonpland, A. (1821), p 69.

[9] The 'Aristotle' Humboldt was referring to was the Aristotle of the *Problemata*, who as noted in Chap. 2 was not the author of that work.

[10] Trinity House is the organisation responsible fro the safety of shipping and sailors in UK waters. Tyndall carried out most of his investigations from a ship lying off the White cliffs of Dover, Kent.

of water and tiny snowflakes reflect and scatter sound, just as molecules of gas in the atmosphere scatter light, thus reducing visibility.[11]

Frustratingly, during the weeks that Tyndal spent aboard boats owned by Trinity House off the White Cliffs of Dover listening to a selection of the foghorns at the South Foreland lighthouse, the atmosphere remained resolutely clear. The first opportunity he had to discover what effect fog has on sound came during a break he took in London. During several days when a fog enveloped the city he carried out a number of experiments at the Serpentine, the lake in Hyde Park. For one of these "I chose three organ-pipes of different lengths, a dog-whistle, a small bell struck by mechanism, and some percussion-caps. These were well heard across the Serpentine near the Watermen's Boathouse. At the same time I could converse with ease with my assistant across the water."[12]After several such experiments he concluded "Fogs have no such power to deaden sound as, since the time of Derham, has been universally ascribed to them."[13]

Tyndal was quite right because the conditions that favour fog, i.e. stable, calm air, also favour sound propagation. But how to explain the many occasions when the lighthouse was clearly visible from the sea but its foghorns were inaudible? These were often warm, sunny days. Tyndall claimed that in these conditions atmospheric turbulence due to rising columns of water vapour caused by evaporation from the earth's surface, which he called "acoustic clouds", scatter and absorb sound, just as Humbolt had surmised decades earlier.

However, in an account of these investigations, while noting Humbolt's observations, Tyndall said nothing about why audibility frequently seems to improve at night.[14] Nor did he have much to say concerning the audibility of distant battles about which he had been informed in the course of his investigations, namely that during some battles of the American Civil War there were occasions when events on the battlefield could be seen but not heard.[15] A witness to the battle of Gaine's Mill (27th June, 1862) had written to Tyndall describing how, from a vantage point on the opposite side of a valley some 2 km wide, he "distinctly saw the musket-fire of both [Confederate & Union armies], the smoke, individual discharges, the flash of the guns. I saw

---

[11] The preferential scattering of blue light by air was established experimentally by Tyndall in 1868 (see: Hoeppe, G. (2007) Why the Sky is Blue: Discovering the Color of Life. Princeton University Press, p 159–165).

[12] Tyndall, J. (1874) On The Atmosphere As A Vehicle Of Sound. Philosophical Transactions, Vol 164, p 183–244, p 209.

[13] Tyndall, J. (1874) p 211.

[14] Tyndall, J. (1874), p 195.

[15] Ross, C.D., (2000) Outdoor Sound Propagation In The US Civil War. Applied Acoustics, 59, p 137–147.

batteries of artillery on both sides come into action and fire rapidly. Several field-batteries on each side were plainly in sight. ...Yet looking for nearly two hours, from about 5 to 7 pm on a midsummer afternoon, at a battle in which at least 50,000 men were actually engaged, and doubtless at least 100 pieces of field-artillery, through an atmosphere optically as limpid as possible, not a single sound of the battle was audible to General Randolph and myself."[16] This was remarkable enough but there was more: the "cannonade of that very battle was distinctly heard at Amhurst Court-house, 100 miles west of Richmond, as I have been most credibly informed."[17] Tyndall ignored this particular detail, possibly because he could not have explained it in terms of the theory he was offering. In fact, the correct explanation for the audibility of sounds at such huge distances was not discovered until well into the twentieth century.

While a turbulent atmosphere does affect the transmission of sound through the air, particularly when the dimensions of the turbulent eddies are similar to the wavelengths of the sounds that encounter them, Tyndall had overlooked two factors that usually far outweigh the effect of turbulence. These are that wind speed and temperature both vary with altitude.

In fact, the year before Tyndall began his investigations on behalf of Trinity House, had spent several months lecturing in the United States where he met Joseph Henry, an American scientist whose research and discoveries in the fields of electricity and magnetism equalled and in some cases exceeded those of Michael Faraday. Henry had completed a study on the audibility of fog signals for the Light House Board of America, so was able to give Tyndall a demonstration of a foghorn he had had installed on the shore of Staten Island. But regarding the audibility of fog signals Henry arrived at a different conclusion: in his view the most important factor that determines the distance at which a fog signal can be heard is that wind speed increases with altitude. Measuring the speed of wind far above ground, however, essential if his explanation was to carry weight, proved difficult, though Henry came up with an ingenious ad hoc method to show that wind speed does indeed increase with altitude:

> During these investigations an attempt was made to ascertain the velocity of the wind in the upper stratum as compared with that in the lower. The only important result however was the fact that the velocity of of a cloud passing over the ground was much greater than that of the air at the surface, the velocity of the latter being determined approximately by running a given

---

[16] Tyndall, J. (1874), p 234–35.
[17] Tyndall, J. (1874), p 235.

distance with such speed that a small flag was at rest along the side of a pole. While this velocity was not perhaps greater than six miles per hour, that of the cloud was apparently equal to that of a horse at full gallop.[18]

Henry was not the first person to realise that a variation in wind speed might affect the distance at which sound can be heard, though he was the first to obtain direct evidence of this fact. In any case, the effect of wind on audibility is a matter of everyday experience, and has been known since antiquity. But its effect on audibility wasn't systematically investigated until 1812, when a Swiss physician, François Delaroche, spent several months with a colleague in the middle of a large field on windy days listening to the sound of hand-held bells being struck by hammers.[19] The experiments established that sounds are always audible at greater distances downwind than they are upwind; but Delaroche offered only measurements, not explanations.[20] Henry knew of Delaroche's research because he had read a short paper on the subject of the propagation of sound by George Gabriel Stokes, the Lucasian Professor of Mathematics at Cambridge University from 1849 to 1903. In 1857 Stokes had suggested that the reason why sounds are heard at greater distances downwind than they are upwind is that an increase in wind speed with altitude deflects sound back towards the ground that would otherwise pass over a listener.[21]

A few years later, exactly the same explanation occurred to Osborne Reynolds, professor of engineering at Owens College in the English city of Manchester.[22] Unaware of Stokes' conjecture, at least to begin with, Reynolds set out to disprove the explanation then widely accepted for the effect that wind has on the distance at which sounds can be heard distinctly. This was essentially that offered by Derham in 1708: a favourable wind accelerates sound, and a contrary wind retards it. But, as Reynolds pointed out, wind

---

[18] Henry, J. (1886) The Scientific Writings of Joseph Henry, Vol 1. The Smithsonian Institution, Washington, p 390.

[19] Delaroche died from typhus brought back to France from Russia by the remnants of Napoleon's Grand Armée.

[20] Delaroche, F. (1816) Sur l'influence que le vent exerce dans la propagation du son, sous le rapport de son intensité". Annales de Chemie et de Physique, Vol 1, p 177–195.

[21] Stokes, G.G. (1857) On The Effect Of Wind On The Intensity Of Sound. Report of the British Association, Dublin, p 22–23.

[22] Now the University of Manchester. Reynolds was one of the first of a generation of professors of engineering to be appointed in England during the latter half of the nineteenth century. His most famous pupil was J.J.Thomson, discoverer of the electron. According to Thomson, Reynolds "was one of the most original and independent of men, and he never did anything or expressed himself like anybody else. The result was that it was very difficult to take notes at his lectures…" (Thomson, Sir J.J. (1936) Recollections and Reflections. Cambridge, p 15).

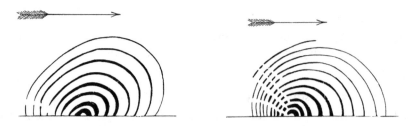

Fig. 4.1 Reynolds' sketches of the pattern of ripples created when drops of water fall into a slow moving stream. The arrow indicates direction of flow

speed is usually a tiny fraction of the speed of sound and so in itself has a negligible effect on the propagation of sound. On the other hand, if wind speed varies with height, being least at ground level due to friction between the air and the surface across which it moves, the resulting velocity gradient would cause the speed of sound to increase ever so slightly with height downwind and to be correspondingly incrementally retarded with height by a contrary wind. This velocity gradient creates what is known as wind shear, which in turn brings about a small but significant distortion in the envelope of sound travelling away from the ground.

A chance observation helped Reynolds to visualise the effect of wind shear on the propagation of sound on a windy day. While walking along a stream he noticed that ripples produced by water dripping from a pipe into the water close to the bank do not spread uniformly. As they expand into the faster moving water away from the bank, the distance between successive ripples increases downstream, distorting their envelope. Had the water been stationary, the ripples would have been uniformly semicircular. It's a simple and entertaining experiment that anyone can perform by dropping tiny pieces of wood into a slow moving stream as close to its edge as possible (Fig. 4.1).[23]

As an engineer, Reynolds knew that the matter could be settled only by quantitative experiments. His apparatus consisted of an electric bell positioned 30 cm above the ground in the middle of a large open field and a hand held anemometer. All he had to do was to measure the speed of the wind at different heights and determine the greatest distance at which the bell could be heard clearly, both downwind and upwind. In every case, the greatest distance at which the bell could be heard was always downwind, i.e. to leeward. At the same time he found that the sound of the bell was never as loud at ground level as it was even a metre or two above, regardless of wind direction, a fact he confirmed by climbing a tree several metres tall to,

---

[23] Even the smallest pebble will create splashes that form secondary ripples that obscure the pattern of the primary ripples.

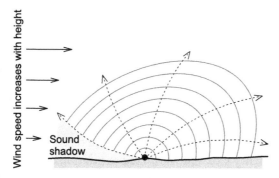

Fig. 4.2 Effect of wind shear on the propagation of sound. Downwind of the source the upper portion of the wavefront topples over towards the ground. Sound is not heard upwind within the sound shadow

as he put it, "recover" the sound. He also found that he could increase the distance at which the bell was audible by raising it well above the ground.[24] Here, then, is the one of the reasons why church bells are usually hung in tall towers, why *muezzin* call the faithful to prayer from the top of a minaret and why songbirds choose the highest trees from which to sing: the higher the source is above the ground, the greater the distance at which it can be heard due to vagaries of wind and weather.

Reynolds concluded that as long as the velocity of wind increases with height, sound is "lifted" (but not destroyed) to windward and "brought down" to leeward. In these circumstances there will be what he called a "sound" to windward (Fig. 4.2).

Having successfully explained the effect of wind on the propagation of sound, Reynolds turned his attention to the audibility of sounds at night. He realised that temperature must affect the passage of sound through the atmosphere in much the same way as wind does because the speed of sound in air is temperature dependent. Moreover, he knew that the temperature of the atmosphere diminishes at a fairly constant rate with height above the ground.

This had been established a century earlier by Horace-Bénédict de Saussure, the Swiss natural philosopher famous for his ascents of Alpine peaks. In 1787 de Saussure organised what was only the third expedition to reach the summit of Mont Blanc, 4,810 m above sea level.[25] During the ascent he made

[24] Reynolds, O. (1874) On The Refraction Of Sound By The Atmosphere. In: Reynolds, O. (1900) Papers on Mechanical and Physical Subjects Vol 1. CUP, p 89–106.

[25] In 1760, De Saussure had offered a reward to the first person to climb Mt. Blanc. The first successful ascent was made in 1785. John Tyndall, who was as well known as an accomplished Alpinist as he was for his scientific work, climbed Mt. Blanc twice, in 1857 and 1858 (See: Tyndall,

measurements with a barometer and thermometer and found that air temperature diminishes at a rate of approximately 0.7 °C per 100 m of elevation.[26] A few years later the French chemist Joseph-Louis Gay-Lussac reached an altitude of 7000 m above sea level during a solo ascent in a hydrogen balloon from Paris. At that altitude his thermometer registered −9.5 °C, almost 40 °C less than it had on the ground.[27]

Gay Lussac's ascent set an altitude record that stood for 58 years. But during the summer of 1862, the English meteorologist James Glaisher, accompanied by Henry Coxwell who acted as pilot, made a series of ascents in a gas-filled balloon, one of which probably reached at least 10,000 m.[28] The exact height of that ascent, which took place on 5th September, is not known because Glaisher became temporarily incapacitated at 8,800 m due to lack of oxygen while the balloon was still rising and so was not able to record the final readings of the barometer which acted as the balloon's altimeter. Coxwell somehow managed to remain conscious and stop the ascent by releasing gas from the balloon.[29] Glaisher regained consciousness and continued to take measurements as they descended. In his account of the flight he noted, in the laconic style that one associates with a stiff upper lipped Victorian Englishman, that "No inconvenience followed my insensibility; and when we [landed] it was in a country where no conveyance of any kind could be obtained, so I had to walk between seven and eight miles."[30] That distance, coincidentally, was approximately the height above ground they had probably reached in the balloon.

At a time when ballooning was considered to be either an amusement or an adventure, Glaisher's balloon flights were unusual in that they all had a scientific purpose. His precise systematic measurements of atmospheric conditions high above the ground were not equalled for several decades. And among these was the discovery that the temperature of the atmosphere varies not just with altitude but that it also depends on the time of day: "…within 100 feet near to the earth we now know there may be a decline of temperature of several degrees during the mid-hours of the day, and that during the

J. (1860) The Glaciers of the Alps, Being a Narrative Of Excursions and Ascents. John Murray, London, Ch 11).

[26] The accepted value for dry air, the so-called "normal lapse rate", is approximately 1 °C per 100 m. In humid air it is 0.65 °C per 100 m.

[27] Gay-Lussac, L. J. (1804) Relation d'un voyage aerostatique. Journal de Physique, 59, p 454–62.

[28] By the time Glaisher began ascending in balloons, hydrogen (expensive to produce and dangerous to use) had been replaced by so called Town Gas (principally methane, seven times denser than hydrogen but cheaper to produce), which is why his balloon flights took off from the vicinity of gas works. In Glaisher's and Coxwells's 1862 ascent, it was Wolverhampton gas works.

[29] Holmes, R. (2013) Falling Upwards. Collins, p 211–19.

[30] Glaisher, J. (1871) Travels in the Air. Richard Bentley & Sons, London, p 54.

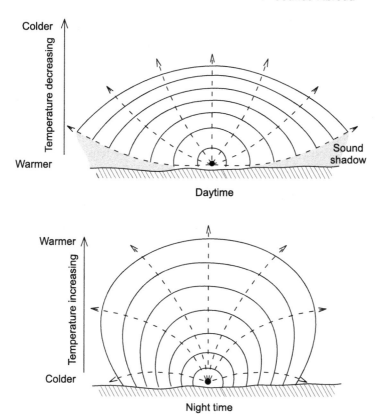

Fig. 4.3 Effect of temperature gradients within the atmosphere on the propagation of sound. Sounds refract away from a region of low velocity to one of higher velocity. On a sunny day sound refracts away from the ground and so is not heard at a distance. At night sound refracts towards the cooler ground and can be heard at a distance

mid-hours of the night there may be, and generally is, an increase of several degrees."[31]

Knowing this, Reynolds surmised that the speed of sound must also vary slightly with altitude due to changes in air temperature. Thus on a clear day, when the atmosphere is warmest at ground level, sound will propagate in a similar fashion to the way it does to windward, i.e. as the envelope of sound expands, its lower part refracts away from the ground, creating a sound shadow. The reverse happens on a clear night: the upper part of sound envelope refracts towards the ground (Fig. 4.3).

---

[31] Glaisher, J. (1871), p 84.

The reason why the temperature of the atmosphere varies with height is that compared to the ground, air is a very poor absorber of the sun's radiant energy. During the hours of daylight on a clear day the ground is warmed by the sun, which in turn warms the air in contact with it. Hence on a sunny day the air just above the ground is always warmer than it is higher up. But on calm, cloudless nights the ground cools by radiating infrared radiation. So does the atmosphere, but at a lesser rate, so that after sunset the temperature of the ground soon falls below that of the air a few tens of metres above it. This cools the air in contact at ground level, creating a positive temperature gradient in the first hundred or so metres of atmosphere, i.e. the temperature the air increases with altitude. On cold, clear winter nights this temperature gradient can be very pronounced: at ground level air temperature might be 0 °C while a few tens of metres higher up it might be as much as 15 °C (Fig. 4.4).

The effect of a positive temperature gradient on sound is to make it audible at greater distances than it would be otherwise because the speed of sound increases in step with temperature. The portion of the sound envelope travelling away from the ground speeds up within the warmer air and

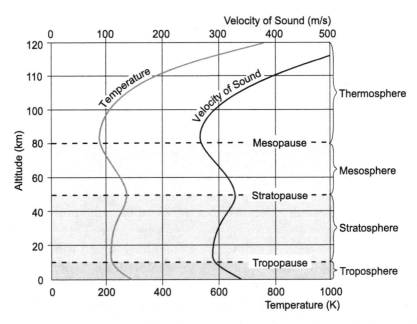

Fig. 4.4 The temperature profile of the Earth's atmosphere. Changes in the speed of sound closely mirror those of the change in atmospheric temperature with altitude above the Earth's surface. Above the troposphere atmospheric temperature increases leading to a gradual increase in the speed of sound within the stratosphere

refracts back towards the ground. Nor is this effect confined to the hours of darkness: during clear, calm days in winter and in polar regions, when the surface temperature is at or below freezing, sounds are sometimes clearly audible at much greater distances than is usual at lower, temperate latitudes. A remarkable instance of this phenomenon was described by William Parry, a pioneering British Arctic explorer, in his account of an expedition to discover the North West passage that he led in 1824.

> The extreme facility with which sounds are heard at a considerable distance, in severely cold weather, has often been a subject of remark; but a circumstance occurred at Port Bowen, which deserves to be noticed as affording a sort of measure of this facility, or at least conveying to others some definite idea of the fact. Lieutenant Foster having occasion to send a man from the observatory to the opposite shore of the harbour, a measured distance of 6,696 feet [2,041m], or about one statute mile and two-tenths, in order to fix a meridian mark, had placed a second person half-way between, to repeat his directions; but he found on trial that this precaution was unnecessary, as he could without difficulty keep up a conversation with the man at the distant station. The thermometer was at this time $-18°$ [$-18°F = -8 °C$], the barometer 30.14 inches, and the weather nearly calm; and quite clear and serene.[32]

However, all else being equal, the effect of temperature alone on the path taken by a sound is usually much less than that of wind, which made it very difficult for Reynolds to obtain clear evidence of temperature-dependent acoustic effects. The slightest wind will usually overwhelm the effect of temperature gradients on the propagation of sound, at least over distances of a few kilometres, and so Reynolds' investigations into the effect of temperature on audibility were not as conclusive as those into the effect of wind. One of the problems he faced is that the effect of temperature on sound is noticeable only over distances of hundreds if not thousands of metres, which made it necessary to replace the electric bell of his earlier experiments with a pistol. And, as with Colladon on Lake Geneva, the distance between the person responsible for creating the sound and the person listening for it made coordinating their roles next to impossible. But the greatest problem was wind: on the days during which these experiments were made, ideal conditions, i.e. calm, cloudless days, were seldom met with. As a result, trials carried out on land were inconclusive. However, those on water were more successful.[33]

---

[32] Parry, W.E. (1826) Journal of a Third Voyage for the Discovery of a North-West Passage from the Atlantic to the Pacific (1824–25). John Murray, p 58.

[33] Reynolds spent several weeks on board a friend's yacht off the North Norfolk coast.

Although he had no way to measure the temperature of the air at different heights in the vicinity of his trials, he was able to tell whether the temperature of the air was increasing or decreasing with height because this affects light in the same way as it does sound. When the air next to the ground is warmer than that higher up, light from the horizon is refracted away from the ground and distant objects disappear from view. This is the cause of those inferior mirages of the sky that look like distant pools of water on warm surfaces such as roads. The reverse happens when the temperature of the air in contact with the ground is colder than that higher up: objects beyond the horizon are raised into view, an effect known to sailors as looming.[34] By keeping an eye on the horizon, Reynolds was able to gauge whether the temperature of the atmosphere was increasing or decreasing with height and correlate it with the maximum distance at which the sound of the pistol shots could be heard.

All the trials were made during daylight hours, so he assumed that the temperature of the air decreased with altitude, which would cause sound to refract away from the ground, making the pistol shots inaudible. But when he conducted his investigations at sea, he found the reverse because cold water cools the air in contact with it, which will create a positive temperature gradient—i.e. temperature increases with altitude. Reynolds wrote that on the day of his most conclusive trial, he heard sounds of guns being fired as far away as eight miles (13 km) as well as the paddles of a steamer that he estimated was even more distant than the guns:

> For the sake of my experiments, what I had been in hope of was a state of the atmosphere which would cause great upward refraction of sound, and I was naturally on the *qui-vive* for any indications of such a state. All morning I had been watching the distant objects to see whether they were lifted or depressed by the refraction of light. They loomed to a remarkable degree, which showed that the upward variation of temperature was the reverse of what I wanted.[35]

There is yet another reason why sounds can travel further across open water than over open ground: the difference in acoustic impedance between air and water is much greater than it is between air and earth and so water is a much better reflector of sound than solid ground. Even a perfectly level, grassy surface absorbs and dissipates sound energy to a far greater degree than

---

[34] When conditions over the sea favour looming, one might see a ship that is beyond the horizon appear to float in the air above it. For more see: https://aty.sdsu.edu/mirages/mirsims/loom/loom.html (accessed 1/3/2021).

[35] Reynolds, O. (1876) On the Refraction of Sound by the Atmosphere. In: Reynolds, O. (1900) Papers on Mechanical and Physical Subjects, Vol 1. CUP, p 162.

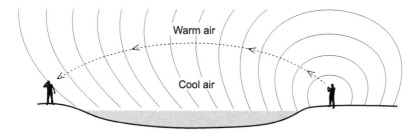

Fig. 4.5 Propagation of sound over water. Sound travels faster in warm air than in cold air causing the upper part of the wavefront to head back to the ground

the smooth surface of calm water. Indeed, combined with a positive temperature gradient, sound can travel across a large expanse of open water in a series of hops: downward refraction in air followed by reflection at the surface of the water until it becomes inaudible due to absorption by the atmosphere. In this situation, sound is largely confined, or ducted, within the layer of air in contact with the water surface. This is why sounds are heard particularly well over lakes at the beginning or the end of a warm and windless day when the layer of air in contact with water is cooler than that higher up (Fig. 4.5).

Another unexpected effect of air temperature on sound is that as long as the temperature of the atmosphere decreases with altitude, in the absence of wind, sounds are more likely to be heard uphill from the source than the other way around. Balloonists often find it impossible to talk with people on the ground because although they can hear the voices of groundlings, those on the ground may be unable to hear what the balloonists are saying to them. This situation occurs when the temperature of the atmosphere decreases with altitude, which in turn causes sound to refract away from the ground. Hence the voice of someone on the ground is carried up to the balloonist. The same, of course, happens to the sounds made by those in a balloon. These also refract away from the ground and so don't reach those below unless the balloon is almost overhead.

The effect of temperature-dependent refraction is also evident with aircraft noise. If you have an unobstructed view of the horizon, you may have noticed that on a clear day you can usually see a low-flying plane as it approaches you long before you hear it. This is because the effect of temperature on sound is much more pronounced than its effect on light. The sound of the turbulence of the hot gasses emitted by the engines, which is the source of almost all of the sound of an aircraft in flight, is refracted away from the ground due to the typical daytime temperature gradient of the atmosphere. It is only when the plane has been visible for some while that its engine noise reaches the ground where one is situated (Fig. 4.6).

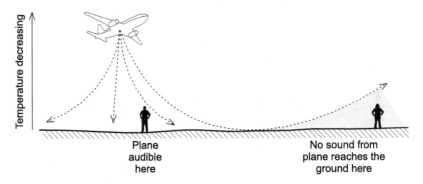

Fig. 4.6 Refraction of sound from an airplane as it travels towards the ground. Sound that grazes the ground is refracted back into the sky because its speed increases within the warmer air

Songbirds appear to be instinctively aware that sound is refracted away from the ground on a warm day because they will usually seek a perch as high above their surroundings as possible from which to broadcast their songs. The distance at which a bird on the ground can be heard is far less than it is if it is several metres higher. But unlike aircraft noise, birdsong is intended to be heard. Indeed, it must be heard by other birds of the same species if it is to be worth the effort of producing it because the purpose of the song is both to attract mates and warn off interlopers.

The finer details of these songs, of course, are gradually lost as they travel through air due to absorption of high frequencies, turbulence and scattering by vegetation. Taken all together, these losses limit the distance at which birdsong is effective to between ten and one hundred metres depending on the species of bird.[36] But there are at least a couple of features of birdsong that partially compensate for these losses. In the first place, many songbirds are remarkably loud given their size. For example, the male Hooded Warbler weights about 15 g, yet it is capable of singing at 80 to 90 decibels (measured at 1 m). Compare that to the loudness of a human voice during a normal conversation: 60 dB at a similar distance from the speaker. The second feature of birdsong that mitigates transmission losses is that most of the acoustic energy of the song is typically confined to a single frequency. Variations in tone are achieved by altering the frequency, just as humans do when whistling a tune. By concentrating energy in a single frequency, the song travels further and stands out from background noise.

---

[36] Wiley, H.R. (2015) Noise Matters, The Evolution of Communication. Harvard University Press, Ch 4.

Given their relative size, it is very surprising that the power of the human voice during normal conversation is less than that of many songbirds. The reason is that when speaking, one has to filter out some of the frequencies due to the vibration of the vocal folds to create the intricate phonemes that are the building blocks of speech. And to achieve this, some of the energy of the vibrations of the vocal folds is dissipated within the vocal tract as heat. In place of vocal folds and a vocal tract, birds have an organ known as a syrinx. The syrinx is lined by two thin membranes that vibrate to produce sound. In comparison with human speech, there is little acoustic filtering within the syrinx so most of the energy developed within it makes it into the air.

The human voice is also limited in the distance at which it can be clearly heard by the same factors as birdsong. Even in ideal conditions, a quiet, windless day, a shouted conversation between two people ceases to be intelligible when they are more than a couple of hundred metres apart. Like any other sound, as one's voice travels through the air it spreads out and becomes fainter. At the same time, the sibilant sounds fail to make it to the listener's ear because their high frequencies are absorbed by the intervening air.

But some communities have found a way to reduce the effects of some of these losses: imitate birds. Don't vocalise, whistle. This is the idea behind so-called whistled languages, which are still in use by some isolated communities living in mountainous countryside or in forests.[37] The best known example of a whistled language is Silbo used in La Gomera, one of the seven islands that make up the archipelago of the Canary Islands.[38] In fact, Silbo is not a distinct language, it is Spanish. The same is true of every other known example of whistled languages: they are all versions of the local spoken language. It is therefore more accurate to class them as whistled speech rather than whistled language.

Whistling has at least two advantages over vocalising. A whistle can be some 20 dB louder than vocalised speech, so can potentially be heard at a greater distance. And all whistled speech employs frequencies between 1000 and 2,500 Hz, frequencies that the ear is very sensitive to. However, restricting frequency to a narrow band means that whistling can't reproduce the full range of complex phonemes necessary for spoken words and so is more limited in what it can express. Whistled speech is perfectly adequate for exchanging greetings, giving warnings or orders and declaring intentions. But it's not a suitable medium in which to engage in an exchange of ideas.

Voiced speech relies a much broader range of frequencies than whistling can achieve, which means that its finer details are lost through absorption

---

[37] Meyer, J. (2015) Whistled Languages: A Worldwide Inquiry on Human Whistled Speech. Springer.
[38] Francesca Phillips (2009) "Written in the Wind", a film about el Silbo of La Gomera.

and scattering much more rapidly than a whistled conversation as it travels through the air. A study carried out in a valley in the French Alps established that whistled communications were intelligible up to 700 m from the source whereas the spoken word was all but inaudible at 40 m.[39] The author of this research claims that in favourable conditions whistled speech can be intelligible up to a distance of 5 km.

Zones of Silence

But what of the cannonade that was heard in Amhurst, a hundred miles distant from the battle of Gaine's Mill? In fact, it had long been known that gunfire and explosions can on occasion be heard across distances of hundreds of kilometres. Derham knew of several accounts of gunfire that had been heard at improbably large distances. He wrote that "… at the [1674] siege of Messina, the report of the great guns reached the ears of the inhabitants of Augusta and Syracuse, at almost 100 Italian miles; and likewise when the French besieged Genoa [in 1684], the firing of their guns was plainly heard as far as Monte Nero, upwards of 90 Italian miles."[40] A century later, Chladni reported hearing the sound of cannon fire during the battle of Jena (14th Oct, 1806) from Wittenberg, his home town—almost 150 km away. He mistakenly attributed it to conduction of sound through the ground.[41]

But gunfire had been heard at an even greater distance, several years before the events described by Derham. Moreover, the account of that earlier event included a curious detail seemingly unknown to him: the gunfire was not heard by people much closer to the guns than those who did hear it.

During a bitterly fought but inconclusive sea battle between the Dutch and English Fleets off Dunkirk that lasted for several days in early June 1666, Samuel Pepys (1633–1703), the naval administrator famous for his diary, recorded that gunfire from the battle was heard in Greenwich Park, London but not at Deal or at Dover. These are towns on the Kent coast on

---

[39] Meyer, J. (2017) Whistled Languages Reveal How the Brain Processes Information. Scientific American, Feb.

[40] Derham, W.I. (1708) Experiments and Observations on the Motion of Sound, &c. By the Rev. Mr. Derham, Rector of Upminster, and F. R.S. In: Hutton C., Shaw, G. Pearson, R. (eds.) The Philosophical Transactions Of The Royal Society Of London, From Their Commencements, In 1665, To The Year 1800; Abridged With Notes And Biographic Illustrations, Vol V from 1703 to 1712, London 1809, p 380–95, p 338.

[41] Ullmann, D. (2007) Life and work of E.F.F. Chladni. Eur. Phys. J. Special Topics, 145, 25–32, p 30.

the other side of the English Channel from Dunkirk and about one third of the distance between Dunkirk and London.

> ...it is a miraculous thing that we all Friday, and Saturday and yesterday, did hear every where most plainly the guns go off, and yet at Deal and Dover to last night they did not hear one word of a fight, nor think they heard one gun...how we should hear it and they not, the same wind that brought it to us being the same that should bring it to them: but so it is.[42]

From Dunkirk to London is about 200 km as the crow flies, and since the guns were not heard on the coast it can't have been a wind that carried the sound of their fire all the way to London, at least not in the sense that Pepys would have understood the matter. We now know that what he heard was an extreme form of the temperature-dependent refraction of sound that is always accompanied by a phenomenon that has come to be known as a *zone of silence* , and which invariably occurs whenever the sound of a very powerful explosion is heard at a great distance, even very occasionally as much as 1000 km from the source. In these circumstances there are always one or more broad zones closer to the explosion and more or less concentric with it, where it is not heard. Incidentally, the next time Londoners heard gunfire from Dunkirk was in May 1940, during the forced evacuation of British and French soldiers from France.

As we have seen, there were several instances where the sound of distant gunfire was accompanied by a zone of silence during the battles of the American Civil War,[43] but it wasn't until the First World War that the phenomenon was studied systematically. Zones of silence between zones of audibility were a frequent occurrence during the artillery barrages on the Western Front. Indeed, the phenomenon was first noticed within weeks of the outbreak of war. Ewoud van Everdingen, a Dutch meteorologist, established that during the siege of Antwerp during the second week of October, 1914, gunfire from the siege cannons was heard by people up to 200 km away in the far north of the Netherlands, but that the guns were not audible between 60 and 100 km from the city.[44]

A memorable example of the sounds of battle from the Western Front being audible in London occurred at the start of the Battle of Messines that

---

[42] Pepys, S. Diary entry for 4th June 1666.

[43] Ross, C. D. (2000) Outdoor Sound Propagation In The US Civil War. Applied Acoustics, 59, p 137–47.

[44] Everdingen, E. Van (1914) De hoorbaarheid in Nederland van het kanongebulder bij Antwerpen op 7–9 October. Hemel en Dampring, 6:1914, p 81–85.

began in the early morning of 7th July, 1917. The British had dug 19 tunnels under German fortifications and filled them with 450 tonnes of explosives. The detonation of the mines, which killed some 10,000 German soldiers, was heard in London. One of those who heard it was the British Prime Minister, David Lloyd George. He was awake in his study at 3 a.m. awaiting news of the start of the battle when he heard the explosion for himself.[45]

A thorough investigation of sounds heard at huge distances from their source, a phenomenon known as anomalous propagation of sound, was made in 1917, following the largest explosion ever to occur in London. This was the accidental detonation of 50 tonnes of TNT at a munitions factory at Silvertown, in the East End of London, that killed 73 people and injured 400. The subsequent inquiry found that the explosion was heard 160 km away in Norfolk and South Lincolnshire as well as in London and the adjacent counties, but not in Cambridge or parts of Essex and Suffolk, which lie between London and Norfolk. Furthermore, in Norfolk and Lincolnshire the explosion was heard not as a single event but as a series of two, three and even four distinct reports, which were said to sound like muffled gunfire.

The need to dispose of large quantities of unused munitions in France, Germany and the Netherlands following the First World War provided scientists with a unique opportunity to carry out controlled investigations into what until then had been largely a matter of anecdote, notwithstanding the inquiry into the Silvertown explosion. Observers at different locations in a vast area around selected munitions dumps were recruited and asked to note what, if anything, they heard when the explosives were detonated. When their reports were collated, they confirmed that twenty to thirty kilometres beyond the explosion there was a zone, that usually extended a further 100–160 km from the site of the explosion, where nothing was heard, while observers beyond this zone reported hearing the sound tens of minutes after the detonation.[46]

At the time when these controlled explosions were carried out, next to nothing was known about the nature the atmosphere beyond 10 km above sea level, the altitude supposedly attained by Glaisher and Coxwell during their 1862 balloon flight. Indeed, until the discovery of oxygen and nitrogen in 1772, air was assumed to be a single substance. Gay-Lussac had collected samples of air at 6500 m during his solo 1804 balloon ascent and had analysed them himself within hours of returning to the ground. He found that the proportion of oxygen in air to be the same in these samples as it is at ground

---

[45] "Mr Lloyd George hears the explosion." The Times, 8 June, 1917.

[46] Richardson, E.G. (1953) Sound. Edward Arnold & Co., London, p 18–21.

level. The only other fact that was known with certainty about the atmosphere is that both temperature and pressure diminish with altitude, which implied that the temperature of the atmosphere must eventually fall to zero at some point.

The very lowest possible temperature, i.e. the point at which matter is devoid of all thermal energy, is −273 °C, known as *absolute zero* on the Kelvin scale of temperature.[47] Taking the normal lapse rate of the atmosphere to be 1 °C per 100 m implies that the temperature of the atmosphere should be −273 °C at an altitude of a mere 27 km above sea level. However, during the last few years of the nineteenth century, a French meteorologist, Léon Philippe Teisserenc de Bort, had discovered that above 10 km, the temperature of the atmosphere appears to be constant. Between 1896 and 1898, employing small, unmanned high-altitude hydrogen balloons of his own design equipped with thermometers, he found that the temperature of the atmosphere does indeed fall steadily with altitude to a height of approximately 10 km. Above this, however, his measurements indicated that the temperature remains constant. He concluded that there are at least two layers in the atmosphere. He named the lower one the *troposphere*, i.e. the sphere of change, from the Greek word "tropos", which means "to turn", because within that layer the atmosphere is churned up by currents of air that rise and descend. The apparent absence of a temperature gradient within the atmosphere just above the troposphere implied an absence of convection currents, which led him to surmise that above 10 km the air forms a stable layer. He named this region the *stratosphere* or "sphere of layers" from *stratus*, the Latin for layer.

Teisserenc de Bort's discovery was later confirmed by a German meteorologist, Richard Assmann. Assman's balloons reached even greater altitudes than de Bort's because they were made of rubber, which expands as the pressure of the surrounding air drops thus providing increased buoyancy in the thinner air of the upper atmosphere.[48] Intriguingly, both men detected a slight increase in temperature at those altitudes, but chose to ignore this as an instrumental error.

Variations in air temperature within the stratosphere play a central role in bringing about zones of silence. But being ignorant of the temperature profile within this region of the atmosphere the early investigators of

---

[47] The Kelvin scale is used in physics and thermodynamics. 1 K represents the same temperature interval as 1 °C, but the zero point of the latter scale is based on the freezing point of water, which is 273 K. The Celsius or centigrade scale is more convenient than the Kelvin scale for everyday use.
[48] Teisserenc de Bort's *L'Aérophile* reached an altitude of 14 km, Assmann's *Cirrus* reached an altitude of 19 km.

the phenomenon were unable to come up with a definitive explanation for absence of sound at ground level within these zones. One suggestion was that the upper stratosphere is composed of hydrogen and helium, which as the two lightest gases in nature would necessarily rise to the top of the atmosphere.[49] The speed of sound in hydrogen is almost four times greater than it is in the mixture of oxygen and nitrogen that are the main constituents of air, which would cause sounds that reach the upper atmosphere to refract back towards the ground. But this explanation was rejected when calculations showed that both hydrogen and helium molecules in the atmosphere can reach velocities that exceed the escape velocity due to the Earth's gravitational field. This, of course, is why there is no hydrogen or helium in the Earth's atmosphere.[50]

The troposphere is between 10 and 20 kms deep, being least at the poles and greatest in the tropics. The stratosphere extends a further 50–60 kms above this. Although the composition of air is almost the same in both these layers, the temperature profile is not. Air temperature decreases with altitude within the troposphere, whereas the reverse happens within the stratosphere. In fact, the very pronounced temperature inversion within the stratosphere was first inferred from studies of zones of silence, as was the upper limit of the stratosphere. Erwin Schrödinger, better known as one of the principle architects of quantum mechanics, had taught meteorology in an Austrian military academy during the First World War. One of the problems he became interested in was that of sound propagation through the atmosphere and in an academic paper on the subject he showed that the distance between molecules in the atmosphere above 60 km is too great to allow audible sound to propagate without considerable loss of energy due to absorption by molecules, though infrasound is much less affected.[51] This, of course, is the explanation for the high degree of attenuation suffered by sounds on Mars: at its surface its atmospheric pressure is a mere 1% of that of the Earth—i.e. similar to that of the Earth's stratosphere.

But a zone of silence is a by-product of the long-distance propagation of sound and can't occur in the absence of powerful sounds. As we have seen, the energy of sounds travelling close to the ground is gradually absorbed and

[49] Von dem Borne, G. (1910) Über die schallverbreitung bei Explosionskatastrophen. Physikalische Zeitschrift XI, p 483–88.

[50] Hydrogen can be obtained by the electrolysis of water (H_2O) and helium is created by radioactive decay of elements within the Earth's crust. When some radioactive substances decay, they emit alpha particles (two protons and two neutrons) which capture two electrons each to form a helium atom. Helium is present in small amounts in underground gas deposits, and this is the only source of that gas on Earth.

[51] Shrödinger, E. (1917) Zur Akustik der Atmosphäre (On the Acoustics of the Atmosphere). Physikalische Zeitschrift, 31 July.

dissipated as heat by the surface over which it travels and within the atmosphere so that even the loudest explosion will usually be inaudible at distances of more than a couple of dozen kilometres. However, powerful sounds travelling away from the ground reach the stratosphere in under a minute, where they begin to speed up in the increasingly warmer air, thus gradually changing direction and eventually returning to the ground a considerable distance from the source. This happens only when the temperature of the air in stratosphere directly above the troposphere is unusually warm. In the absence of a strong temperature inversion within the stratosphere, sound from a powerful explosion is unlikely to refract back to Earth and reach the ground several tens of kilometres from the source. And without a zone of audibility far from the source, there is no corresponding zone of silence.

Coincidentally, the most powerful man-made non-nuclear explosion occurred in 1917, the same year as the Silvertown explosion, when two ships, one of which was fully loaded with explosives destined for France, collided in the harbour of the Canadian port town of Halifax. The resulting explosion destroyed much of the town around the harbour, killed 1500 people outright and injured a further 9000. The explosion was heard in Charlottetown, Prince Edward Island, roughly 215 kms north of Halifax, and in North Cape Breton, 360 kms east. More recently, on 11th December, 2005, an explosion at an oil depot in Buncefield, Hertfordshire, was heard as far away as the Netherlands, some 300 km distant, though nothing was heard on the East Anglian coast. And on 4th August, 2020, an explosion of almost three thousand tons of ammonium nitrate stored in a portside warehouse in Beirut, which killed 215 people and wrecked much of the town, was heard 240 km away in Cyprus.

Although the explanation for the long-distance propagation of sound and its associated zones of silence was eventually discovered during the interwar years, one aspect of the phenomenon remained unresolved. During the First World War gunfire on the Western Front was usually heard in Germany during winter and in England during the summer. The explanation was only discovered after the Second World War when the phenomenon was studied in detail and it was found that up to an altitude of 10 km the prevailing wind in the Northern Hemisphere blows west to east, but that well above this there is a seasonal change. In winter, high altitude stratospheric winds continue to blow west to east, but in summer they blow east to west.[52] The combination of a strong stratospheric temperature inversion and high altitude wind almost guarantees that sound that reaches the stratosphere will return to the ground.

---

[52] These stratospheric winds reach their greatest velocity at altitudes between 40 and 70 km above sea level.

Volcanic Eruptions

The most powerful natural sounds are due to volcanic eruptions and, unsurprisingly, these have on occasions been heard at great distances. But the audible sound of an eruption seldom, if ever, matches the spectacle of thousands of tonnes of rock and ash being hurled hundreds of metres into the air because the acoustic energy of these eruptions is largely inaudible infrasound.

Shortly after sunrise on Sunday, 18th May, 1980, Mount St Helens, a volcano in Washington State, USA, which had been the source of hundreds of small local earthquakes during the preceding weeks, erupted without warning. The eruption reduced the height of the mountain by some 400 m and devastated the surrounding countryside: buildings and trees within an area of 600 square kilometres were flattened and 57 people killed. The death toll would have been far greater had the eruption occurred during a weekday when teams of loggers would have been at work in the vicinity of the mountain.

The person closest to the eruption who lived to tell the tale was almost 15 km away. Extraordinarily, he heard not a whisper, almost certainly because most of the acoustic energy of the eruption was infrasonic. But twenty minutes later a series of very loud bangs, which were described variously as the sound of large guns, sonic booms or thunder, were heard in Vancouver, which is some 300 km due north of the mountain. Similar sounds were heard in Seattle (150 km North) and Victoria (200 km North). But no one within 100 km of the eruption reported hearing anything, whereas it was heard, with varying degrees of intensity, at distances between 150 and 300 km from the volcano, with the loudest sounds being heard in Vancouver. There were even a handful of reports of the eruption been heard up to 900 km away.

Here was a classic example of a zone of silence, created because the infrasonic pulses from the eruption travelled up into the stratosphere where a temperature inversion reversed their direction and sent them back towards the ground. That the infrasonic pulses became audible, and, moreover, were heard as a series of distinct loud sounds, requires explanation. The transformation of infrasound into audible sound in this case has been attributed to the bunching up within the stratosphere as trailing pulses caught up with slower moving leading pulses to create a shock wave that was heard as a sharp explosion by those on the ground. The separation of the original pulse into several distinct sounds can be explained by assuming that the path of the pulses though the upper atmosphere was determined by the direction in which they left the volcano. The steeper their initial inclination, the higher

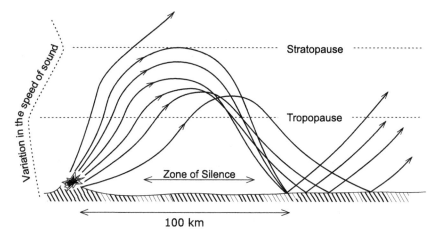

Fig. 4.7 Sound paths in the vicinity of Mt St Helens following its eruption in May 1980. Sound that returns to the ground due to refraction within the stratosphere is reflected back up into the atmosphere resulting in a second zone of silence[55]

they would rise into the stratosphere before being turned back towards the ground. And the longer their path, the later they arrived back at ground level (Fig. 4.7).[53]

The circumstances that bring about a zone of silence provide the most likely explanation for the huge distances at which the eruption of Krakatoa was heard. The report on the event compiled by the Royal Society recorded that "The sounds of the eruption were heard with great distinctness over the most distant parts of Java and Sumatra throughout the morning of the 27th, but it is very remarkable that at many places in the more immediate neighbourhood of the volcano they ceased to be heard after 10am, although it is known that the explosions continued with great intensity for some time longer."[54] Krakatoa lies in the straights between these two large islands, i.e. midway between the farthest points at which the eruption was heard on Sumatra and Java. Hence a zone of silence must have extended over much of each island.

Although no zones of silence beyond Java and Sumatra came to light during the subsequent investigation conducted by the Royal Society, that is almost certainly because the area across which the eruption was heard is mostly ocean, so that there would have been few if any witnesses within

[53] Dewey, JM (1985) The Propagation Of Sound From The Eruption Of Mount St Helens Of 18 May 1980. Northwest Science 59, 79 (92).

[54] The Eruption of Krakatoa and Subsequent Phenomena. In: Symons, J. (ed) Report of the Krakatoa Committee of the Royal Society. Trübner & Co., London, 1888, p 80.

[55] Dewey, JM (1985) ibid.

these zones. However, for the sounds of the eruption to be audible at the huge distances it was heard, zones of silence must have been present because, as we have seen, audible sound cannot usually travel through the lower atmosphere for more than a couple of dozen of kilometres due to the conversion of its energy into heat. And refraction of sound within the stratosphere tends to bring sound back to the ground within a couple of hundred kilometres of the source. Hence the sound of the eruption must have reached the island of Rodriguez by means of a series of hops across the intervening ocean, each consisting of refraction within the stratosphere followed by reflection from the sea.

Wherever the eruption was heard, it was invariably taken to be distant gunfire. The Chief of Police on Rodriguez noted that "Several times during the night (26th–27th August) reports were heard coming from the eastwards, like the distant roars of heavy guns."[56]

On Diego Garcia, 3,600 km west of Krakatoa, "…the sounds were very distinctly heard, and were supposed to be those of guns fired by a vessel in distress; a belief which likewise prevailed at Port Blair in the Andaman Islands, and at several places less remote from Krakatoa. In Sri Lanka [then known as Ceylon], and also in Australia, the sounds were heard at many different places far removed from one another."[57]

The inaudible infrasonic pulses from the eruption travelled around the world several times and were detected only because their passage caused minute changes in air pressure recorded by recently installed barographs in the meteorological stations of several countries. It was, in fact, the first time infrasound had been detected, if only indirectly, which led to the realisation of the existence of aerial vibrations below the threshold of human hearing. Moreover, it showed that these very low frequency vibrations can travel huge distances with little loss of energy. Mersenne would, doubtless, have been thrilled to have one of his more fanciful conjectures vindicated: a sound powerful enough to circle the Earth (Fig. 4.8).

Brontides

A few years after the eruption of Krakatoa, on several occasions people in Bengal reported hearing what they took to be the sound of distant gunfire. The earliest published account of these mysterious sounds appeared in the journal of the Asiatic Society of Bengal.

---

[56] Symonds, J. (1888), p 79.
[57] Symonds, J. (1888), p 80.

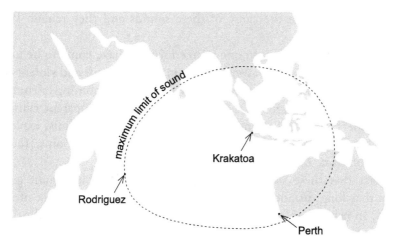

Fig. 4.8 Map showing the area within which the eruption of Krakatoa was heard

The sounds resemble the booming of distant cannonade, and they are usually heard during the months from April to September in a lull after a squall, or after a shower of rain, or when the clouds begin to break up. Barisal Guns they are called because at Barisal the explosions happened to be first noticed, but the area is vast over which such noises are heard.[58]

The members of the Society decided that the phenomenon was worth further investigation. This revealed that these mysterious sounds always came from the south or southeast, i.e. from the Bay of Bengal, that sometimes several sounds were heard in rapid succession, and that they varied in loudness from one occasion to the other.[59]

The phenomenon attracted widespread interest, and a detailed study by an amateur Belgian geologist and naturalist, Ernest van den Broeck, during the 1890s found that similar unexplained booming sounds had been heard elsewhere around the globe and that in some locations they had been heard with sufficient frequency to have been given specific names by the locals: "mistpoeffers" in Belgium and the Netherlands, "brontidi" and "marina" in Italy, "lake guns" in North America, "uminari" in Japan and "retumbos" in the Philippines. He also found that these booming sounds usually seemed to come from the direction of the sea. Mistpoeffers, for example, always appeared to have their origin in or beyond the North Sea. However, he

---

[58] Babu Gaurdas Bysack (1888) On The Barisal Guns. Proceedings, Asiatic Society of Bengal, March, p 97.

[59] Lt.-Col. J. Waterhouse, (1888), p 110.

was unable to establish the cause of these sounds and they remain largely unexplained to this day.[60]

In all probability there is no one source for what have come to be known collectively as brontides. To powerful, man-made explosions and violent eruptions, which as we have seen are well-known sources of booming or rumbling sounds that may appear to the uninitiated to have no apparent cause, we can add others: powerful thunderstorms, earthquakes, tsunamis, exploding meteors, booming sand dunes and, in our own day, sonic booms. You will find more on some of these phenomena in Chap. 6.

The fact that these unexplained sounds often resemble distant gunfire suggests that long distance propagation and zones of silence must often be involved. In other words, the source may sometimes lie a hundred or more kilometres from where it is heard. But in such cases it is difficult to establish a direct correlation between cause and effect without first-hand information about the source, which is why there is as yet no definitive explanation for Barisal Guns on any of the occasions when they have been heard.

It may be that in some instances brontides are due to earthquakes. It is well known that earthquakes on land can give rise to loud explosive sounds in addition to the usual series of rumbles. Moreover, humans are less sensitive to movements due to low frequency earth tremors than they are to low frequency airborne sounds, so it is possible to hear an earthquake without feeling it. So one might well hear faint pops, rumbles and booms without being aware that the cause is directly beneath one's feet.[61] Bearing in mind that the sound of brontides usually appear to come from the direction of the sea it is possible that they are due to undersea earthquakes. An undersea earthquake can cause the surface of the sea to rise slightly and rapidly and this can create a shockwave within the air above that may be heard at some distance as a boom.

Perhaps the best-known example of earthquake-induced sounds that have been mistaken for gunfire or explosions are the so-called Moodus noises. These have been recorded intermittently in and around the small town of Moodus in Connecticut, USA, since at least the middle of the seventeenth century, when English settlers first purchased the land from local tribes. But they were almost certainly heard long before that because the native name for the area was "Machimoodus", which means "the place of noises" in English.

---

[60] Van Den Broeck, E. (1895–7) Un phenomene mysterieux de la physique du globe. Ciel et Terre, 1895–7, 16, p 447, 479, 515, 535, 601; 17, p4, 37, 99, 148, 183, 208, 348, 399.

[61] Gold, T., Soter, S. (1979) Brontides: Natural Explosive Noises. Science, Vol 204, 27 April, p 371–74.

An investigation with sensitive seismometers during the late 1970's found that these hitherto unexplained airborne sounds were associated with earthquake tremors that had their origin a few kilometres below the surface and which were far too small to be felt even by someone directly above the epicentre. The resulting sounds have been described as being on some occasions like that of a cork popping out of a bottle and on others like a sonic boom.

It has since been established that audible airborne sounds can be produced when the surface of the ground moves less than a millimetre as long as the movement occurs very rapidly, i.e. within milliseconds. If it takes longer than this, the resulting aerial vibration is infrasonic, which is, of course, inaudible to humans.

These days, unfortunately, it has become increasingly difficult to distinguish what have been termed brontides by some scientists and "USAs" (unidentified sounds in the air) by others from the fog of man-made noise. If we hear a loud boom we probably shrug our shoulders and mutter "sonic boom" and leave it at that.[62]

Underwater Sounds

Most of us spend comparatively little time in water and even less time fully immersed in it. In any case, if you are scuba diving you are more likely to be in search of sights, not listening for sounds. So unless you are prepared to plunge your head underwater and make a point of listening it is by no means obvious that sound not only can travel through water, it does so far more efficiently than it does through air. Nevertheless, the fact that sound can travel through water was known to natural philosophers in antiquity. Aristotle noted that "…sound is heard both in air and in water, though less distinctly in the latter."[63] He may have learned this from divers who would have heard underwater noises as they searched the seabed for sponges, shellfish and pearls, or when salvaging cargo from sunken ships. He also knew from his own careful study of fish that not only are they extremely sensitive to noise but also that several species produce distinctive sounds, something they would not do unless they could be heard by other fish. Fishermen, he noted, were acutely aware that the splash of their oars would scare away their

---

[62] Businger, J. A. (1968) Rabelais' Frozen Words And Other Unidentified Sounds Of The Air. Weather, 23, p 497–504.

[63] Aristotle, 350 BC, On the Soul, Book 2, part 8. http://classics.mit.edu/Aristotle/soul.2.ii.html (accessed 18/03/2021).

quarry and took great care to lower their nets into the water as quietly as possible.[64]

A couple of millennia later Leonardo da Vinci noted in passing that "If you cause your ship to stop, and place the head of a long tube in the water, and place the other extremity to your ear, you will hear ships at a great distance from you. You can also do the same by placing the head of the tube upon the ground, and you will then hear anyone passing at a distance from you."[65] Leonardo's various manuscripts were not properly gathered together and edited into the now famous Notebooks until the nineteenth century, which may explain why this particular observation seems to have escaped the attention of natural philosophers of later generations.[66]

Nevertheless, Leonardo's discovery was probably already known to fishermen far from European shores. The fishermen of Northern Fukien, a province on the Chinese mainland opposite Taiwan, are said to have located shoals of fish using a length of bamboo 5 cm in diameter and 1.5 m long, which they immersed to a depth of 1 m while placing an ear against the upper end of the tube. The noise of a shoal of fish can, it was said, be heard up to 1.5 km away and was described as resembling a distant rumble of thunder. Apparently the same technique was used in Malayan waters.[67]

It wasn't until the eighteenth century, however, that natural philosophers in Europe turned their attention to the propagation of sound through water. The earliest of these investigations was performed by Francis Hauksbee, Curator of Experiments at the Royal Society. Hauksbee modified the bell-in-a-vacuum experiment by placing a small bell in a large, sealed jar filled with air and immersing this in a tank filled with water. He found that he was able to hear the bell ringing "seemingly, very little less, in respect to its Audibility, and grave at least two or three Notes deeper than it was before…".[68]

Several decades later the phenomenon was revisited by Jean-Antoine Nollet, a French natural philosopher. Abbé Nollet, as he was known to his contemporaries, had abandoned a promising career in the Church for one in the sciences soon after being been ordained. He became France's leading

---

[64] Aristotle, 350 BC, History of Animals, Book 4, part 8. http://classics.mit.edu/Aristotle/history_anim.4.iv.html (accessed 18/03/ 2020).

[65] The Notebooks of Leonardo Da Vinci, Arranged, rendered into English and Introduced by Edward MacCurdy, Volume 1, Jonathan Cape, 1938, p 284.

[66] Leonardo was not mentioned by Chladni in his review of the history of the investigations into the propagation of sound in water. (See: Chladni, E.F.F. (2015) Treatise on Acoustics: Sect. 212, p 163).

[67] Needham, J., Ling, W., Robinson, K.G. (1962) Science and Civilisation in China, Vol 4, Part 1: Physics & Physical Technology. OUP, p 210.

[68] Hauksbee, F. (1708) An Account of an Experiment touching the Propagation of Sound through Water. Philosophical Transactions, 26, p 371–72.

advocate of the use of experiments to reveal the workings of nature, and his lecture demonstrations became so popular that they were even attended by "duchesses, peers and lovely ladies".[69] He was also one of Europe's foremost "electricians" and the author of an influential theory of the nature of electricity. Nollet's electrical ideas were not universally accepted, however, his chief opponent being Benjamin Franklin, who had his own ideas on the subject. The issue led to a long-running and bitter dispute between them that divided the natural philosophers of the day.

During the late 1730's Nollet turned his attention briefly to the question of whether or not fish can hear. Seemingly unacquainted with either Aristotle's writings on the subject or Leonardo's underwater listening tube, he and his contemporaries were unsure of the answer because fish don't have external ears. Nollet was aware of anecdotal evidence that they could hear. According to Robert Boyle, fish could be trained to feed at the tinkling of a bell, and travellers returning from the Far East reported that a gong was sometimes used for the same purpose in China. Nearer home, the fishermen of Brittany were known to use the sound of drums to drive fish into their nets. Nollet was not satisfied with anecdotal evidence, however, and set out to discover the truth of the matter for himself. With impeccable French logic he reasoned that nature would have provided fish with the organs of hearing only if sound can travel through water. And the only way to settle the issue was to immerse himself fully and listen. Easier said than done because, as is apparent from his account of the investigation, the Abbé could not swim.

> During the summer of 1740 I took advantage of the warmest days to carry out these experiments in the Seine; I chose a very deep spot by an island where there were no currents in the water, and I drove a stake in that I could use to immerse myself conveniently, and I accustomed myself little by little to remain under water without breathing, in such a manner that after a few days I could remain submerged for 12 seconds without any ill effect.[70]

With his head submerged just below the surface for those few precious seconds he arranged for an assistant on the bank of the river to make a variety of noises: to speak, blow a whistle, ring a small bell and fire a pistol. Nollet heard them all, albeit considerably muted, and concluded that sound can travel through water and consequently that fish must indeed be able to hear.

---

[69] Sutton, G.V. (1995) Science for a Polite Society: Gender, Culture, and the Demonstration of Enlightenment. Westview Press, p 225.

[70] Abbé Nollet (1743) Mémoire Sur L'ouie Des Poisons Et Sur La Transmission Des Sons Dans L'eau. Memoires de L'Academie Royale des Science, p 199–244, p 205.

A year or two later he performed another experiment that established a more significant fact: water is a better medium for sound than air. He submerged himself in a large barrel full of water and bashed two stones together as hard as he could. Not only was the sound "unbearable", the shock of the impact on his body was "like the sensation that is produced when a solid body held between the teeth is struck by another solid body".[71] One shudders to think what prior experience had supplied him with this information.

On this issue, Franklin agreed with Nollet, though, unlike him, he had no qualms about submerging his head in water. In 1762 he wrote to a friend on the subject of the best medium for the propagation of sound, suggesting that air is not as good as water: "It is a well-known experiment, that the scratching of a pin at one end of a long piece of timber, may be heard by an ear applied near the other end, though it could not be heard at the same distance through the air. And two stones being struck smartly together under water, the stroke may be heard at a greater distance by an ear also placed under water in the same river, than it can be heard through the air. I think I have heard it near a mile; how much further it may be heard, I know not; but suppose a great deal farther, because the sound did not seem faint, as if at a distance, like distant sounds through air, but smart and strong, as if present just at the ear..."[72] On reading this, an eminent physicist remarked admiringly: "Is there anything Benjamin Franklin didn't try?".

Why water should be a better medium than air for sound propagation was for a long while attributed to the fact that the speed of sound in water is greater than it is in air. However, the actual reason is that the absorption of sound energy by water is a tiny fraction of what it is in air.[73]

Direct measurement of the speed of sound in water proved to be very much more difficult than in air. As we saw in Chap. 2, the first successful determination was made by Daniel Colladon in 1826, some two centuries after the earliest measurements of the speed of sound in air. The difficulty he faced was not only that of devising and setting up the necessary apparatus, it was also that sound travels over four times faster in water than it does in air. Hence if the time of travel was to be accurately determined, the distance over which the measurement had to be made was correspondingly greater, which, in turn, required a very large body of still water. In fact, Colladon already

---

[71] Abbé Nollet (1743) ibid, p 222.

[72] Franklin, B. (1762) Letter XLIV To a Friend. In: Franklin, B. (1769) Experiments And Observations On Electricity Made At Philadelphia In America By Benjamin Franklin LLD & FRS To Which Are Added Letter And Papers On Philosophical Subjects. London, p 435–7, p 435.

[73] Attenuation of sound in air is some six million times more than it is in water. Saltwater attenuates sound to a slightly greater degree than fresh water.

knew the speed of sound in water, having calculated it from measurements of the compressibility of water that he had obtained in Paris the previous year. The purpose of the Lake Geneva experiment was merely to confirm the work he had carried out in Paris, which, he was relieved to find, it did.

Incidentally, in his account of this experiment Colladon mentions that at first he couldn't think of a way of hearing the sound of the distant underwater bell without immersing his head in water, which suggests that he, like Abbé Nolett, wasn't aware of Leonardo's observation. But he managed to improvise an off-the-shelf solution involving a watering can, which led him to design and commission the horn that he used in the final trials.

> I then thought that a metallic vase, closed at its base and immersed by means of a weight might perhaps serve to transmit the sound from the water to the air in the vase and that one could then hear it outside. To try this I was able to take a watering pot ballasted so as to sink in the water. I was so impatient to see what result I would obtain that at 5 o'clock in the morning I aroused Alphonse de Candolle [a Swiss botanist] and told him of my idea. We dressed in a hurry; we asked the gardener to remain at the dock with the bell, while with de Candolle, I went across the lake, which in this part is about 1500 meters wide. Having arrived at our destination, I immersed the watering pot and gave the signal to strike the bell. Without putting my head in the water I immediately heard in the watering pot the sound from the bell.[74]

The effect of the compressibility and density of a liquid on the speed of sound was already known in Colladon's time: the more incompressible a liquid, the more rapidly sound travels through it, whereas the greater its density, the less the speed. But compressibility is also affected by pressure: with increasing pressure a liquid becomes less compressible because its molecules grow closer to one another. Consequently, the speed of sound in water is slightly greater at the bottom of a deep ocean than it is nearer the surface. Moreover, density is affected by temperature because liquids shrink as they cool, becoming denser and less buoyant. Hence, in the absence of convection currents, a large body of liquid is always coldest at its greatest depth. But water is a notable exception to this rule because it expands when it's temperature drops below 4 °C, which causes its density to *decrease*. As a result, water at a temperature of less than 4 °C floats above water that is warmer than 4 °C. Hence in a body of

[74] Colladon, J-D (1893) Souvenirs et Memoires—Autobiographie de Jean-Daniel Colladon. In: Lindsay, R. B. (1973) Acoustics: Historical and Philosophical Development Benchmark Papers in Acoustics, Vol 1. Dowden, Hutchinson & Ross, p 196–201.

very cold water that is not frozen, the warmest, i.e. the densest, water lies at the greatest depths.

Taking the combined effects of pressure and temperature into account we find that sound travels faster in warm water than it does in cold water and that its speed increases with depth. In shallow bodies of water such as ponds or small lakes, neither temperature nor pressure change significantly with depth so, to all intents and purposes, the speed of sound is the same whatever the depth. But in deep oceans, below a shallow surface layer in which the temperature remains more or less constant due to mixing brought about by the action of wave and wind at the surface, and which is seldom more than a few tens of metres deep, the temperature of water falls from whatever it may be at the surface to a minimum of about 2 °C to 4 °C at depths of 500 m to 1000 m and thereafter remains fairly constant however deep the water. This drop in temperature, known as a *thermocline*, causes the speed of sound to decrease with depth. At depths of more than 1000 m, however, the increasing pressure of water reverses the decrease in speed due to falling temperature so that at a depth of several kilometres the speed of sound is slightly greater than it is just below the surface (Fig. 4.9).

As you would expect, variations in the speed of sound in water has a similar effect on the propagation of sound as it does in the atmosphere: sound refracts

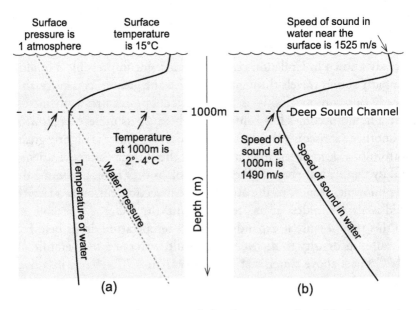

Fig. 4.9 a Temperature and pressure of the deep sea varies with depth as shown. **b** The speed of sound in water varies with depth because it is affected by both temperature and pressure

away from regions where its speed is greatest towards regions where its speed is least. But unlike the atmosphere, which is constantly agitated by winds, below the churning surface layer ocean waters are usually very still. This can lead to ducting, that is to say to underwater sounds being largely confined to a layer within which there is a marked reversal in the speed of sound. The most spectacular example of ducting occurs within a layer hundreds of metres below the ocean surface where the effect of increasing pressure reverses the effect of decreasing temperature on the speed of sound. This layer is known as the Deep Sound Channel (DSC).

Evidence for the DSC was discovered in 1937 by an American geophysicist, Maurice Ewing, who was studying underwater refraction in the North Atlantic using small explosions as a sound source. Low frequency sounds within the DSC can travel for thousands of kilometres because they spread in only two dimensions rather than three as sounds do in unconstrained spaces. There is evidence that blue whales exploit the DSC in order to communicate with one another over huge distances. Submariners, on the other hand, aim to keep well away from the DSC to avoid been detected at long ranges by the enemy, always supposing that they can dive to such great depths without being crushed.

One of the more unexpected aspects of ducting within the DSC is that it transforms a short-lived sound into a drawn out rising crescendo that ends abruptly. The cause is that the speed of sound is least at the centre of the duct. Hence the sound travelling directly along the axis arrives *after* sounds that have travelled along other paths, even though it has travelled the least distance. Figure 4.10 shows a few of the many possible paths taken by a sound that has its origin at the centre of the DSC, i.e. the depth at which the speed of sound is at a minimum. The paths are not symmetrical about the axis of the sound channel because the rate at which the speed of sound changes with depth is not uniform: it is greater above the axis than it is below it. Moreover, sound is not entirely confined to a shallow layer because the DSC does not have impenetrable upper and lower boundaries. A sound directed at a large angle to the axis will reach the surface of the sea, where most of it will be reflected back into the depths.

When the Second World War gave way to the Cold War, Ewing's discovery of the DSC was exploited by the US Navy. It secretly developed an underwater sound surveillance system known as SOSUS (SOund SUrveillance System) that eventually covered the North Pacific and North Atlantic oceans and the Mediterranean Sea. Each of these systems consisted of a vast array of thousands of underwater hydrophones that could detect and track the submarines of the Soviet Union at distances of a thousand kilometres and

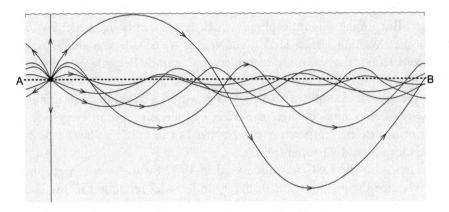

Fig. 4.10 Possible paths of sound rays that have their origin within the Deep Sound Channel. The dotted line, AB, is the channel's axis. The speed of sound in seawater just below the sea's surface is 1525 m/s. This drops to 1490 m/s at the centre of the DSC, and rises thereafter with increasing depth. See Fig. 4.9 for details of how and why the speed of sound varies with depth

more. And to increase the effectiveness of SOSUS, both the British and the Americans built ultra quiet hunter-killer submarines that were used not just to find and shadow Soviet submarines and surface ships, but also to listen to and record the distinctive sounds produced by the propellers and engines of particular vessels. This would enable particular vessels to be identified when they came within range of SOSUS microphones. Until the late 1960's Soviet submarines were inherently noisier than those of the British and Americans. However, SOSUS was able to provide only the bearing of a submarine, not its exact position and eventually Soviet submarines came up with ways of masking and reducing their acoustic signature. According to an assessment of SOSUS by the Royal Navy "At best it can only be taken as a good guide and at worst it was totally misleading."[75]

Ewing was also inadvertently responsible for the long running conspiracy theory that the US authorities are concealing evidence that a flying saucer had landed near Roswell, New Mexico. For a few years after WW2, the US Air Force worked on an atmospheric version of SOSUS. In 1945 Ewing persuaded General Carl Spaatz, Chief of Staff of the Army Airforce, that a sound channel within the tropopause similar to that in the ocean would duct powerful sounds over large distances making it possible to detect any above ground nuclear tests that the USSR might carry out in the future.

---

[75] Hennessy, P., Jinks, J. (2015) The Silent Deep; The Royal Nay Submarine Service Since 1945. Allen Lane, p 328.

The top-secret enterprise was called Project Mogul. To detect the explosions, microphones had to be lofted high into the atmosphere using huge polyethylene balloons filled with helium. Their size made them visible from the ground, though, of course, those who saw them didn't know what to make of them, hence the suggestion that they might be UFO's. One of these balloons came down near Roswell, New Mexico, in the summer of 1947. Unsurprisingly, the military authorities refused to release any information about what had really happened, and was happy to go along with the idea that they were concealing an alien spacecraft and the bodies of its occupants. Project Mogul never delivered the on its promise and was quietly shelved.

Keeping an Ear to the Ground

If you have ever been bothered by the unwelcome sounds of a neighbour's music coming through a shared wall, you will be only too well aware that sound can travel through solids. And should the wall be thick and dense enough to make the music barely audible, it is possible to hear it more distinctly by pressing your ear against the wall. In fact, it is sometimes claimed that pressing an ear to the ground was a method favoured by the native Indians of the American West in order to hear the sound of the hooves of distant galloping horses or stampeding buffalos, though there is scant evidence that they actually did so. That's not to say that sounds can't be heard through the ground, however, as Leonardo discovered when listening to sounds through a tube. In *Scouting for Boys* there is paragraph on how to improve hearing sounds transmitted through the ground, the last sentence of which must surely have led to the occasional bloody mishap:

> If you put your ear to the ground or place it against a stick, or especially against a drum, which is touching the ground, you will hear the shake of horses' hoofs or the thud of a man's footfall a long way off. Another way is to open a knife with blade at each end; stick one blade into the ground, hold the other between your teeth and you will hear all the better.[76]

Methods such as these have been used since ancient times during sieges, when defenders devised ways of detecting the sound of digging to locate the tunnels of the attacking army.

---

[76] Baden Powell, R. (1908) Scouting for Boys. H. Cox, London.

Military tunnelling, or mining as it came to be known, was widely employed by both sides on the Western Front during the First World War. Its purpose was primarily to get beneath the enemy's defences and blow them up. And, as with the sieges of earlier wars, the only way to locate an enemy tunnel was to listen for sounds of digging. In the early years of the war, the methods used were not very different from those recommended to Scouts. The most straightforward of these was to drive a narrow wooden stake into the ground and bite on the other end firmly with one's teeth, which would feel the vibrations from within the ground. Another method was to sink a large, empty oil drum into the earth and fill it with water. Immersing one's head in the water improved one's chances of hearing the enemy miners because water reduced the difference in acoustic impedance between the ground and the ear.

Below ground there was a deadly cat and mouse game between miners on opposing sides as they dug tunnels to intercept and destroy those of the enemy. These countermines might be filled with explosives to destroy the enemy tunnel or used to break into it and attack the digging party within. To pin point underground sounds, the French developed a device that consisted of two mechanical microphones, known as geophones, each linked by a rubber tube to the listener's ears, rather like a stethoscope. The operator placed the phones some 50 cm apart on the floor of the tunnel and, keeping one stationary, slowly moved the other in a circle around it. When the sound from each geophone was equally loud the listener would know that the direction of the sound source was at right angles to the line joining the phones. Another bearing would be taken from further along the tunnel, and the point where the two bearings intersected would give a fairly precise position of the source of the sound. The sound from these early geophones was not amplified electronically and so even in the most favourable geological conditions, i.e. in limestone, digging was audible only within 100 m of the listener.

Benjamin Franklin's example that "the scratching of a pin at one end of a long piece of timber, may be heard by an ear applied near the other end" is a special case of the conduction of sound in a solid because sound is largely confined within the narrow batten by reflection from the inner surfaces of the timber and so does not fall in intensity as it would in a much larger solid body, where, of course, it would spread out in all directions. And this offers an explanation for another of those snippets of wild-west folklore: placing one's ear against the iron track of a railway to hear the approach of a train well before one can see it. This should work not because sounds travels through iron faster than it does through air but because the sound of the train's wheels is ducted through the rail with little loss of energy. However, rails in those days were laid with gaps between them to allow for thermal expansion, gaps which would interrupt the transmission of sound, so it's unlikely that it would be possible to hear the sound of a distant train in this way.

Measuring the speed of sound in a solid is even more difficult than it is in either gases or liquids because the speed of sound in solids is usually far greater than it is in either of those media. So much so that until the advent of electronic apparatus capable of measuring fractions of a second it was not possible to measure the speed of sound in a solid directly, as it is for sounds in gases and liquids. Nevertheless, as we have seen, Chladni was able to measure the speed of sound in various solids indirectly by determining the frequency of sounds created when a narrow rod made of the solid under in investigation is stroked with a resin-coated cloth. These days the speed can be determined directly, i.e. by measuring the time taken to travel the length of a sample of the solid medium.

But it is, in principle, a simple matter to demonstrate that sound in a solid travels a great deal faster than it does in air. Indeed, Biot's determination of the speed of sound in cast iron relied on hearing the sound of a bell through the iron before that through the air. Thomas Young had earlier noted that.

> The velocity with which impulses are transmitted by solids, is in general considerably greater than that with which they are conveyed by air … I have also found that the blow of a hammer on a wall, at the upper part of a high house, is heard as if double by a person standing near it on the ground, the first sound descending through the wall, the second through the air[77]

Sound in solids are attenuated or weakened just as they are in liquids or gases. But whatever the medium, high frequencies are attenuated to a far greater degree than low frequencies. Hence the characteristic rumbles and thuds of subterranean sounds such as those that often accompany earthquakes, which, along with thunderstorms, are a source of loud natural sounds.

Earthquakes are due to extremely powerful vibrations created when there is sudden movement between large bodies of solid rock beneath the surface of the planet. This produces several types of very low frequency seismic waves that are felt as tremors at the surface. The fastest moving of these waves is longitudinal, like sound in air, and is known as a primary or P-wave because it reaches the surface before the others and is the cause of the low-frequency rumble that is often heard during an earthquake. This can give the impression

---

[77] Young, T. (1845) Young's Course of Lectures Vol 1. Taylor and Walton, London Lecture XXXI: On the Propagation of Sound, p 292. It's worth finding a long brick wall somewhere quiet so that you and a companion can repeat Young's experiment.

that the source of the sound is far away, possibly because we usually assume that distant rumbles are due to thunderstorms. Moreover, it is impossible to locate the source of earthquake-induced rumbles because there are no high frequencies to aid the ear's direction-finding system. Incidentally, the tremors of the ground and the resulting damage to structures are principally caused to the slower moving secondary or S-waves, which are transverse, i.e. side-to-side. S-waves are silent, which explains why the rumble is usually heard just before the ground begins to shake. There is more on these so-called *stick and slip* sounds that are produced when surfaces rub together in Chap. 6.

Sound Shadows

Have you ever stopped to wonder why it is possible to hear sounds but not see lights around corners? As we saw in the previous chapter, Newton did and concluded that this is the reason why light cannot be a wave. He pointed out that "All motion propagated through a fluid diverges from a rectilinear progress into the unmoved spaces". This, he added, is a matter of observation, not conjecture: "That these things are so, anyone may find by making the experiment in still water ... And we find the same by experience also in sounds which are heard through a mountain interposed; and, if they come into a chamber through the window, dilate themselves into all the parts of the room, and are heard in every corner; and not as reflected from the opposite walls, but directly propagated from the window, as far as our sense can judge."[78] Newton was right about sound but drew the wrong conclusion concerning light.

As we saw in Chap. 2, the tendency of waves to wrap around an object or spread out from an opening is known as diffraction. And the degree to which a wave diffracts depends on the ratio of its wavelength to the size of the obstacle or opening it encounters. If this is small then very little diffraction occurs and vice versa: for a given wavelength, diffraction increases as the size of the object or opening decreases. You can confirm these things for yourself by taking a leaf out of Newton's book and tossing a small pebble into the still waters of a pond and noting what happens when the resulting ripples encounter solid objects that protrude above the surface of the water such as a narrow upright pole or a floating log (Fig. 4.11).[79]

---

[78] Newton, I. (1687) Philosophiæ Naturalis Principia Mathematica, Prop 42, Theorem 33.

[79] Alternatively, use the interactive iPad app "*Ripple*" by Paul Falstad, which gives you a much greater degree of control over the parameters of a situation in which a wave encounters an object or opening than can be achieved with a physical ripple tank.

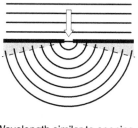 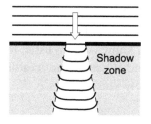

Wavelength similar to opening   Wavelength less than opening

Fig. 4.11 Diffraction of waves after passing though an opening. The degree of spreading depends on the relative sizes of wavelength and opening. The wave spreads less when its wavelength is smaller than the width of the opening

The diffraction of sound is much greater than that of light due to the enormous difference in their respective wavelengths compared to the objects they ordinarily encounter. The wavelength of visible light is approximately a million times less that that of audible sound. Moreover, many audible sounds are composed of wavelengths that are comparable to the dimension of objects we encounter in daily life such as buildings, buses and boulders, whereas visible light has wavelengths that are a thousand times less than the smallest objects that we can discern unambiguously with the naked eye (i.e. objects less than 1 mm across). This is why in most circumstances sound doesn't cast distinct shadows as light does, a fact that was at the heart of Newton's thinking when he rejected the possibility that light is a wave.[80]

To hear the effects of sound diffraction listen to broadband sounds near the edge of a wall or a fence, as well as at edge of an open doorway or window. Anywhere, in fact, where there is an edge around which sound can diffract. But what should one listen for? Most sounds consist of a broad range of frequencies, so even in the absence of reflections we will be able to hear lower frequency sounds from a broadband source that is shielded from direct view. The wavelengths of low-pitched sounds are a metre or more, while those of high-pitched sounds are ten centimetres or less. This is why low-pitched sounds diffract around objects to a far greater degree than high pitched ones. If a sound source that is out of sight around a corner consists of a broad range of frequencies, as most sounds do, the bass notes will dominate what you hear until you turn the corner and can see the source directly. It's the same with an opening: the treble notes in music heard through a doorway or window are distinctly audible only if you are more or less directly in front of

---

[80] The fuzzy edge of a shadow, known as its penumbra, is not due to diffraction but to the fact that most sources of light, such as the sun or a light bulb, are extended, not points like a star. A point source casts a pin-sharp shadow.

the opening. But if the opening is very large, then even the low frequencies become inaudible unless you are opposite the opening.

Once you know what to listen for, you will realise that the effect of diffraction on sound is usually noticeable as an absence of sound, as the following account amply illustrates.

> When the pitch of the note is high and the obstacle large, the sound shadow may be very marked. The writer has met with a striking instance of this on Pilling Moss in North Lancashire. In the Spring the sea-gulls resort in large numbers to the Moss to lay their eggs, and when the young birds are able to fly, the air is filled with their shrill screams. There is a road at a little distance from the nests, and by the side of the road there is sometimes a row of stacks of peat. The length of one of these stacks is many times as great as the wavelength of the screams of the birds, and consequently a good sound shadow is formed. As one walks along the road the alternations of sound and silence are very marked. Opposite the gap between two stacks the sound is unpleasantly loud; opposite the stack itself there is almost complete silence, and the change from sound to silence is quite sudden.[81]

And here is another example of the absence of diffraction of low frequency sounds which occurred in circumstances that, fortunately, few of us are likely to encounter:

> Some few years since a powder hulk exploded on the river Mersey. Just opposite the spot there is an opening of some size in the high ground which forms the watershed between the Mersey and the Dee. The noise of the explosion was heard through this opening for many miles, and great damage was done. Places quite close to the hulk, but behind the low hills through which the opening passes, were completely protected, the noise was hardly heard, and no damage to glass and such like happened. The opening was large compared with the wave-length of the sound.[82]

The author of this account gave another example of the effects of diffraction, one that can be experienced at leisure and in circumstances that are not difficult to seek out.

> A very well marked sound shadow is often formed by a mountain ridge. If, for example, there is a stream running at the foot of the ridge the rush of water is

---

[81] Capstick, J.W. (1913) Sound: An Elementary Text-Book For Schools And Colleges. CUP, p 103.
[82] Glazebrook, R. T. (1883) Physical Optics. Longmans, p 149.

heard so long as you can see the stream. On descending sometimes only a few inches below the ridge to a position from which the stream cannot be seen, the noise, too, will appear to stop abruptly. The change from the full rush of the waters to almost complete silence is most marked. A perfect sound shadow is formed.[83]

Flowing water, even when turbulent, is not necessarily noisy. The next time you are by the banks of a babbling brook look and listen: noisy water is invariably bubbly. As we shall see in Chap. 6, the sound we associate with flowing water in a river or stream is due to those bubbles, though in a way that may surprise you. Moreover, the sounds due to bubbles in these circumstances are composed of high frequencies, which diffract hardly at all at an edge.

An absence of high frequencies is very evident near the edge of a cliff overlooking the sea. When very close to the edge the full spectrum of frequencies in the sound of waves breaking on the beach below is audible. But move even a metre or two away from the edge and only the lower frequencies are audible.[84]

In fact, you can notice the absence of higher frequencies because they don't diffract as much as low frequencies simply by holding an open palmed hand ten or fifteen centimetres from one or other of your ears.[85] The source of sound has to be composed of a broad range of frequencies, such as that of road traffic. Stand so that your ear is facing a road and as a car drives past alternately cover and uncover your ear with your hand as you listen to the overall timbre of its sound. You should notice that when your hand covers your ear the high frequency hiss of its tyres is much less evident. At the same time the overall sound is less loud, something that you may attribute to sound that is blocked by your hand. But listen carefully and you will realise that the reduction in loudness is actually due to the loss of the higher frequencies.

It's also worth repeating this experiment while listening to water being boiled in an electric kettle. During the initial phase the timbre of the hissing kettle is altered noticeable when you hold your open hand close to your ear. But when the water comes a rolling boil you will find that there is no difference in the timbre of the sound because higher frequencies which are the source of the hiss you hear during as the water is heating up are absent in the sound of steadily boiling water. In both cases, reflections and other sounds will mask the effect you are listening for, so it is probably best to place the kettle on a table away from the walls of the room you are in.

---

[83] Glazebrook, R. T. (1883) *Physical Optics*, Longmans, p 149–50.

[84] I noticed this when walking along the cliff top path on the Seven Sisters on the Sussex coast.

[85] Alternatively use an A5 sized sheet of stiff card held 15 or 20 cm from your ear.

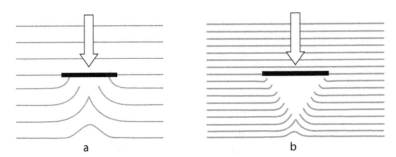

Fig. 4.12 Waves diffract around an object. **a** Longer wavelengths wrap around the object. **b** shorter wavelengths create a shadow beyond the object

We saw in Chap. 3 that diffraction plays an important role in sound location by land based vertebrates because the head prevents high frequency sounds from diffracting around it to reach the ear furthest from the source. Whispering provides a very obvious example of how one's head blocks high frequency sounds. A whisper that can be heard when the speaker faces you becomes all but inaudible when he or she turns their back on you. A whisper consists principally of high frequencies, which is why it is difficult to hear people whisper when they have their back to you because high frequencies can't diffract around the whisperer's head to reach your ears (Fig. 4.12).

The Doppler Effect

When a noisy car races past you, you may notice that the pitch of its engine falls slightly as it draws away. The Doppler effect or shift, as this change in pitch is known, is very obvious when a police car or ambulance is sounding its siren as long as the siren emits a constant frequency, but with practice it is something that can be noticed in the sound of a car, a noisy motorbike, a train or a plane, as long as these are in motion. In the case of a low flying plane, the Doppler shift is very evident in the whine of its engine's turbines as it flies overhead: their pitch falls noticeably as it flies away from you.

The explanation for this phenomenon predates the advent of fast moving vehicles. In 1842 a Viennese physicist, Christian Doppler, published a paper entitled "On The Coloured Light Of The Double Stars And Certain Other Stars Of The Heavens" in which he wrote: "We know from general experience that a ship of moderately deep draught which is steering toward the oncoming waves has to receive, in the same amount of time, more waves with a greater impact than one which is not moving or is even moving along in the same direction of the waves. If this is valid for the waves of water, then why should it not also apply with necessary modifications to air and ether waves?"[86] Doppler's "air waves" are sound, his "ether waves" are light and the phenomenon has since come to be known as the "Doppler Effect".

As the title of his paper suggests, Doppler was primarily interested in the effect that he believed motion should have on the colour of stars. He showed mathematically that when a source of light is moving relative to an observer, there will be a measurable change in its perceived frequency. As we saw in Chap. 2, by 1830 the idea that light is a wave, and that its colour is determined by its frequency, was accepted by the majority of physicists. Doppler claimed that the perceived change in frequency due to their motion relative to the earth is the reason why stars appear to have different colours. Even to the naked eye, it is obvious that while most stars appear to be white, a few are reddish, others yellowish and some have a faint bluish cast. Assuming that all stars emit only white light, he argued, a star moving away from the Earth should appear reddish whereas one moving towards the Earth should appear bluish. Doppler was, in fact, describing a phenomenon that later came to be known among astronomers as "red shift".[87]

"If a radiant object…approaches the…observer at a velocity that is comparable with the speed of light [its colour] will pass from white to green at increasing velocities, from there to blue, and finally to violet…upon moving away the white light will gradually shift to yellow, orange and finally to red."[88] Doppler calculated that the minimum relative velocity necessary to obtain a perceptible shift in colour is 250,000 m/s, which is slightly less than one tenth the speed of light. In fact, the colour of a star is not due to its motion but to its surface temperature; as stars go, Betelguese is relatively cool and Rigel is extremely hot. Even assuming that all stars are intrinsically white, which they are not, for a star to appear noticeably red (e.g. Betelguese) or blue (e.g. Rigel) its velocity relative to a stationary observer would have to approach the speed of light, i.e. 300,000,000 m/s, some ten times greater than Doppler's theoretical value.

Another flaw in Doppler's reasoning was that he ignored the fact that the visible spectrum is itself part of a much larger spectrum of radiation, the so-called electromagnetic spectrum. The red end of the visible spectrum gives way to infrared radiation and the violet end to ultraviolet radiation.

---

[86] Doppler, C. (1843) Über das farbige Licht der Doppelsterne (On The Coloured Light Of The Double Stars And Certain Other Stars Of The Heavens). Proceedings of the Bohemian Society of Science, p 108–9.

[87] The fact that stars are not stationary was discovered by Edmund Halley in 1718 when he realized that the position of some stars in the sky in his day differed by up to half a degree from that recorded by Greek astronomer Hipparchus c.127 B.C.

[88] Doppler. C. (1843), see Sect. 6.

Both these types of radiant energy had long been known to science in Doppler's day. Infrared was discovered by the Anglo-German astronomer William Herschel in 1800 and ultraviolet by Johann Ritter, a leading German Naturphilosoph, shortly after.[89] A very hot body such as a star emits all these radiations (and more) and all of them are subject to a Doppler shift due to the motion of the star relative to an observer. Consequently, when a star is moving away from an observer, the part of the ultraviolet spectrum closest to the violet end of the visible spectrum is shifted into the violet end of the spectrum, 'filling in' for the violet that has been shifted to blue, while the red end shifts into the near infrared. The reverse happens if the star approaches the observer: infrared is shifted into the red end of the visible spectrum and violet into ultraviolet. In other words, the overall visible spectrum perceived by the observer remains unchanged whether a star is receding or approaching and its intrinsic colour remains unchanged, however fast it is moving.

The most famous example of the application of the Doppler effect in astronomy was the discovery in 1929 by Edwin Hubble that most galaxies exhibit a redshift, which he interpreted as evidence that they moving away from one another. In fact, the redshift that Hubble discovered is due to the expansion of space itself, not to galactic motion. As the universe expands, the distance between galaxies increases and wavelengths of radiant energy such as light are stretched resulting in what is known as "cosmological redshift". Hence the evidence for the expansion of the universe is cosmological redshift, not Doppler redshift. And to detect the shift in the spectrum of those distant bodies, Hubble used the spectral signatures of elements present on the surface of a star. The position of these in the visible spectrum are displaced slightly towards one end or the other of the visible spectrum in line with Doppler's original prediction.[90]

---

[89] *Naturphilosophie* was a diverse school of thought that was particularly influential in Germany and which was part of the widespread Romantic reaction to the perceived narrowness and materialism of the 18th Century science and philosophy. One of their ideas was that natural forces are paired opposites (hot v. cold, light v. dark etc.). Ritter believed that Herschel' s warm radiations beyond the red end of the visible spectrum must be balanced by cold radiations beyond the violet end.

[90] These visible spectral signatures of elements were first observed by William Wollaston in 1802, but he thought they marked the boundaries between colours. They were independently discovered in 1814 by Joseph Fraunhofer who made a systematic study of them. They are known as spectral lines because the source is observed through a very narrow slit, so the resulting image is a line rather than, say, a spot. Charles Wheatstone, a British scientist, established that they could be used to identify different metal elements in 1835. And in 1864 the British astronomers William and Mary Huggins succeeded in identifying elements present in in a distant nebula, thus issuing in the era of stellar spectroscopy.

Doppler was unable to offer any experimental evidence for his theory and most astronomers and physicists of his day doubted that there was such an effect. One sceptic was a young Dutch meteorologist, C.H.D. Buys Ballot, and in 1845 he devised an experiment to test Doppler's theory using sound rather than light. Buys Ballot's idea was to have a musician play a steady note of known pitch on a French horn while riding on a railway wagon at a steady speed. A second musician standing by the tracks would estimate the pitch of the note as the wagon approached and as it receded. The musicians would then swap roles: the stationary one would sound the horn and the one on the moving train would estimate the pitch. If Doppler was right, the pitch perceived by both the stationary and moving musician should be higher that that being played as they approached one another and lower as they moved apart.

Buys Ballot employed musicians because there were no instruments in those days with which to measure the frequency of a sound directly. A trained ear was the most straightforward method available to him to establish the pitch of a note.

For the experiment, he assembled four teams, each of which consisted of two musicians and an observer. One of the musicians was equipped with a horn while the other had the job of estimating and recording the pitch of note being played. The task of the third person was to coordinate proceedings. One team rode on an open railway carriage, which was drawn by a steam locomotive. The three remaining teams were placed at the side of the railway track at intervals of 400 m and asked to estimate the pitch of the horn being played as it approached and receded. This arrangement provided data of the change in pitch that occurs when the observer is stationary and the source is in motion. The teams then switched roles. The team on the moving wagon had to estimate the pitch of the horn played by the stationary team (i.e. source stationary, observer in motion). When the results were collated and analysed they confirmed that the pitch of the horn always rises when source and observer are approaching one another and drops when the move apart. In the experiment the change in pitch was a semitone, which even an untrained ear can perceive, and Doppler's idea was vindicated, if only qualitatively.[91]

The standard textbook treatment of the Doppler effect is somewhat misleading because it does not take fully into account the actual circumstances in which we ordinarily experience it. In the first place, it assumes that the both the speed of the source and its pitch is constant. In these

---

[91] The experiment was recreated in 2017 by Charles Hazlewood for BBC Radio 4 and broadcast on 15 Aug, 2017. The program, "The Doppler Effect with Charles Hazlewood", is available on iPlayer. (accessed 1/05/2021).

idealised circumstances, as the source approaches a stationary observer, its forward motion reduces the interval between successive compressions and rarefactions, which results in decrease in wavelength and a corresponding an increase in the rate at which these variations reach the ear. As a result the perceived pitch of the source is increased. The reverse happens as the source moves away from the observer: the interval between successive compressions and rarefactions increases and the perceived pitch decreases. If the source is travelling faster than the speed of sound, the compressions and rarefactions pile up in front of the of the source and form what is known as a shock front, which as we shall see in Chap. 6 is the cause of the sonic boom that accompanies any vehicle moving faster than sound.

Tyre noise, the major source of the noise produced by a fast moving wheeled vehicle travelling on tarmac, is not subject to a Doppler shift because the source (the interaction between tyre and road) does not move with the vehicle, i.e. it is effectively stationary.[92] Nor will there be a Doppler shift if the source is emitting white noise because, depending on whether it is approaching or receding, ultrasonic and infrasonic frequencies are shifted into the audible spectrum, just as happens with the spectrum of a moving star. To hear a Doppler shift, the source must have a distinct pitch, even though this will usually be made up of several frequencies.

Another assumption made in elementary accounts of the Doppler shift is that the source approaches the observer head on. Were it possible to hear the shift from a vantage point directly facing the moving source, the change in frequency would be instantaneous and would occur the moment when the source draws level with the observer. But in the real world this situation is highly unlikely, unless the observer is replaced by a microphone on a boom. Stand in the path of an ambulance or fire engine and you won't survive to tell the tale of that instantaneous change in frequency (Fig. 4.13).

In practice the observer is usually to one side or the other of the path of the moving source and experiences the change in frequency as a diminishing glissando rather than the abrupt change implied in the elementary explanation of the phenomenon. The duration of the glissando depends on the distance of the observer from the path of the source. It increases the further the observer is from the vehicle's path, and is particularly evident in the whine of the turbines of a low flying aircraft as it passes overhead. If you record the sound of the plane with an audio spectrometer app the glissando stands out

---

[92] At high speed, electric vehicles are potentially as noisy as petrol & diesel vehicles because tyre noise is the dominant source of sound from vehicles when their speed exceeds some 50 km/hr. Motorways will be noisy long after the last fossil fuel car has been towed to a scrap yard.

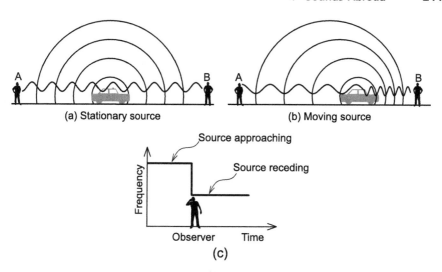

Fig. 4.13 **a** Doppler shift in ideal conditions. When the source is stationary both A and B hear the same pitch. **b** When the source is moving towards B, B receives more vibrations per second than A and hears a slightly higher pitch whereas A receives fewer vibrations per second and hears a lower pitch. **c** An observer directly in the path of the source will experience an instantaneous change in frequency as it sweeps past

clearly from the overall noise of the exhaust gases emerging from the engines (Fig. 4.14).

At the other end of the scale, a Doppler shift should in principle be just about audible in the buzz of a bee or a fly as it whizzes past you. And should you want concrete proof that the buzz is Doppler shifted, deploy an audio spectrometer app on your phone or tablet on a day when bees are busy searching for nectar from your garden's flowers. The drop in the frequency of the buzz will show up as a brief downtick on the display. I think I have sometimes heard the drop in frequency, but I'm not sure my ears are up to it, so I can't be sure.

The Doppler effect has many applications. Blood flow in a patient's arteries can be determined with a Doppler echocardiogram which employs ultrasound. Radar, which uses microwaves in place of sound, is used by meteorologists to monitor the movement of clouds. Police use portable radar units to measure the speed of cars. In all these cases the direction of motion and speed of the moving object is determined by comparing the frequency of the ultrasound or microwaves emitted by the transmitter with the frequency of the reflected signal. The Doppler shift of light emitted by the elements and compounds in a moving source enables astronomers to determine the motion of stars and their planets within our galaxy. And as we saw in the

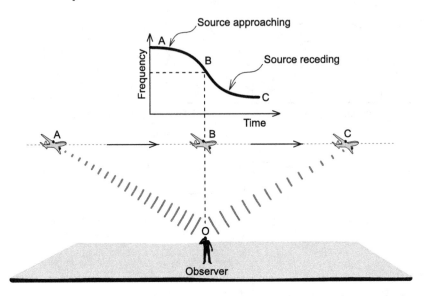

Fig. 4.14 Doppler shift of the noise from the turbines of an aircraft engine is heard as a gradually diminishing glissando

previous chapter, bats and toothed whales that employ echolocation to hunt make use of the Doppler effect to detect the direction of movement of their prey. Moreover, they have been doing so for some fifty million years. It seems that once again, science and engineering were late to the party (though they have made up for lost time).

5

Sounds Returned

Abstract Chapter 5 is about the reflection of sound and the many ways in which that affects what we hear. The symbolism and mythology of echoes. Echoes and whispering galleries of various sorts are described and explained, as is reverberation and its consequences. Other topics include echolocation in war and peace.

The Sound of Silence

The quietest place in the world is reputed to be the anechoic chamber at Orfield Laboratories in Minneapolis, USA.[1] There are anechoic chambers in universities and research establishments across the globe, so whether or not this claim is correct is a moot point. In any case, it's very unlikely that you would be able to notice the difference between one such chamber and another by ear alone.

These rooms, or *chambers* as they are more usually called, are not intended to be silent spaces to which one can retire to escape the hurly-burly of everyday life. They are laboratories used by engineers to study noise emissions and acoustic performance of devices such as engines, microphones, electric motors and loudspeakers and by psychologists to investigate aspects of the perception of sound.

---

[1] Within the Orfield chamber a sensitive microphone can detect sounds of −9.4 dB, which is well below the threshold of human hearing (defined as 0 dB for a healthy human ear).

© The Author(s), under exclusive license to Springer Nature
Switzerland AG 2021
J. Naylor, *Now Hear This*,
https://doi.org/10.1007/978-3-030-89877-9_5

What makes an anechoic chamber especially quiet is that it is completely isolated from its surroundings by being suspended on springs within a building with thick concrete walls so that it is unaffected by external vibrations. And to ensure that the sounds made by the devices being tested within the chamber are not affected by reflections of those sounds, all its surfaces are lined with hundreds of large wedges of porous fibreglass that scatter and absorb sound as completely as possible. In the quietest type of anechoic chamber, the floor is a stiff, open mesh that allows sound to pass though to an array of fibreglass wedges below.

Setting aside its primary practical function, such a room offers a fascinating and instructive insight into how we normally experience sounds. On entering an anechoic chamber you very quickly become aware of some of the consequences of a total absence of acoustic reflections: your ears feel as if they are bunged up and every sound is abruptly curtailed. Moreover, in the absence of the reflections that boost the loudness of sounds within an enclosed space, voices must be raised to be heard clearly. And should a speaker within the chamber turn his or her back to you, their voice will become fainter and noticeably deeper because high frequencies don't spread out as much as low frequency ones as they issue from a speaker's mouth and so can't wrap around their head and reach your ears. After a few minutes, in the absence of extraneous noises, you become acutely aware of the rustle of your clothing and of sounds from within your body—air entering and leaving your lungs, the beat of your heart, and even the noise of blood flowing through your arteries near your ears.

Those bodily noises famously inspired John Cage to compose his best-known and most controversial work, the tersely titled 4′33″. The occasion was a brief visit he made to an anechoic chamber at Harvard University in 1951. Cage had expected that he would experience complete silence, but was nonplussed when he became aware two sounds, one high pitched and the other low pitched. When he asked the engineer in charge about them, he was told that the high one was due to the activity of his nervous system and the low one was blood coursing through his veins and arteries. The experience led him to conclude that absolute silence is unattainable. And to convey this insight, the resulting composition requires the performer to sit at a piano in complete silence for four minutes and thirty three seconds during which the audience becomes conscious of what they would otherwise dismiss as extraneous and intrusive noises: people coughing, clearing their throats, shuffling, etc. The point of the piece is to get the audience to accept that all sounds are music because "Wherever we are what we hear is mostly noise. When we ignore it, it disturbs us. When we listen to it, we find it fascinating."[2] To judge from audience reactions over the years, it would appear that the work has never had many fans.[3]

A cynic might conclude that the title of the composition represents the amount of time Cage spent in the anechoic chamber.[4] Had he stayed longer he might have begun to experience some of the unpleasant symptoms such as spatial disorientation and mild nausea that some first-time visitors to these rooms report feeling. One possible reason they do so is that within these chambers one's senses are in conflict: our eyes inform us that we are in a dark and poky room, our ears infer that the absence of reflected sounds means we are in a boundless space.[5]

So when we say that we prize silence, we almost certainly don't have in mind the somewhat oppressive stillness of an anechoic chamber. If anything, spending more than a few minutes on one's own in such a room can make one yearn for the background murmur of daily life. It also makes one aware of the degree to which reflections affect and alter sounds. Usually, we become aware of reflected only in particularly reverberant spaces. In an anechoic chamber you realise that reflections give sounds a fullness and depth that would be absent were the world completely anechoic, something that becomes apparent as you emerge from the chamber and ambient sounds acquire a fizz and liveliness of which you had been unaware before you entered the chamber.

Indeed, in the absence of any acoustic reflections not only would the world sound odd, it would sound unnatural because the hearing system evolved within environments in which reflections are an integral aspect of the soundscape. And nothing makes one more aware of this fact than a total absence of reflections, a situation that we encounter in its most extreme form within an anechoic chamber. Moreover, as we shall find out in this chapter, reflections are the source of a surprisingly large number of intriguing and arresting acoustic effects. And although Nature can never match the stillness of an anechoic chamber, there are places and situations where reflections are reduced to a minimum. In the absence of wind or rain, the silence of a large open field or a mountain summit, particularly when covered in a thick blanket of freshly fallen snow, is due in large part to an dearth of

---

[2] Cage, J. (1961) The Future Of Music: Credo. In: Silence: Lectures and Writings. Wesleyan University Press, p 3.

[3] Ross, A. (2010) Searching for Silence, John Cage's art of noise. The New Yorker, October 4.

[4] I have found no evidence for the claim that the duration of the work, 273 s, is related to the fact that −273 °C is absolute zero, the temperature at which atoms cease to have any energy.

[5] Anechoic chambers are dark even when brightly illuminated because the wedges that line its surfaces absorb light, though not as efficiently as they absorb sound.

reflecting surfaces. Due to its porosity, a thick covering of freshly fallen snow is a particularly good absorber of sound.

One of the things that surprised de Saussure when he reached the snowy summit of Mont Blanc was that every sound was much fainter than expected. A pistol shot, he noted, was no louder than a small firecracker: "un coup de pistolet n'y fit pas plus de bruit qu'un petit petard de la Chine n'en fait dans une chambre [a pistol shot made no more noise than a small Chinese firecracker makes in a room]"[6] He knew that reflections contribute to the loudness of sounds and could see that on the open summit there were few features to reflect sounds, but he mistakenly attributed faintness of the pistol shot to the thinness of the air, an explanation that was accepted for much of the following century.

Four years before de Saussure reached the summit of Mont Blanc, a French natural philosopher, Jacques Charles, designed and constructed the first man-carrying hydrogen balloon. During his second ascent on 1st December, 1783, Charles managed to reach the not inconsiderable altitude of 3000 m. Had he not experienced acute pains in his ears due to the drop in air pressure, which forced him to return to the ground, he too might have noticed an unusual quality of the aerial soundscape in which he found himself because when floating far above the ground in a balloon one is in an almost perfectly anechoic environment. With the exception of the envelope of the balloon and its basket, there are no reflecting surfaces in this situation and every sound seemingly vanishes into thin air. The effect of these conditions was often remarked upon by pioneering balloonists during the 19th Century, an era when all balloons were gas-filled and therefore totally silent, unlike modern hot air balloons. James Glaisher, who made several balloon ascents between 1862 an 1866, had this to say about the unusual quality of sounds made while handling a gas balloon at altitude:

> But this sound in that solitary region, amid a silence so profound that no silence on earth is equal to it; a drum-like sound meeting the ear from above, from whence we usually do not hear sounds, strikes one forcibly. It is, however, one sound only; there is no reverberation, no reflection; and this is character-istic of all sounds in the balloon, one clear sound, continuing during its own vibrations, then gone in a moment. No sound ever reaches the ear a second time. But though the sound from the closing of the valve in those silent regions is striking, it is also cheering, it is reassuring, it proves all to be right; that the

---

[6] de Saussure, H-B (1852) Voyages dans les Alpes Partie Pittoresque des Ouvrages de H.-B, De Saussure, 2nd edition. Paris, p 288.

balloon is sound, and that the colder regions have not frozen tight the outlet for gas.[7]

Echoes

The paradigm of reflected sound is, of course, an echo. Look up "echo" in a dictionary, however, and you will find that it is invariably defined as a reflection of sound. But whoever came up with that definition omitted an essential feature: in an echo both the original sound and its reflection are each distinctly audible.

You can, of course, hear identifiable reflections of sounds without hearing their source. It all depends on the lie of the land and the location of the sound source. In a town, buildings can both block and reflect sounds, making it difficult to locate the source of a sound and leading to some compelling and often entertaining auditory illusions.

I live near a main road that is lined on both sides with an almost continuous row of houses. When walking on the western side of the road I occasionally hear the sound of bells of a nearby church coming from the east whereas the church itself is actually several hundred metres to the west. The sound of the bells can't reach me directly because it is blocked by the houses on western side of the road so what I hear is their reflection from the houses on the eastern side of the road. About half way down there is a break in the row of houses on the western side due to a side road, and as I cross it the sound of the bells instantly switches to its true direction and masks its reflection. The same thing happens when traffic on a highway about a kilometre to the west of this road is particularly heavy and there is a strong temperature inversion or a continuous westerly wind: the sound of the traffic appears to come from the east until I reach that side road.

When I am in my north-facing garden I often hear the sound of trains travelling along the railway line 500 m to the south of my house. But the sound invariably appears to come from the north because it is reflected by the houses behind my garden while the direct sound is blocked by my house. It's the same with overflying airplanes: they usually cross the northern sky heading east above my garden, but when I am in the front room of the house it always seems that the flight path has shifted south because sounds from the aircraft are reflected by the houses on the opposite side of the street.

---

[7] Turnor, C.H. (1865) Astra Castra: Experiments and Adventures in the Atmosphere. Chapman and Hall, London, p 388.

Another situation in which you will hear reflected sounds is as you drive at speed past a wall, a fence or an upright post. The source of sound is mainly the hiss of the tyres. Its reflection is particularly noticeable if there is a succession of narrow upright structures such as posts or bollards because the reflections are separated by brief intervals of silence. If the width of the reflecting structures varies, the result is rather like a barcode of sound. If you are a passenger, close your eyes and you may be able perceive the relative dimensions of these structures aurally, and even their relative distance from the car due to a change in their loudness and timbre.

In all these situations, the listener is much nearer to the reflecting surface than to the source. To hear an echo—i.e. the original sound and its reflection—the situation has to be reversed: you have to be nearer to the source than to the reflecting surface so that the original sound reaches your ears before it reflection does. At the same time, you have to be sufficiently far from the reflecting surface for the original sound to cease before all or part of its reflection reaches one's ears because otherwise it will mask the reflection.

The limiting factor here is physiological because the human ear is unable to distinguish clearly between individual sounds that are very closely spaced in time. It is often claimed that our hearing system requires at least one tenth of a second between sounds if it is to perceive them as distinct events. Given that sound travels at 340 m/s, in a tenth of a second a sound will have travelled 340 m/s × 0.1 s = 34 m. So to hear a distinct echo of a sound you have made the distance to the reflecting surface has to be at least half this distance, i.e. 17 m. However, you will find that you can hear a distinct echo from a surface that is much closer than this if the original sound is very brief, such as a clap. I can usually hear a distinct echo of a clap when I am 10 m from a wall, which means that the interval between the clap and its reflection is 0.06 s ((10 × 2)/340 = 0.06 s). If the reflecting surface is closer than this, the echo merges with the incident sound and the reflection makes it appear as if the original sound lasts slightly longer than it actually does. We'll consider the consequences of such situations later in this chapter when we take a look at reverberation.[8]

As you might expect, the further you are from the reflecting surface the more you should be able hear of the reflection of the original sound. For example, to hear an echo of *all* three syllables of "peekaboo" distinctly the first syllable of the reflection (pee) has to reach your ears *after* you have finished uttering the final syllable (boo). Assuming that it takes a fifth of a second to utter each syllable, which was Mersenne's estimate, and that the syllables

---

[8] Echolocating bats can detect a distinct echo from objects at distances of a few centimetres.

of a polysyllabic word can be said slightly quicker than this, you can say "peekaboo" in half a second, during which time sound travels 160 m. Hence the reflecting surface must be at least 80 m away for the reflection of the entire word to be heard distinctly.

The more distant the reflecting surface, however, the fainter the echo because sound spreads out in all directions, both as it makes its way to the reflecting surface and after it has been reflected. At the same time, a sound becomes fainter because it loses energy due to absorption as it travels though the air. In fact, you hear only the sound that is reflected *directly* back at you, which is a tiny fraction of the original sound. Sometimes the loss of higher frequencies due to absorption by the reflecting surface is very obvious in the timbre of reflection (Fig. 5.1).

The ideal echoing surface is hard, vertical and high, such as a cliff or a solid wall. But it need not be perfectly smooth because the wavelengths present in the majority of sounds vary from tens of centimetres to several metres. Hence a slightly irregular surface can be a far better reflector of sound than a smooth mirror is of light. The wavelength of light is less than a thousandth of a centimetre so even barely perceptible irregularities in the surface of a mirror will spoil it as a reflector whereas the rough surface of a cliff can reflect sound almost as well as a smooth wall. Trees on the edge of a dense wood can return surprisingly distinct echoes of sounds such as a barking dog.

Due to diffraction, however, a wave can't be reflected by an object that is smaller than its wavelength because the wave will wrap around the object rather than being reflected by it. The reason we can see a tiny object such as a single grain of sand is that the wavelength of light is at least 2000 times less than the average size of a grain. Even with the most powerful optical

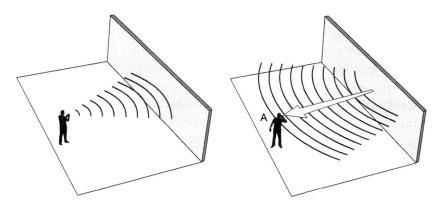

Fig. 5.1 A simple echo. Only a fraction of the reflected sound reaches your ears at A so the echo is always much fainter than the original sound

microscope, however, we can't see atoms because the diameter of an atom is some four times smaller that the shortest wavelength of visible light (which appears violet to the human eye). The shortest wavelength of audible sound is one hundred thousand times *greater* than that of violet light, which is why the objects capable of reflecting sounds have to be at the very least several centimetres across. Fortunately, the world is full of objects that are more than large enough to reflect sounds of every wavelength. Were this not so the world would in effect be a vast anechoic chamber.

Some sounds produce a better echo than others: brief sounds such as claps, explosions and gunshots produce echoes that are more distinct than sounds that are drawn out, such as a sibilant hiss. The same applies to words: sibilant words are poorly echoed compared to short, sharp ones. That advocate of unbridled empiricism, Francis Bacon, noted that "There are certain letters that an echo will hardly express; as *S* for one, especially being principal in a word. I remember well, that when I went to the echo at Pont-Charenton, there was an old Parisian, that took it to be work of spirits. For, said he, call *Satan*, and the echo will not deliver back the devils name; but will say, *va t'en*; which is as much in French as *apage*, or avoid. And thereby I did hap to find, that an echo would not return S, being but a hissing and interior sound."[9]

A couple of centuries later, the British physicist, John Strutt, more commonly known by his title, Baron Rayleigh, came to the same conclusion: "I suppose it must have been noticed before now that the *s* sound is badly returned by an echo. Standing at a distance of about 150 yards from a large wall, I found that there was scarcely any response to even the most powerful hiss. *Sh* was heard a little better; *m, k, p, g* pretty well; *r* very well; *h* badly; *t* badly; *b* seemed half converted into *p* by the echo. The failure of the hiss seems to be the fault of the air rather than of the wall, for a powerful hiss heard directly at a distance of 200 yards had very little *s* left in it."[10]

In fact, if you want to hear a distinct echo, a clap is usually the best sound to employ because not only is it loud and brief, it is also composed of a broad spectrum of frequencies that can produce interesting and unusual acoustic effects in certain situations. And while on the subject, there is an art to clapping loudly, which flamenco has perfected. The trick is to maximise the of contact between the hands. This requires that the fingers of one hand strike the palm of the other. Clapping with palms alone is much quieter;

[9] Bacon, F. (1627) Sylva Sylvarum. In: Bacon, F. (1826) The Works of Francis Bacon, Vol 1. London, p 331.
[10] Strutt, J.W. (1877) Acoustical Observations: Audibility of Consonants. In: Strutt, J.W. (1899) Scientific Papers Vol 1, 1869–1881. Cambridge, p 317.

try it.[11] Other sources of loud, brief, broad-spectrum sounds include bursting a balloon or banging two wooden blocks together.

Finally, unless the source sound is loud, its echo may be too faint to hear it in a noisy environment. About the only time when silence is more or less guaranteed in an urban environment is in the hours after midnight when human activity has largely ceased and the world is as quiet as it ever gets. But deliberately summoning up echoes in the dead of night is an un-neighbourly act and so you should take advantage of suitable moments and situations during daylight hours when you are unlikely to disturb or alarm other people. As for bursting balloons, the danger these days is that they may be mistaken for explosions.[12]

From Myth to Mathematics

There's no doubt that echoes have fascinated and beguiled people since ancient times. Archaeologists have often noted that many of the cliffs and canyon walls decorated in prehistoric times with petroglyphs and painted images also return strong echoes. In fact, myths and legends from around the world suggest that the belief that an echo is the voice of a spirit that inhabits rocks is universal, although the detail of how the spirit came to be there varies from place to place. Such a belief could explain why people ignorant of the nature of sound might regard an echoing cliff an auspicious place to decorate with hunting scenes. Moreover, some anthropologists have suggested that towards the end of the Paleolithic era, people exploited both resonance and reverberation within caves when choosing where to depict certain types of animals.

Inevitably, given their remoteness in time and space, we know more about some of these myths than others. To the ancient Maya of Mexico, an echo was the voice of one of the lords of the underworld, *Tepeyollotl*, or "Heart of the Mountain". *Tepeyollotl* was also responsible for earthquakes, which are a source of subterranean sounds, and was associated with the jaguar, possibly on account of its growl.

According to ancient Norse lore, echoes are the voices of mischievous dwarves who live in rocks and delight in teasing mortals by repeating the last words of overheard conversations. Indeed, the word for an echo in Old Norse is *dvergmál*, which means 'dwarf talk' and is a combination of *dvergr*

---

[11] Fletcher, N. H. (2013) Shock Waves And The Sound Of A Hand-Clap — A Simple Model. Acoustics Australia, Vol. 41, No. 2, p 165–8.

[12] In these days of heightened security, the sound of a busting balloon is likely to cause alarm, so should be reserved for times and places far from people.

for dwarf and *mál* for talk.[13] The ancient Irish referred to an echo as *mac alla*, or son of the rock or cliff.

A much more detailed account of the link between echoes and spirits is found in the creation mythology of the South Pacific islanders where an echo is the voice of *Tumuteanaoa*, a female spirit that lives among rocks and gorges. According to legend, she was discovered by Rangi, the Polynesian sky god, while he was exploring the lands he had just created. On approaching a rocky gorge he called out "Hello, there!" and was surprised to receive the same words in answer. All attempts to discover the source of the voice were met in the same way: his every question was thrown back at him. Exasperated, Rangi began to curse, only to be cursed in return. As he made his way up the gorge, however, he eventually caught sight of the creature that had taunted him and only then was she willing to divulge her name and tell him that she was the mother of the rats that infest her domain.

And long ago the artless folk of ancient Greece personified these disembodied sounds as a nymph who repeats all she hears. From such lore, Greek and Roman poets fashioned several whimsical versions of the story of Echo, whose grisly fate was to be unable to use her voice other than to repeat involuntarily any sound she heard.

In the romance of *Daphnis and Chloe* by Longus, a second century AD Greek poet, Echo is a virginal wood-nymph, taught by the muses to sing and to play the flute and lyre. Pan, the amorous, half-human, half-animal god, grows jealous of her singing and when she resists his advances he incites a group of shepherds, whose patron god he is, to tear her limb from limb and scatter her remains far and wide. But, to Pan's dismay, his murderous plot fails to silence her because even the smallest fragments of her dismembered body continue to sing and are everywhere the source of mournful echoes.

Another and far better known account of how the unfortunate nymph met her fate was concocted by the Roman poet, Ovid in his *Metamorphoses*, a lively retelling of ancient myths and tales that all have the theme of transformation from one state of being to another—hence the title of the work. In Ovid's version Echo was stripped of the power of speaking her own words and condemned to repeat those of others by Hera, the long-suffering wife of Zeus, because she had seemingly connived in Zeus's secret trysts with other nymphs by engaging Hera in needless conversation while the nymphs made good their escape.

One day, as the mute nymph wandered across a remote meadow, she caught sight of the beautiful adolescent Narcissus and was overwhelmed with

---

[13] In Norse mythology dwarves were renowned as skilled blacksmiths who lived in underground.

desire for him. Narcissus was searching for his companions, from whom he had become separated. When he called out "Is anyone here?", Echo replied "Here". "Come to me", he shouted back and heard in reply: "Come to me." The dialogue continued like this until at last Narcissus urged her to show herself, saying "Here, let us meet together", and was answered with a single word: "Together". But when she went to embrace him, he rebuffed her. Heartbroken, she took to hiding in caves and pined away until only her voice remained, her bones having turned to stone. Thus it is that we sometimes hear her reply when we call out. Fittingly, Narcissus suffered an equally poignant end: he glanced into a pool and fell so in love with his reflection that he wasted away when it would not answer him. The gods took pity on the youth and turned his remains into the flower that now bears his name.[14]

But centuries before Ovid penned *Metamorphoses*, hard-headed materialists had already taken a matter-of-fact approach to the phenomenon. For Aristotle "an echo occurs, when … the air originally struck by the impinging body and set in movement by it rebounds … like a ball from a wall."[15] And some fifty years before *Metamorphoses* was written, another Roman poet, Lucretius, gave the same explanation in *De Rerum Natura* and mocked those gullible yokels for whom echoes are the voices of imaginary satyrs, nymphs and fauns.[16]

We saw in chapter two how that matter-of-fact approach to nature re-emerged with a vengeance during the seventeenth century, when increasing numbers of natural philosophers turned their backs on the accumulated wisdom of the ages and brought about the transformation in thought and practice that in the twentieth century came to be known as the scientific revolution. And where Echo is concerned, her principle nemesis was Marin Mersenne. You will recall that Mersenne employed echoes of short polysyllabic phrases to measure the speed of sound and so was well aware of the conditions necessary to hear an echo, such as the least distance between the speaker and the reflecting surface at which a distinct echo can be heard. He was also the first person to cast a sceptical eye on reports of strange and unusual echoes. Indeed, he proposed "a new science of Sounds, which one will call, if one wishes, Echometrie, or the measurement of Sounds" on the grounds that much could be learned about the nature of sound in general through the study of echoes in particular.[17] Not only can echoes be used

---

[14] Ovid, *Metamorphoses* Book III, 359–401.

[15] Aristotle, 350 BC, De Anima, Book 2.8. http://classics.mit.edu/Aristotle/soul.2.ii.html (accessed 20/03/2020).

[16] Lucretius, De Rerum Natura, Book IV, lines 576–581.

[17] Mersenne, M. (1636) Harmonie Universelle. Premier Livre De La Nature Des Sons, Prop XXVI, p 50.

to measure the speed of sound, they also provide evidence that sound, like light, obeys the law of reflection.[18] Hence to simplify things, ever since the seventeenth century scientist have often used rays in place of wavefronts to determine the path that sound takes as it travels through the atmosphere, a technique known as ray tracing. As you will have seen, many of the diagrams in this book use sound rays in place of waves.

Even the most committed materialist can't fail to be intrigued by the possibility of a dialogue with a disembodied voice, however, and in Mersenne's day there were numerous accounts of fabulous echoes that, it was claimed, would respond intelligently to one's questions. One such improbable echo was said to reply in French when spoken to in Spanish. Mersenne realised that this is impossible, but wondered if there are words or phrases that might persuade someone that the echo is indeed in a different language and came up with several phrases in Greek in which an echo of the final few words would sound as if it were Latin.[19]

Another improbable echo involved a tower near the Aventine Hill in Rome. It was said that the echo from the tower could repeat the entire first verse of Virgil's *Aeneid* eight times.[20] Mersenne estimated that these eight repetitions would require at least 32 s and therefore that the reflecting surface would have to be half a league (i.e. 2 km) from the speaker. Nor, he pointed out, is the human voice loud enough to be heard at that distance, let alone its reflection. He didn't, however, ask himself how the tower was supposed to be able to repeat the phrase over and over again.[21] Perhaps the anonymous author of this account of the fabulous echo had in mind what we now call a flutter echo when adding this detail to the description of the echo.

Those lengthy phrases favoured by some of the seventeenth century savants fascinated by echoes seem excessive when all that is necessary to produce an echo is a loud percussive noise such as a clap. However, should a clap strike you as unduly prosaic, a dialogue with an echo is possible in the right circumstances. Having found a suitable reflecting surface and bearing in mind that some words are more clearly echoed than others, choose phrases for which the echo of the last word or syllable would sound like a sensible reply to a statement or a question.

---

[18] Mersenne, M. (1636) Harmonie Universelle. Premier Livre De La Nature Des Sons, Prop XXVI, p 49.

[19] Mersenne, M. (1636) Harmonie Universelle. Livre Troisiesme Des Mouvemens, Prop XXI, Corollary VII, p 219.

[20] *Arma virumque cano, Troiae qui primus ab oris…* (Arms, and the man I sing, who, forced by fate etc.).

[21] Mersenne, M. (1636) Harmonie Universelle. Livre Troisiesme Des Mouvemens, Prop XXI, p 214–15.

Just such an echo is to be heard from the north wall of the Château in Chinon, France, with the necessary phrases supplied by the local tourist office, should that be required. Walk up the aptly named Rue de Echo until you reach a raised platform on the right hand side of the road.[22] Climb onto it, face the castle wall and shout as loudly as you can: "Les femmes de Chinon, sont-elles fidèles?", with an emphasis on the final syllable. A faint echo comes back: "Elles?" and you reply loudly: "Oui, Les femmes de Chinon". The echo replies: "Non". The walls of the Château are approximately 150 m from the platform and are hidden from view by trees, though this doesn't affect the echo in the least. However, you may find that you have to alter the rate at which you say the sentences in order to hear unambiguous replies. And, of course, you can have the exactly the same conversation with any wall you wish as long as it is sufficiently far away from you.

Athansius Kircher was, if possible, even more fascinated by echoes than Mersenne, but in his case enthusiasm sometimes got the better of him. He devoted sixty pages of his *Musurgia Universallis*[23] to the phenomenon and returned to the subject twenty years later in another of his books, *Phonurgia Nova*,[24] without, however, adding much to what he had written in the earlier work. He devised a Latin terminology for echoes that was widely employed well into the eighteenth century. The reflecting surface was named the *centrum phonocampticum* and the optimal position from which to hear a distinct echo was the *centrum phonicum*.[25] The distance between the *centrum phonicum* and the *centrum phonicampticum* is not fixed, it depends on what the listener wishes to hear echoed. Where polysyllabic words are concerned, Kircher claimed that to hear the final syllable in its echo the distance must be 100 paces. For two syllables it is 190 paces, for three it is 270 paces and up to 600 paces for seven syllables.[26]

Kircher was particularly interested in multiple echoes, and claimed to have come across an exceptionally remarkable example in Avignon. He said that it was caused by a series of seven equally spaced buttresses supporting a long wall, from each of which one could hear a distinct echo. This led him to speculate about how to engineer an echo that would produce a complicated

---

[22] Ask for directions at the Chinon tourist office or look it up on Google Maps.

[23] Kircher, A. (1650) Musurgia universalis sive Ars Magna Consoni et Dissoni, Vol II, Book IX. Rome, p 247–308.

[24] Kircher, A. (1673) Phonurgia nova, sive conjugium mechanico-physicum artis & naturae paranympha phonosophia. Campidonae.

[25] Kircher, A. (1650) Musurgia universalis sive Ars Magna Consoni et Dissoni, Vol II, Book IX. Rome, p 238.

[26] Kircher, A. (1650) Musurgia universalis sive Ars Magna Consoni et Dissoni, Vol II, Book IX. Rome, p 264.

but sensible reply to a question. The speaker would address the Almighty loudly in Latin: "Tibi vero gratia agam quo clamore?" [How shall I cry out my thanks to thee?] and would hear the following echoes: "Amore, More, Ore, re.", i.e. "with thy love, thy wont, thy words, thy deeds."[27] This would require five reflecting surfaces arranged at different distances from the speaker, each of which would reflect the entire question. The distance to the wall *nearest* the speaker would be such that he hears all three syllables of the last word CLAMORE in the echo. A fainter echo from the next wall arrives slightly later and so is almost entirely masked by the echo from the first wall except for AMORE. And so on, each echo being fainter than the previous one. Allowing 1/5 s per syllable, the distance to the furthest reflecting surface, from which one would hear a faint "re", would be some 250 m. Not that Kircher, or indeed anyone else, ever put this madcap idea into practice.

Indeed, such was Kircher's preoccupation with unusual echoes that he even took to devising ways of tricking the unwary into believing that an echo could provide them with intelligible answers. He relates how he once had fun at his friends' expense, who were bemused to hear an echo reply "Constantius" to the question "Quod tibi nomen?" (What is your name?). The "echo" was an actor hidden out of sight behind a wall.[28]

Kircher's schemes notwithstanding, it seems that no one has gone to the trouble of deliberately designing and erecting a structure for the sole purpose of creating an echo. If there is an echo within a building, or from its external walls, then this is invariably a happy accident; and happily there are lots of those, as the following pages will confirm. Not that the idea never crossed anyone's mind. Gilbert White, the eighteenth century English naturalist, once wrote to a friend suggesting that if one was going to the expense of having a building erected, it would be worth doing so in such a way as to contrive an echo: "Should any gentleman of fortune think an echo in his park or outlet a pleasing incident, he might build one at little or no expense. For whenever he had occasion for a new barn, stable, dog-kennel, or the like structure, it would be only needful to erect this building on the gentle declivity [i.e. a downward slope] of an hill, with a like rising opposite to it, at a few hundred yards distance; and perhaps success might be the easier ensured could some canal, lake, or stream, intervene. From a seat at the *centrum phonicum* he and his friends might amuse themselves sometimes of an evening with the prattle of this loquacious nymph; of whose complacency and decent reserve more

---

[27] Kircher, A. (1650) Musurgia Universalis sive Ars Magna Consoni et Dissoni, Vol II, Book IX. Rome, p 267.

[28] Kircher, A. (1650) Musurgia Universalis sive Ars Magna Consoni et Dissoni, Vol II, Book IX. Rome, p 268.

may be said than can with truth of every individual of her sex; since she is
… *quae nec reticere loquenti, Nec prior ipsa loqui didicit resonabilis echo* [who
could neither hold her peace when others spoke, nor yet begin to speak till
others had addressed her]."[29] The quotation is from Ovid's story of Echo and
Narcissus. But as we shall see, not only is it unnecessary to go to the trouble
of engineering an echo, coming across places by accident where they can be
heard adds to their interest and charm.

Mersenne's and Kircher's interest in echoes was enthusiastically taken up
later in the seventeenth century by the members of the newly established
Royal Society of London Accounts of echoes frequently featured in the
minutes of the Societies meetings.[30]

Multiple Echoes

Fortunately, Kircher didn't expend all his prodigious curiosity and ingenuity
on these fanciful acoustic diversions. He also gathered information, much
of it second hand, on the unusual acoustics of several buildings, ancient and
modern. He was the first person to describe and explain a remarkable multiple
echo that could be heard in the courtyard of the Villa Gongoza-Simonetta, a
sixteenth century palace in Milan.[31] The Villa had two wings at the back of
the main building and the echo was produced in the courtyard between the
wings. The wings were each 17 m long and 34 m apart. Kircher never visited
the Villa and relied instead on reports sent to him by a fellow Jesuit, Matthäus
Storr, from whom he learned that the echo repeated between 24 and 30 times.
Kircher explained this by showing how the original sound would reflect back
and forth between the smooth outer walls of the upper floors, becoming
fainter with each repetition until it finally becomes inaudible.

The Villa itself was an architectural marvel, but its fame seems to have
rested principally on that multiple echo, which became one of the attractions
on the Grand Tour of Europe undertaken by wealthy young gentlemen in the
seventeenth and eighteenth century.[32] Indeed, although there is no evidence

[29] White, G. (1778) Letter XXXVIII. In: White, G. (1778) The Natural History of Selborne. Cassell and Co., London, p 100.

[30] Gouk, P. (1982) Acoustics in the Early Royal Society 1660–1680. Notes and Records of the Royal Society of London, Vol. 36, No. 2, p 155- 175, p 161–63.

[31] Kircher, A. (1650) Musurgia Universalis sive Ars Magna Consoni et Dissoni, Vol II, Book IX. Rome, p 289–91.

[32] Famous visitors to Villa Simonetta included Thomas Jefferson and James Boswell, who fired a pistol from an upper story opposite a wall and claimed to have counted 58 repetitions of its echo. Later, the echo was studied by the French mathematician and ardent Bonapartist, Gaspard Monge and by the Swiss mathematician, Daniel Bernoulli.

that its architect deliberately designed the courtyard to produce the echo, the Villa became far better known for its echo than for its architecture. In fact, he left the Villa unfinished, so the echo "materialised" decades later when the building was finally completed. In all likelihood it was first noticed by the builders who completed the Villa because they made a small window in the middle of the otherwise windowless wall of the upper story of one of the wings from which a visitor could summon the echo. That the echo put the Villa on the map is confirmed by the fact that in the earliest image of the Villa, reproduced in *Musurgia Universalis*, the window is prominently shown and marked with a star.[33]

Mark Twain was among its many famous visitors and included an account of his visit to the Villa in *Innocents Abroad*, his account of his travels through Europe and the Middle East in 1869.

"We arrived at a tumble-down old rookery called the Palazzo Simonetti—a massive hewn-stone affair occupied by a family of ragged Italians. A good-looking young girl conducted us to a window on the second floor which looked out on a court walled on three sides by tall buildings. She put her head out at the window and shouted. The echo answered more times than we could count. She took a speaking trumpet and through it she shouted, sharp and quick, a single "Ha!" The echo answered:

> "Ha!–ha!—ha!–ha!–ha!-ha! ha! h-a-a-a-a-a!" and finally went off into a rollicking convulsion of the jolliest laughter that could be imagined. It was so joyful-so long continued-so perfectly cordial and hearty, that every body was forced to join in. There was no resisting it.

Then the girl took a gun and fired it. We stood ready to count the astonishing clatter of reverberations. We could not say one, two, three, fast enough, but we could dot our notebooks with our pencil points almost rapidly enough to take down a sort of short-hand report of the result. My page revealed the following account. I could not keep up, but I did as well as I could.

I set down fifty-two distinct repetitions, and then the echo got the advantage of me. [My companion] set down sixty-four, and thenceforth the echo moved too fast for him, also. After the separate concussions could no longer be noted, the reverberations dwindled to a wild, long-sustained clatter of sounds such as a watchman's rattle produces. It is likely that this is the most remarkable echo in the world."[34]

---

[33] Kircher, A. (1650) Musurgia Universalis sive Ars Magna Consoni et Dissoni, Vol II, Book IX. Rome. Image of the Villa is between p 282 & 283.

[34] Twain, M. (1869) Innocents Abroad. Collins, p 126–7.

Twain's account is puzzling because he says that the echoes eventually "moved too fast" to be distinctly heard. In fact, the time between echoes in this case would remain the same because it depends only on the distance between the reflecting surfaces, so it would have been their reduction in loudness that made them difficult to follow.

Villa Simonetta was badly damaged by Allied bombing towards the end of the Second World War. Although the building was repaired, the wings were not rebuilt and so, alas, its multiple echo is no more (Fig. 5.2).

The Villa's multiple echo, however, was not and is not unique; it was a just particularly striking example of a flutter echo. This form of echo occurs when sound is reflected over and over again between two parallel surfaces that are sufficiently far apart for distinct reflections to be heard. The result has been compared to a ricochet that rapidly fades into silence. There are well known examples of this type of echo in Iran (Shah Abbas Mosque in Isfahan), India (the "clapping portico" in the Fortress of Golkonda in Hyderabad, and Mexico (between the massive walls of Great Ball Court at Chichen Itza). In the case of the first two, the flutter echo can be set off just by striking a stiff sheet of paper with a finger because the sound is focused back to the floor by a domed roof. Such echoes were said to be tautological because they

Fig. 5.2 Villa Gongoza-Simonetta. The multiple echo was due to repeated reflections between the two wings of the Villa. The source of the echo was produced at the window marked RS (upper right wing)[35]

[35] Kircher, A. (1650) Musurgia universalis sive Ars Magna Consoni et Dissoni, Vol II, Book IX. Rome. Between p 282 & 283 (Image in Public Domain, Wikimedia Commons).

"return a clap with the Hands, or a Stamp with the Foot, eight, nine or ten times distinctly; the Noise dying as it were, and melting by degrees, becomes constantly weaker and weaker."[36]

Multiple echoes can also be produced within the large, solid semi-circular arch of a bridge or viaduct as long as the surface under the arch at either end of the gap it spans is more or less horizontal and made from a good reflector of sound such as concrete, asphalt, brick, stone or water.

> An interesting multiple echo is found at Echo Bridge over the Charles River at Newton Upper Falls, Massachusetts. Just beneath the arch of the bridge there is a platform which is easily accessible. If an observer stands on this platform and claps his hands the echo comes back to him about a dozen times before it dies out.[37]

This bridge, which carries an aqueduct, was built in 1870 and its arch has a span of 40 m.

Multiple echoes can be heard under the western span of Maidenhead Railway Bridge that carries the Great Western Railway across the river Thames. Indeed, the echoes are so pronounced that the western span of the bridge is known as the "Sounding Arch". The bridge was designed and built by Isambard Kingdom Brunel and opened in 1839. It was widened in 1890. It has several claims to fame, quite apart from its echoes. Its two arches are the widest and flattest brick arches in the world (39 m wide, 9 m tall), and was immortalised in 1844 by William Turner in his dramatic painting "Rain, Steam and Speed."[38] The Thames Tow Path goes under the western end of the span and provides a platform from which one can summon its multiple echoes. The sound of one's voice, a clap or the stamp of a foot when one is on the tow path under the bridge is reflected back and forth several times between the hard surface of the path and the surface of the water at the far end of the arch (Fig. 5.3).

In fact, multiple echoes are a feature of wide arches as long as long as one can get under them. I first came across such an echo was on the riverside path under the westernmost arch of the Pont de Vinade, which crosses the Charante River near the village of Bassac.[39] Ever since then I always

---

[36] Clare, M. (1737) The Motion of Fluids, Natural and Artificial; In particular that of the Air and Water. London, p 350.

[37] Jones, T.J. (1937) Sound: A Textbook. Chapman and Hall, London, p 87.

[38] The viewpoint of the scene depicted in "Rain Steam and Speed" is from the eastern side of the bridge. The sounding arch in the painting is thus furthest from the viewer. The painting hangs in the National Gallery in London, U.K.

[39] Pont de Vinade was built in 1842 and has four arches each with a span of approximately 20 m.

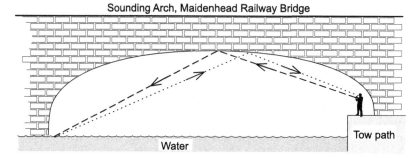

Fig. 5.3 Multiple echoes within an arch. The sound of a clap is reflected back and forth several times between the surface of the river at the far side of the arch and the tow path

clap whenever I find myself under an arch and have found that more often than not there is a flutter echo, though none in my experience have matched the one I heard within the "Sounding Arch" under the Maidenhead Railway Bridge.

Flutter echoes can be a problem in halls with high ceilings because a pronounced multiple echo can make it difficult to hear music or speech clearly, but the phenomenon is so entertaining that is worth clapping should you find yourself inside a tall, domed room, assuming this is permitted. And should you find yourself between two tall buildings or in a large, open space flanked by tall, parallel walls you can produce a flutter echo by standing close to one of the walls and clapping or popping a balloon. You have to be close to one of the walls to maximise the distance that the sound of the clap travels before reaching your ears; standing between the walls reduces this time and make the echo less distinct.[40] The distance between the walls also plays a part because a distinct echo is not audible if they are less than 4 or 5 m apart. But don't take my word for it: find a couple of parallel walls and experiment.

Echoes in Nature

Before there were large man-made structures, or indeed any structures at all, echoes were, of course, heard only from bluffs and cliffs or in gorges and valleys. And given that, by and large, much of the world's topography has not changed significantly over the centuries, there are places that have long been renowned for their echoes. Some of these echoes are sufficiently remarkable to

---

[40] The optical equivalent of a flutter echo is the multiple reflections seen in two mirrors that face one another by someone standing between them.

be worth seeking out, if only for the fun of retracing the steps of travellers in earlier ages to confirm their accounts. Accounts of remarkable echoes often used to be included in guide-books, especially those published during the nineteenth century, when tourism became both affordable and fashionable among the middle classes.

Canyons and gorges, wherever they are to be found are always a reliable source of echoes. The Gap of Dunloe in County Kerry, Ireland, was once celebrated for its echoes, though nowadays tourists drive through the gap seemingly unaware of this. Horseshoe Canyon in the Canyonlands National Park in Utah not only echoes magnificently, there also are life-sized petroglyphs of human figures and animals on some of its cliffs that may well be evidence of the age-old association between echoes and rocks.

A huge rocky bluff, know as the Lorelei, on the eastern bank of the Rhine near St Goarshausen, is associated with a famous echo, though these days human activity makes it all but impossible to hear it. The etymology of its name suggests that the echo has been known since antiquity: *Luren* (murmuring) + *ley* (rock). But the legend of Lorelei, a beautiful maiden who sits atop the bluff combing her golden hair and singing, thereby unwittingly causing boats to be wrecked on the rocks below by distracting boatmen, is a nineteenth century invention by Heinrich Heine, the German poet.[41]

The echoes from the cliffs surrounding Lake Königsee in Bavaria are a well-established attraction, though it is necessary to take a boat trip on the lake to hear them. At suitable points in the tour a guide in traditional costume will sing folk songs or play a trumpet or horn to demonstrate the amazing echo that can heard from the rocky cliffs that surround the lake. The surface of the lake enhances the echo because water reflects and does not absorb sound.

Echo Point at Katoomba, New South Wales, Australia offers a vantage point from which one looks out on the spectacular panorama of the Blue Mountains beyond the town. The cliffs to the west and east of Echo Point are close enough to return echoes but few people who visit Katoomba seem to take advantage and call out loudly to raise an echo. In any case, were they to do so, they might well not hear the echo because the babbling mass of visitors will drown it out.

Other examples of dramatic vistas that return echoes include the caldera of Vesuvius, the notorious volcano near Naples, and the cliffs around Masada in Israel.

Dr Robert Plot, a seventeenth century English naturalist, the first professor of Chemistry at the University of Oxford and the first keeper of Oxford's

41 Heine, H. (1824) Die Lorelei.

Ashmolean Museum, included several examples of unusual echoes in his *Natural History of Oxford*. One of these, in Woodstock Park, in Woodstock, near Oxford, was one of the Park's attractions.[42] This echo is still there, so to speak. To hear it, enter from the Park Gate at the end of the town, walk down the drive towards the Palace, leave the road on the right and go a little way down the slope to face the bridge, and clap or call out as loudly as you dare.[43]

Musical Echoes

Should you find yourself at the foot of a broad flight of steps, turn to face it and clap your hands: you may be rewarded by an unusual echo that sounds like a chirp and has come to known as a chirped echo. To hear the chirp clearly, however, it is necessary that the staircase is in the open air and not flanked by buildings or nearby vertical surfaces that might reflect the sound of the clap and mask the chirp. If the flight of steps is sufficiently long, the pitch of the chirp diminishes noticeably towards the end of the echo. And because it has an identifiable pitch, the chirp is also known as a musical echo.

The chirp, known technically as a repetition pitch, is the result of successive reflections from each of the vertical risers. Its pitch is determined by the time interval between successive reflections, which in turn depends on the depth of each step, as shown in Fig. 5.4. In other words, it is the rate at which the reflections arrive at the ear that determines the pitch of what you hear in this situation. The reason the pitch of the chirp diminishes slightly towards the end if the upper steps are much higher than the height of the listener is that the distance each reflection travels from a step to the listener increases towards the top of the stairway. This increases the time interval between successive reflections, leading to a slight drop in the repetition pitch.

A chirped echo can also be elicited from a long row of equally spaced palings or a wall with evenly spaced vertical corrugations such as those in the steel panels used for fencing and the walls of industrial buildings. In this case, face the reflecting surface, clap and listen for the chirp. How, if at all, does the distance to the reflecting surface affect what you hear?[44]

It is very likely that chirped echoes were noticed in antiquity because they can be heard within the semi-circular stepped theatres of ancient Greece and

---

[42] Plot, R. (1677) Natural History Of Oxford Being An Essay Towards The Natural History Of England. Oxford, p 7–18, p 7–10.

[43] Personal communication.

[44] Crawford, F.S. (1970) Chirped Handclaps. American Journal of Physics, 38, p 378.

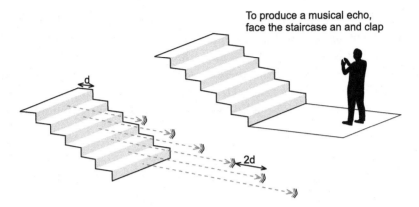

Fig. 5.4 Musical echoes from a staircase are due to the rapid succession of reflections that reach the ear from the vertical surface of each step

Rome, such as the fourth century BC amphitheatre in Epidauros in the Peloponnese. The original amphitheatre had thirty four rows of banked seats; the Romans added a further twenty one, which has had the effect of increasing the duration of the chirp. A chirp can also be elicited from the stairway of El Castillo, the Mayan pyramid of Kulkan at Chichen Itza, in present day Yucatan. Indeed, the chirp produced by the steps of this pyramid has become an attraction in its own right: visitors are invited to stand close to the foot of the stairway, clap their hands and listen to the resulting reflection.

El Castillo's chirp is so pronounced that some archaeologists have speculated that it was deliberately engineered to recreate the call of the Mayan sacred bird, the colourfully plumed quetzal. But the likelihood that the Maya understood the phenomenon well enough to design and build a structure capable to reproducing the call of the quetzal is vanishingly small. Nor can one take it for granted that they ever noticed the unusual quality of the echo once the pyramid had been constructed. Even though a chirped echo, wherever it is produced, can sound remarkably birdlike, we have no evidence that the Maya, nor indeed any ancient peoples, understood the nature of sound sufficiently well to design structures having specific acoustic properties. One has only to consider how incomplete and imperfect was the ancient Greek's knowledge of acoustics despite their tradition of natural philosophy to realise a less scientifically developed civilization is most unlikely to have understood these things any better.[45]

---

[45] Without doubt, reverberation within the large spaces of Gothic cathedrals adds to their spiritual atmosphere, if only to encourage silence. Did the builders of these cathedrals take into account the effect that prolonged reverberation might have on a congregation?

In fact, the musical echo and its cause was discovered by accident in the latter half of the seventeenth century by Christiaan Huygens, who, as we saw in chapter two, was one of the leading natural philosophers of his day and the author of the first mathematical account of the propagation of a physical pulse through a medium. He was also an accomplished musician, which enabled him to understand the nature of this particular echo. Huygens spent fifteen years in France as a leading member of the *Académie Royale des Sciences*. It was while he was residing in France that he came across a musical echo in the grounds of the Chateau de Chantilly.[46] In a letter to a fellow academician, he described that he had been standing between a broad flight of steps and a noisy fountain when he noticed "a certain musical tone that lasted as long as the fountain continued to spout."[47]

Puzzled, he sought its cause, and by listening attentively he quickly deduced that the sound was due to the interval between the reflections of the sound of the fountain from the individual steps of the staircase, "For any sound, or rather any noise, that is repeated with equal, small interstices, gives a musical tone…"[48] He surmised that the pitch of a musical echo is determined by the depth of the tread, and to confirm this he there and then rolled up a sheet of paper into a cylinder the same length as the depth of a single step—which he measured and found to be 17 inches (45 cm). Blowing over one of the open ends of the paper cylinder he found that it produced a note that had the same pitch as that of the chirped echo. Huygens could have given the pitch of the sound in terms of its frequency, something that was first achieved by Mersenne in 1636. But the measurement of pitch in terms of frequency rather than as a musical note was not common in the seventeenth century, quite apart from the practical difficulties of doing so given that the tuning fork, which could have provided him with a known pitch, was not invented until 1711. Incidentally, the fundamental frequency of an organ pipe 45 cm long and open at both ends is approximately 380 Hz.

Although the Chateau has undergone alterations since Huygens' visit, the flight of steps and fountain are as they were in his day, so it is still possible to hear the musical echo just as he did. Several years ago a Dutch physicist, Franz Bilsen, visited the Chateau to hear the echo for himself and to check Huygens' conclusions and calculations.[49] But rather than listening to the sounds of the

---

[46] Chateau de Chantilly is some 30 km north of Paris.

[47] Letter to P de la Hire. In: Huygens, C. (1905) Oeuvres Complètes, 10, Correspondance 1691–1695, p 570–71.

[48] Huygens, C. (1905) ibid.

[49] Bilsen, F.A. (1969/70) Repetition Pitch and its Implication for Hearing Theory. Acustica, 22, p 63–73.

tinkling fountain, he clapped his hands and recorded the ensuing echo.[50] Analysis of the recording confirmed Huygens' estimate of the frequency of the chirp.

The pitch of the musical echo at El Castillo differs from that at Chantilly in one significant respect: the drop in pitch from the former's staircase is far more noticeable. The reason is that there are many more steps at El Castillo: 92 to the 50 at Chantilly. The average depth and height of each step at El Castillo is approximately 26 cm, so the topmost step is 24 m above the ground. Hence the distance that each individual echo travels to reach someone at ground level increases markedly towards the top of the pyramid. The initial pitch of the chirp from the steps at the bottom of the pyramid is 920 Hz and falls to 654 Hz from those at the top of the staircase.

Lord Rayleigh noticed a musical echo in his home in Terling, Essex. "At Terling there is a flight of about 20 steps which returns an echo of a clap of the hands as a note resembling the chirp of a sparrow."[51] And there is a very pronounced musical echo from the staircase on The Mound in the gardens of New College in Oxford[52] Having clapped my hands at the foot of this staircase, I would describe the sound it produces as a "meow" rather than a chirp.

Steps are not the only structures that produce a musical echo. There is an interesting and unexpected example of such an echo at Ursinus College in Philadelphia where there is a maze in the form of a concentric paved path. The edge of the path is marked out by slightly raised, bevelled pavers. Shortly after it had been constructed, students discovered that if one stands at the centre of the labyrinth and claps, a faint squeak similar to that produced when a rubber-soled shoe is scuffed on a wood floor can be heard. The squeak is due to the succession of reflections that return to the centre from each of the concentric rows of bevelled pavers.

Reflection Tones

A musical echo requires a rapid succession of several consecutive reflections, but even a single reflection can produce a clearly audible effect when combined with the original sound. The earliest account of this phenomenon

---

[50] Listen to Bilsen's musical echo at Chantilly: https://fabilsen.home.xs4all.nl/Chantillytrap.mov.

[51] Strutt, J.W., Baron Rayleigh, Sc., F.R.S. (1896) The Theory of Sound, Vol.2. Macmillan, p 453.

[52] In the courtyard that one has to cross to reach the gardens of New College, you can produce a very distinct and impressive flutter echo. It was mentioned by Dr Plot in his 1677 Natural History of Oxford, p 16.

was given in a lecture by Nicolas Savart, brother of the more famous Felix Savart, to the French Academy of Sciences in 1838.

Savart had noticed that "If, while a noise is heard, one approaches a reflecting surface, we can hear a distinct note in the midst of this noise, and the note thus produced varies with the distance between the ear and the reflecting body: its pitch becomes lower when this distance increases and higher when the distance decreases."[53]

He listed several situations in which this effect can be heard: "The sound of a coach travelling on a pavement; the sound of a waterfall, the noise made by steam escaping through a hole with force; a drum roll; the sound of one or more trees whose branches and leaves are agitated by the wind, the noise made by a combination of all the sounds that are in a big city; the sound of the sea, which provides a sound of very remarkable intensity." What all these examples have in common is that they all are sources of broadband sounds.

These days, Savart's phenomenon is variously known as a reflection tone, flanging or comb filtering. In the days of steam trains it was sometimes possible to hear it in the hiss of steam issuing from the boiler of a stationary locomotive. An alert passenger walking up and down a platform would hear the pitch of the hiss rise as he approached the locomotive and fall as he walked away from it.[54] These days, given the right conditions, it is possible to hear the effect in the sound of an aircraft engine as it flies overhead.

Both the hiss of steam and the roar of a jet engine are reasonable approximations of white noise, i.e. a broad spectrum of audible frequencies in which no particular note is dominant. But when the reflection of such a sound is combined with the sound reaching one's ears directly from the source, it can lead to destructive interference of particular frequencies. The absence of those frequencies alters the timbre of the original noise due to the extra time it takes for the reflection to reach the ear compared to the direct sound: as the interval increases, the overall perceived pitch drops and vice versa.[55]

To hear the change in pitch as an aircraft flies overhead you have to alter the distance between your ears and the ground, which you can do by alternately crouching and straightening up when outdoors on a large horizontal surface such as a paved area or a lawn. The pitch rises as you crouch (the distance between your ears and the ground decreases) and drops as you stand up (the distance between your ears and the ground increases). To ensure that you hear

---

[53] M.N.Savart (1838) Quelques faits resultant de la réflexion des ondes sonores. Comptes rendus hebdomadaires des seánces de l'Académie des Science, p 1068–79.

[54] Minnaert, M. (1939) Muzikaal geruis, door interferentie ontstaand. In De natuurkunde vanthe vrije velt, Deel II, Geluid, warmte elektriciteit. W.J. Thieme, Zutphen, p 33.

[55] Frequency of repetition pitch, f given by $f = 1/T$ (T = delay between successive reflections in seconds).

the effect clearly, keep moving up and down while the aircraft is overhead so that the change in pitch occurs several times.

A reflection tone is also audible in the sound from an airplane if you are in an open field or a large paved area such as a parking lot. In this case you don't have to move because as the plane approaches the delay between the direct and reflected sound increases, which in turn causes a perceptible drop in pitch. But as the plane draws away, the pitch rises, and so is easily distinguished from the drop in pitch of the whine of the engine due to a Doppler shift. You should also be able to hear a similar change in pitch in the sound of a moving steam locomotive or a car (Fig. 5.5).

Indeed, as Savart found, with practice the change in pitch is audible in many situations in which broadband sound is reflected. Try the exercise mentioned in chapter one: clench your teeth, draw back your lips as much as possible and hiss. The hiss is composed of a broad range of frequencies from 1000 Hz and 10,000 Hz, in other words a good approximation of white noise. Hold a sheet of card in front of your mouth and move it away slowly until it is at arms length and listen to the change in the tone of the hiss. Moving the card back and forth makes the change more noticeable. You can hear the same thing as you approach or move away from a hissing kettle as it comes to the boil. In fact, as we noted in chapter one, reflection tones are employed by some blind people to sense the presence of obstacles.

Reflection tones can be particularly noticeable and troublesome beneath the sea because in calm conditions underwater sounds are strongly reflected at the interface between water and air. As in air, the reflection combines

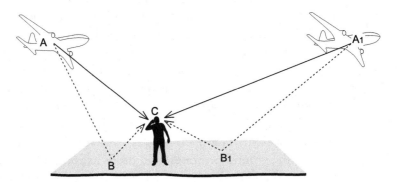

Fig. 5.5 Reflection tones. Sound from the engine reaches the listener directly (AC) and indirectly by reflection (ABC). Interference between direct and reflected sound is responsible for the so-called "comb-filter" effect or Lloyd 's Mirror effect that is heard as a flutter. Path difference becomes smaller as the distance to the sound source increases, i.e. (ABC–AC) is greater than (A_1B_1C–A_1C). Hence what is heard changes continuously as the plane approaches and then recedes

destructively and constructively with the direct sound, which in turn causes a problem for sonar detection. Interestingly, in the field of underwater acoustics the phenomenon is usually referred to as Lloyd's Mirror. Lloyd's Mirror is the name given to an optical experiment performed in 1832 by Humphrey Lloyd, an Irish mathematician and academic, to investigate aspects of the wave nature of light by creating a pattern of light and dark bands, similar to those due to Young's slits.[56]

Echo Tubes and Culvert Whistlers

In science museums that have hands-on exhibits you may come across the somewhat misleadingly named echo tube. This consists of a long narrow tube some 10 to 20 m long and 30 to 40 cm wide that is open at both ends and which can be either straight or coiled (to save space). Stand before one of the open ends and clap your hands once and you will hear an eerie high-pitched "meow" that persists for a second or two. If you listen carefully you will notice that the pitch of the sound drops noticeably towards the end. Resist the impulse to speak into the tube or make any noise other than a clap. Although your voice will echo in the tube, it won't produce the "meow".

That "meow" is due to a rapid succession of reflections of the clap from the far end of the tube. Even though it is open to the air, some sound is reflected back into the tube due to the impedance mismatch between the air within the tube and that outside it. Multiple echoes are produced because the sound of the clap doesn't just travel directly up and down the tube, sound is also reflected over and over again from the sides of the tube so that its path is a zigzag. And, of course, the more times a sound is reflected from the sides, the further it must travel before reaching the far end and returning. You hear the succession of echoes as a prolonged "meow" because the different frequencies that make up the sound of the clap are separated as they make their way along the tube.

The interval between successive reflections is due to the wave properties of sound: not all of the multiple reflections result in an audible echo. Clapping produces a broad range of frequencies and some of these are reinforced when successive reflections remain in step with previous ones as they make their way up and down the tube; those frequencies whose successive reflections are out of step cancel themselves out and are not heard. The mathematical analysis of the progress of sounds that are reflected multiple times as they travel up

---

[56] Thomas Young devised the double slit experiment in 1801 to demonstrate that light has wavelike properties.

and down the tube shows that to produce audible echoes, lower frequencies have to be reflected more often than higher ones. This, in turn, means that the overall distance travelled by lower frequencies is significantly greater than that travelled by higher ones. And because all frequencies travel at the same speed, i.e. at the speed of sound, the higher ones will arrive back at the open end of the tube before the lower ones. Hence the original brief sound of the clap is drawn out into a descending "meow".

Moreover, the lowest pitch that can be echoed in this way is determined by the width of the tube and not by the lowest frequency in the sound of the clap. The reason is that the distance travelled between successive reflections can never be less than the tube diameter. The difference, of course, should be evident if you are able to compare the echo from tubes of different diameters: the wider the tube the lower the final reflected frequency (Fig. 5.6).

The separation of a complex wave into its component frequencies is known as dispersion. We are all familiar with the consequences of the dispersion of white light because it is the cause of the eye-catching spectrum of colours seen in a rainbow or reflections from the surface of a compact disc or DVD. In the case of the rainbow, dispersion occurs because the speed of light through water depends on its frequency. The speed of violet light in water is fractionally greater than that of red light, which causes different colours to follow slightly different paths as they travel through a raindrop. In the case of a CD or a DVD, dispersion is due to constructive and destructive interference, the same process that is the cause of the eerie sound of an echo tube.

You don't have to visit a museum to hear this echo, however. It can also be heard from culverts, pipes and tubes and some boreholes that are long

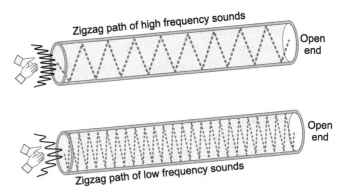

Fig. 5.6 Echo Tube. Low frequency sounds are reflected more often as they travel though the tube and so return to the listener later than high frequency sounds. The result is a glissando that drops in pitch

compared to their width.[57] In fact, the phenomenon first became widely known as the result of a chance encounter. On a May morning in 1970, Frank Crawford, an American physicist, was playing with his children on a beach when his attention was caught by the unusual echoes within a long concrete culvert under the dunes (he estimated that was approximately 1.2 m wide and some 60 m long.) When he clapped his hands at the open end "to my astonishment I heard, not resonances, but a loud descending "zroom" that commenced at high pitch at the same time as the reflected sharp hand-clap and descended within a few tenths of a second to a rather low pitch where it continued to sound for several seconds as it gradually faded away."[58] He christened the phenomenon a "culvert whistler" because of its similarity to the whistles and squeaks that are sometimes picked up by shortwave radios when lighting stokes produce electromagnetic waves that travel back and forth between the magnetic poles of the Earth's magnetic field and which are known as "electromagnetic whistlers". And back at his desk he came up with a mathematical explanation of how the phenomenon comes about.

In fact, as we shall see in chapter six, Crawford's "culvert whistlers" are merely one example of the dispersion of sound, known collectively as "acoustic whistlers". And it seems unlikely that Crawford was the first person to have heard this particular form of acoustic whistler. But it appears that he was first to take notice and have the necessary scientific training to make sense of this eerie sound. A century an half earlier, Biot had concluded his measurement of the speed of sound in an iron pipe by firing a pistol at the open end. He noted that at the far end "The air was discharged out of the pipe with enough force to produce a very strong shock on the hand, to throw more than twenty metres out of the pipe of the light bodies which were put there, and to put out the lights, though they were 950 m away from the place where the shot had gone."[59] Not a word about the acoustic whistler that he would almost certainly have heard a second or so after the pistol shot, even though he had earlier noticed the echo of his voice from the far end. But then, as we have noted several times, it is often very difficult to perceive phenomena with which we are not in some way already familiar. Crawford lived in an age of the wireless and, as a physicist, he knew about electromagnetic whistlers. In Biot's day the battery had only recently been invented and its use was still confined to the laboratory.[60]

[57] Ice borehole sound: https://www.youtube.com/watch?v=q0hYi7Kl1u8 (accessed 12/07/2021).
[58] Crawford, F.S., 1988, Culvert Whistlers. American Journal of Physics, 39, p 610–15.
[59] Biot, M., 1809, "Expériences sur la propagation du son à travers les corps solides et à travers l'air, dans des tuyaux très-alongés", Mémoires de Physique et de Chimie de la Société D'arcueil, p 405–23.
[60] The electric battery was invented by Alexandro Volta in 1800.

Harmonic Echoes

Sound is scattered by large objects in a similar manner to the way in which light is scattered by very small ones. The best-known example of scattering in the case of light is the colour of a clear sky. Scattering by the molecules of the gases that make up the atmosphere cause it to appear overwhelmingly blue because very small particles scatter the shorter wavelengths of sunlight to a greater degree than they do the longer ones. Short wavelengths of light are perceived as violet and blue, the longest ones as red. Scattering by molecules is also responsible for the rosy hues of sunrise and sunset because the light that reaches us when the sun is at the horizon has to pass through a far greater mass of air than when it is high in the sky. As a consequence at sunset sunlight is shorn of much of its shorter wavelengths, wavelengths that are the source of blue skies that lie beyond one's western horizon.[61]

The cause of the sky's colour remained a puzzle until the last decades of the nineteenth century when Rayleigh showed mathematically that it is due to the scattering of light by particles that are comparable in size to the range of wavelengths that make up visible light. Discovery, it is often said, favours the prepared mind, so it is unsurprising that Rayleigh believed that he had discovered the same effect for sound. Having established the role that scattering plays in the colour of a clear sky, he realised a similar effect might be noticeable in the case of a sound when it encounters objects that have dimensions that are close to its wavelength. In the right circumstances it should produce what he called an harmonic echo:

> My attention was first drawn to the subject [of harmonic echoes] by an echo at Bedgeburry Park, the country residence of Mr Beresford Hope. The sound of a woman's voice was returned from a plantation of firs, situated across the valley, with the pitch raised an octave. The phenomenon was unmistakable, though the original sound required to be loud and rather high. With a man's voice we did not succeed in obtaining the effect.[62]

According to Rayleigh, the tree trunks scattered the higher frequencies (i.e. shorter wavelengths) of the woman's voice while the lower frequencies pass between them. Hence a weak echo with a distinctly higher frequency than that of the original sound was heard. A man's voice, he assumed, lacks the high frequencies necessary for such scattering.

---

[61] When the sun is at the horizon the air mass through which its light must pass to reach us is approximately 35 times greater than when it is directly overhead, hence the degree to which short wavelengths are scattered is commensurately greater than when the sun is overhead.

[62] Strutt, J.W., Baron Rayleigh (1873) Harmonic Echoes. Nature, 21 August, p 319–20.

Rayleigh was not only the leading expert on acoustics of his day, he also had a particularly acute ear, so it is very likely that there was something unusual in the echo of the woman's voice. But recently his explanation for the possibility of an apparent increase in the frequency of an echo in terms of preferential scattering of higher frequencies has been questioned. An alternative explanation has been offered based on destructive interference between the reflections from an array of regularly spaced trees.[63] It appears that in Rayleigh's day there was a plantation of regularly spaced trees at Bedgeburry Park. Given that a voiced sound consist of a series of well defined and harmonically related frequencies, depending on the distance between the trees, in certain directions some reflected frequencies should cancel one another out due to destructive interference, so that they are absent in the echo.

This might explain why the echo heard by Rayleigh was an octave higher than the original sound and why the man's voice did not produce a similar effect. His voice would have consisted of a different set of frequencies, which did not lead to destructive interference given the location of Rayleigh and his companions relative to the plantation.

Unfortunately, Rayleigh never followed the matter up; nor, it seems, has anyone else. So the possibility of harmonic echoes has yet to be confirmed. If the more recent explanation of the phenomenon is correct, and should you wish to investigate it yourself, then you should seek out a *regular* array of closely spaced trees of similar girth or, indeed, free standing columns.

In fact, such an array has been built, though its purpose was aesthetic rather than scientific. In 2010 an architect and a sound artist designed and constructed a sound sculpture that consisted of an array of vertical perspex pipes and named it "Organ of Corti". The intention was to make people aware of ambient sounds by comparing the effect on what they heard when inside the array with what they heard when outside the array. "Organ of Corti is an experimental instrument that recycles noise from the environment. It does not make any sound of its own, but rather it attempts to draw our attention to the sounds already present by framing them in a new way."[64] The sculpture was erected in several locations where the ambient sounds were broadband sounds either from distant road traffic or flowing water. People were told to listen to sound that had passed through the array rather than to reflections from the side of the array facing the source. And, indeed people did notice an audible difference because the interactions between the

---

[63] Heller, E.J. (2013) Why You Hear What You Hear. Princeton UP, p 572–77.

[64] *Organ of Corti* (2010–2011) designed by Frances Crow and Francis Prior. See: http://www.liminal.org.uk/organ-of-corti/ (Accessed 4/03/2021).

broadband sound and the array enhances the loudness some frequencies and diminishes those of others due to constructive and destructive interference. It's a pity that the people who created the sculpture did not know of Rayleigh's visit to Bedgeburry Park.

Whispering Galleries

In some buildings there are halls or galleries in which faint sounds can be heard with surprising clarity at a considerable distance from the source. In the seventeenth and eighteenth century such places were known as whispering places. These days they are called whispering galleries. Given the possibility of overhearing secret conversations between plotters or lovers one would expect that they would be of interest to mistrustful princes or jealous spouses. But with one possible exception—the whispering gallery in the Missouri State Capitol building—none has ever been deliberately engineered. Nor is there any suggestion that the Missouri gallery was designed with any nefarious purpose in mind.

The earliest account of a whispering gallery dates from the early decades of the seventeenth century when Francis Bacon wrote that "There is a church at Gloucester, and, as I have heard, the like is in some other places, where, if you speak against a wall softly, another shall hear your voice better a good way off, than near at hand. Inquire more particularly into the frame of the place".[65]

The church is Gloucester Cathedral, and it has been suggested that it was the inspiration for the best-known whispering gallery of them all, that in St Paul's Cathedral in London.[66] However, the whispering gallery at Gloucester Cathedral is simply a narrow corridor on the first floor above the altar, 22.5 m long and approximately 1 m wide, consisting of five straight sections arranged to make half an octagon. A whisper at one end is reflected by each of the angled walls of the corridor so that it can be heard by someone at the far end. St Paul's whispering gallery, on the other hand, is an open circular gallery, 1.5 m wide and a diameter of 34 m (hence its circumference is approximately 110 m), at the base of a tall circular drum that supports the cathedral's dome.

St Paul's was designed and built by Sir Christopher Wren to replace the old St Paul's Cathedral that had been destroyed by the Great Fire of London in 1666. Wren's cathedral was completed in 1710, thirty five years after the

---

[65] Bacon, F. (1627) Sylva Sylvarum Century 2, §148 (p302). In: Bacon, F. (1826) The Works of Francis Bacon, Vol 1. London, p 394.

[66] Another of Gloucester Cathedral's claim to fame is that it was used in the Harry Potter films.

first stone was laid, but exactly when its whispering gallery became one of its attractions is not known for certain. Daniel Defoe mentioned it in passing in his account of his travels around Great Britain published in 1724.

> The whispering place in [Gloucester] Cathedral, has for many Years pass'd for a kind of Wonder; but since, experience has taught us the easily comprehensible Reason of the Thing: And since there is now the like in the Church of St. Pauls, the Wonder is much abated.

A 1741 guide book to the cathedral informed the visitor that "the greatest curiosity of all is the whispering-place; where leaning your head against the wall, you may easily hear all that is said, though it be ever so low, and at the most distant place from you in the gallery: which affords great matter of surprise and innocent diversion to all young persons who come to amuse themselves with this curiosity."[67] The wording suggests that people were whispering to one another years before the guide was written.[68] Moreover, for a fee of tuppence the gallery's warden would direct solitary visitors to the far side and utter a whisper for them to hear. These days there is still a warden, but you must provide your own whisperer or listen to someone else's.

Wren based the design of his dome on that of St Peter's Basilica in Rome, which also has a circular gallery at its base. St Peter's was completed almost a century before St Paul's and its gallery was known to be a whispering gallery in the seventeenth century. A description of its unusual acoustics together with a somewhat questionable explanation of its cause was given by Athanasius Kircher in his *Musurgia Universallis*, which was published in 1673 and which drew on some thirty years of studying buildings with unusual acoustics.[69] So the possibility that the circular gallery in St Paul's would also convey faint sounds from one side of the gallery to the other would have been known in principle from the outset, even though it was not until much later that its unusual acoustics were mentioned in guide books to the cathedral. There is, disappointingly, no direct evidence that Wren was aware of this unusual acoustic property of his gallery, though as an active member of the

---

[67] Boreman, T. (1741) The History and Description of the Famous Cathedral of St Paul's. London, ch 10.

[68] Defoe, D. (1724) Tour thro' the Whole Island of Great Britain, Letter 6, Part 2: Oxford Bristol and Gloucester.

[69] Kircher, A. (1650) Musurgia universalis sive Ars Magna Consoni et Dissoni, Vol II, Book IX. Rome, p 275–76.

Royal Society he would almost certainly have know about Gloucester Cathedral's whispering gallery.[70] His father, Dean Christopher Wren, possesed a copy of Bacon's Sylvia Silvarum and had made a note beside the account of Gloucester's "whispering place", which his son may have seen.

Despite Bacon's exhortation to "Inquire more particularly into the frame of the place", no serious attempt to explain the acoustics of the gallery at St Paul's was made until well into the nineteenth century when George Biddell Airy, who was Astronomer Royal at the time, concluded without any evidence that the dome acts as a curved mirror that focuses sound from one side of the gallery to the other.[71] Kircher's much earlier explanation of the acoustics of the whispering gallery in St Peter's was closer to the mark because he identified the curved wall of gallery and not the dome as the cause. However, Kircher insisted that a whisper could be heard only by someone diametrically across the gallery from the speaker. In 1828, John Herschel, one of the most illustrious astronomers and natural philosophers of his day, and who should have been able to improve on Kircher's explanation, not least because he actually visited the gallery, wrote "In the whispering Gallery of St Paul's, London the faintest sound is faithfully conveyed from one side to the other of the dome, but is not heard at any intermediate point."[72] Herschel's description is still in circulation today even though it was long ago shown to be incorrect.[73]

The secret of the whispers that had for so long surprised and delighted visitors to St Paul's was eventually discovered by Rayleigh when he visited the gallery in 1878 and found that it is not necessary to be directly opposite a source of a whisper in order to hear it. It is, in fact, possible to hear a whisper anywhere along the gallery as long as one's ear is close to the wall. He realised that this implied that sound from the speaker must make its way around the wall of the gallery by a series of reflections: "The whisper seems to creep round [the wall of the gallery] horizontally…".[74] Further investigations revealed several other important facts. In the first place, the speaker should direct the whisper parallel to the wall in the direction of the listener. Second, the whisper is audible only to someone very close to the wall. This is the

[70] On 5 November, 1662, "Mr Powle's description of the whispering-place in the cathedral of Gloucester was brought in by Mr OLDENBURG, and read" at a meeting of the Royal Society. In: Birch, T. (1757) The History of the Royal Society, London, Vol. 1, p 120–23.

[71] Airy, G. B. (1871) On Sound and Atmospheric Vibrations, with the Mathematical Elements of Music. Macmillan and Co, p 145.

[72] Herschel, J.F.W. (1830) Sound, art 32. In: Smedly, E. (ed) (1830) Encyclopaedia Metropolitana, Vol II. Baldwin and Cradock, London, p 752.

[73] Sir John Herschel was the only son of the astronomer, William Herschel.

[74] Strutt, J.W., Baron Rayleigh, Sc., F.R.S. (1896) The Theory of Sound, Vol. 2. Macmillan, p 127.

reason why a faint sound can be heard clearly even by someone on the far side of the gallery: being confined to a shallow band close to the wall of the gallery, sound does not spread out and diminish in intensity as it would if it was propagated across the open space at the centre of the gallery, where it would, of course, spread out in all directions and become inaudible.

In addition to its circularity, there are three other features that make St Paul's whispering gallery so effective: a stone seat runs around the base of gallery wall, the wall of the gallery slopes inward slightly and is topped by a lip. The seat and lip prevent sound from spreading vertically, which maintains its intensity as it makes its way around the gallery. In the case of the gallery of Missouri State Capital, the wall of the gallery is slightly concave which is even more effective in confining sound to a narrow beam (Fig. 5.7).

But what causes a whisper to be confined to a shallow layer no more than a few centimetres deep? As Rayleigh and others later realised, a full explanation of a whispering gallery requires that the wave properties of sound be taken into consideration. Rayleigh pointed out that, being a wave, due to diffraction sound will spread out as it passes into the air through a narrow opening such as an open mouth. As we saw in the last chapter, the degree to which sound spreads beyond an opening depends on its wavelength: lower frequencies spread out more than higher ones. A whisper, which consists primarily of frequencies between 1000 and 3000 Hz, is thus confined to a comparatively narrow cone as it issues forth from an open mouth. This sound is confined to a shallow layer close to the wall as it makes its way around the gallery. The layer is deeper for lower frequencies such as those of voiced speech.[75] This explains why a whispering gallery does not work quite as well for voiced sound: the energy of voiced speech is spread over a larger volume and so its intensity is diminished to a greater degree than that of a whisper as it travels around the gallery wall. In any case, a whisper is a much more effective and arresting demonstration of the unusual acoustics of a whispering gallery than voiced speech because one doesn't expect to hear a whisper at such a distance. Under ideal conditions, when only two people are whispering to one another, you can hear the whispered words so distinctly that it can seem that the whisperer is standing next to you, not several tens on metres away.

There are other subtle features due to the effects of interference that can be heard in a whispering gallery, though only under carefully controlled conditions. In 1920, Sir C.V. Rahman, the distinguished Indian physicist and Nobel laureate, made a systematic study of whispering galleries and discovered that when a faint sound is confined to a single frequency (say a muted

---

[75] You have only to compare the sound of the question "Can you hear me?" when voiced and when whispered to notice how different their timbres are.

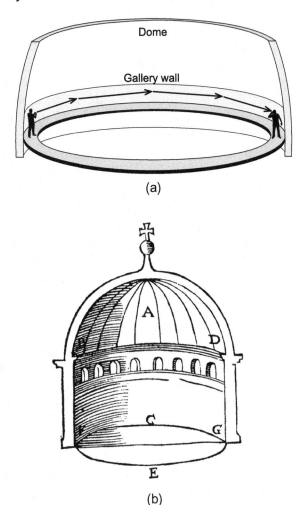

Fig. 5.7 Whispering Galleries. **a** Sounds are reflected around the surface of the gallery's wall. **b** Earliest image of a whispering gallery (St Peter's, Rome).[76]

whistle) then regions of audibility interspersed by regions of silence along the circumference of the gallery are set up. He suggested that this is due to alternating regions of constructive and destructive interference between waves travelling in opposite directions around the wall of the gallery.[77]

---

[76] Kircher, A. (1673) Phonurgia nova, sive conjugium mechanico-physicum artis & naturae paranympha phonosophia. Campidonae, p 69.

[77] Raman, C.V., Sutherland, G.A. (1921) On the Whispering-Gallery Phenomenon. Proc. Roy. Soc. A, 100, p 424–28.

Rahman also found that the acoustic surprises of these galleries are not confined to whispers. He drew attention to what appears to be multiple echo that can be heard when someone claps loudly in a whispering gallery: a single handclap will repeat several times. The repetition is not due to reflection from the dome but to the sound of the clap travelling several times around the wall of the gallery until it becomes too faint to be heard. The person who has clapped hears it every time it passes him, while another person standing in the gallery directly opposite the clapper hears two "echoes" that reach him from opposite directions.

Rahman discovered and investigated several whispering galleries in India, among which are the galleries at the base of the dome of the Gol Gumbaz at Bijapur (completed 1656), The General Post Office (1864) and the Victoria Memorial (1921), both of which are in Kolkata (formerly Calcutta).[78] In fact, there are whispering galleries all over the world: an internet search will throw up several dozen, though they are not all of the same type as that in St Paul's.

Nor is it necessary for a whispering gallery to be within a building. The circular wall that surrounds the octagonal tower known as the Imperial Vault of Heaven in Tiantan Park, Beijing, is well-known as a whispering gallery; the height of the wall is 3.72 m and its circumference is 193.2 m. Unfortunately, a recently installed railing prevents visitors from approaching the wall closely, which greatly reduces its effectiveness as a whispering gallery.

Neither does the reflecting wall have to be a closed circle. Any smooth, curved vertical surface will do, and the best examples are sometimes known as whispering walls. The parabolic surface of the barrage of the Barossa Reservoir in South Australia makes it possible for a whisper to be heard across a distance of 140 m. And in the absence of crowds of visitors and the occasional acoustic installation that masks the sound of a whisper, the curved wall of the central semicircular oil tank in the basement of the Tate Modern art gallery in London is very effective whispering gallery.

It is also possible for people sitting at opposite ends of the same step of a semicircular Greek or Roman theatre to communicate by whispering against the vertical riser of the step. A whispered conversation can be held across a distance of more than 80 m in the Roman theatre at Orange in Southern France. And there are any number of so-called whispering benches, i.e. curved stone benches that have a high, smooth backs. The "Whisper Bench" in New York's Central Park was built in 1936, and there's a very long whispering bench in West Fairmount Park, Philadelphia, USA. The Alameda Park in Santiago de Compostela, Spain has a "banco acústico", the Spanish

---

[78] Raman, C.V. (1922) On Whispering Galleries. Bull. Indian Association for the Cultivation of Science, 7, p 159–172.

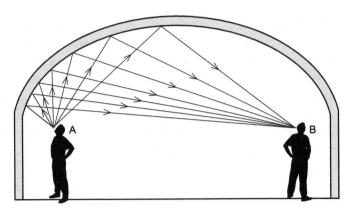

Fig. 5.8 Kircher's ellipsoidal ceiling. The heads of speaker A and listener B are at the respective focuses of the ellipse

for whispering bench. It is known, perhaps inevitably, as the "banco de los enamorados de Santiago de Compostela".[79] Of course, the performance of open-air whispering walls and benches are affected by ambient sounds. High levels of ambient noise due to people, wind and rain will hamper a whispered conversation, as will someone sitting on the bench between the whisperers.

Not all whispering galleries work in the same way as the one in St Paul's. Kircher described a simpler form of whispering gallery in which two people at opposite ends of a room could communicate by reflecting their voices from its ceiling.[80] He suggested that if the surface of the ceiling is ellipsoidal, then by standing at the opposing foci of the ellipse they would be able to communicate in whispers. In fact, as long as the ceiling is curved, all you have to do is stand close to the wall with your back to the room and your whispers will be audible to someone on the far side who also faces the wall; it is not necessary to stand at one of the focal points of the ceiling, even supposing it has an elliptical crossesection. Although no such room existed in his day, a working example is to be found in the "Salle de l'Echo" at the Abbaye de la Chaise-Dieu in the Haute Loire town of the same name (Fig. 5.8).

There is another such whispering gallery in the Statuary Hall in the Capitol at Washington, DC. This used to be particularly effective until it was renovated following a fire in 1901. The ceiling of the hall, which was originally made of wood and covered in paper so that it surface was perfectly smooth, was rebuilt in steel and plaster with recessed panels and ribs. As a result, it is now a far less effective reflector of sound than it was, though it is still possible

---

[79] "enamorados" is Spanish for "lovers".

[80] Kircher, A. (1650) Musurgia universalis sive Ars Magna Consoni et Dissoni, Vol II, Book IX. Rome, p 300–1.

to whisper from one side to the other by standing in the right spot. The guides in the Hall are only too happy to lay on a demonstration for visitors.

Nor is it necessary to be within a room as you will find by visiting the vaulted concourse in Grand Central Station, New York, and the whispering arches of the Loggia dei Mercanti that forms the ground floor of Palazzo della Ragione in Milan.

The Ear of Dionysius, a large cave in Syracuse, Sicily, is mistakenly said to be a whispering gallery, but it is really more like a gigantic ear trumpet. The mouth of the cave is huge and shaped like a tall, irregular triangle, the base of which is a fraction of its height. The cave narrows as it winds into the rock. At the far end there is a small opening through which it is possible to hear sounds within the cave. The idea is that conversations within the cave, which was once used as a dungeon, could be overheard by the jailers. The reality is that reverberation within the cave makes it almost impossible to hear individual sounds distinctly.

A final word of advice: there is no point in visiting a whispering gallery (or siting on a whispering bench) on your own: you will need a companion if you are to explore its unusual acoustics. Take a friend or two and spend a while trying out different sounds and positions along the wall, listening to the results and comparing notes. What is the result when more than two people whisper at the same time?

Echolocation in War and Peace

We saw in the previous chapter that Leonardo da Vinci noticed that it is possible to hear sounds made by a ship under sail by inserting one end of a hollow tube in water and putting one's ear to the other. Can this cursory observation be added to the long list of those of his ideas that finally bore fruit in the twentieth century? If so, then to airplanes, helicopters and tanks we might add sonar, which was developed during the twentieth century as a countermeasure against submarines. But to do so would be a bit of a stretch because there is no evidence that the engineers and physicists responsible for developing sonar were aware of his discovery.

Sonar, an acronym coined by American scientists during WW2 as the acoustic equivalent to "radar", stands for "SOund Navigation And Ranging". Sonar can be passive or active. Passive sonar depends on listening for the sounds made by a moving vessel, be it a submarine or a surface ship. Leonardo's tube was an early example of passive sonar. The technique was revived during WW1 by British, French and American scientists in search of ways of countering the U-boat threat.

Indeed, until the outbreak of WW1 very little was known about underwater acoustics. And had it not been for the submarine, it is possible that there would have been little incentive at that time to investigate the properties of underwater sounds. In fact, the need to discover methods to detect the sounds of submerged submarines led to the formation of the earliest government funded scientific research teams in Britain. This was the Admiralty Board of Invention and Research (BIR), which was set up in 1915. One of the men appointed to its panel of scientists was Ernest Rutherford, who was asked to set aside his research into atoms and radioactivity to work on methods of submarine detection. He concluded that the best method available at the time was to listen to the sounds made by a submarine.

Initially some of the research conducted by these highly trained physicists and engineers were almost as crude as those of Abbe Nollet and Benjamin Franklin. In 1916, Sir Richard Paget, an eminent barrister and an accomplished musician, suggested to Rutherford that if propeller noise was going to be used to locate enemy submarines it was necessary to establish the acoustic signature of their propellers. Paget tackled the problem head on: he arranged to be rowed out to sea in a small boat while a submerged Royal Navy submarine circled around it at some distance. With a sailor holding his legs, he leaned over the side of the boat, submerged his head and listened. After a short while he sat up and sang a scale of notes until he found one that matched that of the dominant pitch of submarine's propeller. Whether this particular piece of research advanced the development of anti submarine warfare is not recorded.[81]

The drawback of Leonardo's tube is that it gives no indication of the direction of the source. Moreover, it depends on sounds made by the target vessel, so it can't detect a stationary vessel if its engines are switched off. In fact, the major source of sound produced by a propeller-driven vessel is the propeller itself because as it rotates it produces bubbles through a process known as cavitation, a phenomenon we shall consider in chapter six. As we noted in chapter four, bubbly water is potentially very noisy. A further limitation of passive sonar is that the propeller sounds of the attacking vessel can mask those of the submarine and so it too must be stationary if it is to detect the enemy. A stationary vessel is, of course, a sitting duck.

Early versions of passive sonar employed two metal boxes attached to the ship's hull and known as "Broca tubes", each of which was separately connected to the operator's ears by rubber tubing.[82] Like the geophones used by the military tunnelers of the Western Front, the pair of Broca tubes was rotated until the intensity of sound in each tube heard by the operator was the same, which under ideal conditions gave an approximate indication of where the sound was coming from. But these tubes were soon replaced by electrically driven microphones adapted to work under water and known as hydrophones. These early hydrophones were far more sensitive to faint underwater sounds and, more importantly, could determine the direction of the source because one side of the hydrophone was covered by a metal baffle. The baffle prevented sound from reaching the hydrophone from that side.[83] During WW1, hydrophones were installed in both surface ships and submarines, and under favourable conditions were able to pick up the sound of a moving vessel at distances up to 80 km.

Active sonar is based on echolocation. Locating objects underwater using echolocation was first proposed in 1912 by Hiram Maxim, an American who invented the modern machine gun, following the sinking of the Titanic. He suggested that the reflection of low frequency pulses of sound could be used to find the submerged portion of an iceberg. The first device capable of doing this was demonstrated in 1914, just before the outbreak of WW1; it detected an iceberg at a distance of 2 km. Unfortunately, this particular device operated at 1000 Hz, which made it unsuitable for direction finding because a low frequency signal cannot be confined to a narrow beam.[84] Hence the later military version of active sonar employed a brief but powerful pulse of high frequency sound emitted from the ship searching for a submarine. Reflections from the submarine's hull enabled a skilled operator of the apparatus to determine its distance and position relative to the attacking ship.

Active sonar was pioneered principally by the British, who called the method ASDIC, an acronym that is said to stand for Allied Submarine Detection Investigation Committee, though no one seems to know the actual source of the name[85] The device itself was invented in 1916 by a French physicist, Paul Langevin, and based on his earlier discovery that a quartz crystal subjected to a high frequency alternating current will vibrate at ultrasonic frequencies.[86] Ultrasound is far more strongly absorbed by water than audible frequencies but it has the advantage that it is highly directional due

---

[82] Manstan, R. R. (2018) The Listeners: U-boat Hunters During the Great War. Wesleyan University Press, p 107.

[83] Bragg, Sir W. (1922) The World of Sound. G.Bell & Sons Ltd, p 175–6.

[84] This device was invented by R. A. Fessenden, a Canadian inventor.

[85] After WW2, ASDIC was renamed *sonar*.

[86] This was the first practical application of the piezoelectric effect discovered by Pierre Curie and his brother, Paul-Jacques Curie, in 1880. Langevin was one of Pierre Curie's PhD students.

to its high frequency and so can be used to pin-point a target.[87] However, the war ended before the device was fully ready for use and so its effectiveness in active service was never properly tested.

Research and development of ASDIC continued during the interwar years, and was considered by the Royal Navy to have tipped the odds against the submarine as an offensive weapon. So confident were the British that ASDIC could locate a submerged submarine that they ignored the painfully learned lessons of WW1, foremost of which was that the German navy—the Kriegsmarine—used their U-boats primarily not to attack warships but to sink merchant ships with the aim of starving Britain into surrender. As long as merchant ships sailed independently of one another, they were easy prey because there were never enough warships to escort them. The Royal Navy eventually accepted that sailing merchant ships in convoys protected by escort vessels, as had been the practice during WW1, was the most effective way to deal with the threat posed by German submarines. A convoy was difficult to locate in the vastness of the Atlantic, and, if found, its escort vessels could offer immediate, concentrated protection. But such was the Royal Navy's initial faith in ASDIC that this lesson had to be learned afresh during what came to known as the Battle of the Atlantic. In any case, as in WW1, U-boats preferred to attack at night while on the surface, circumstances in which ASDIC was of little use.[88] It came into its own only after a U-boat had made its attack and was making its escape underwater.

Nevertheless, along with hydrophones, ASDIC was the only effective method available to the allied navies to detect and locate submerged submarines, and by the outbreak of WW2 in 1939 the Royal Navy had some 160 escort vessels equipped with ASDIC, far too few to cope with the initial U-boat threat as things turned out. By the end of the war the numbers of ASDIC equipped vessels had increased to several hundred. By itself, ASDIC was never as effective as the British or, indeed, some German submariners believed it to be. The problem was the difficulty of maintaining ASDIC contact with a moving submarine and using it to judge the best moment to drop depth charges. On occasions an antisubmarine vessel might spend up to 30 h pursuing a contact without making a "kill". And when an attack did not succeed in destroying the submarine, which was the usual outcome, the

---

[87] Ultrasonic frequencies had to be converted into audible frequencies using a process invented by Fessenden in 1901, and known as "heterodyning".

[88] U-boats were not true submarines because they had limited underwater range due to the need to surface and recharge their batteries using their diesel engines. They were really submergible torpedo boats. However, by 1945, German engineers had begun to build the first generation of true submarines, i.e. vessels that could remain submerged more or less indefinitely. These were designated Type XXI Electroboot.

underwater turbulence due to the explosions rendered ASDIC temporarily ineffective, giving an experienced U-boat skipper time to make his escape. Despite this, Karl Dönitz, who commanded the U-boat fleet throughout WW2, considered it necessary to constantly reassure his crews that ASDIC did not signal the end of the submarine as an effective weapon.

The Germans devised several countermeasures to sonar in its various forms. One of these was the "Pillenwerfer", which the British called a "Submarine Bubble Target" or SBT. This was a small canister filled with calcium hydride that was released from a submerged submarine under attack. The calcium hydride reacted with seawater to create a huge plume of hydrogen bubbles that would reflect ASDIC's ultrasound pulses and confuse its operator. The weak point of the device was that the plume of bubbles remained stationary. Experienced operators had already learned to listen for a Doppler shift in the return echo of a submarine, which was the tell tale sign that is was on the move. There was no Doppler shift from the plume of bubbles created by the "Pillenwerfer".

Clearly, successful deployment of sonar depended as much on the skill of the operator as on the apparatus itself, particularly so where hydrophones are concerned. Much to the surprise and consternation of both submariners and sailors, the sea proved to be a very noisy environment. Operators had to be trained to identify and distinguish the cavitation sounds of propellers from the songs of whales, crashing waves and rain-lashed seas. They also had to learn how to tell the difference between the echo from steel hull of a submarine and that from a fleshy flank of a whale and to employ the Doppler shift in the return echo to anticipate the movements of the target.

For all that, the ocean continued to spring acoustic surprises. On one occasion an American submarine was cruising underwater in the Macassar Straights between Borneo and the Celebes, when its hydrophones picked up an unidentified continuous crackling, which the crew assumed was due to a hitherto unknown anti-submarine weapon deployed by Japanese Navy to mask the sounds of its ships. It turned out to be the sound of a vast multitude of snapping shrimps, whose powerful claws produce an astonishing 180 dB or more when they snap shut. The purpose of this action is to create a shock wave that stuns its prey. So much for the hitherto naively held belief that "There is no sound, no echo of sound, in the deserts of the deep".[89]

At the same time, refraction could play havoc with both active and passive sonar. As we saw in the last chapter, sound is strongly refracted as it travels through layers of water at different temperatures. As with refraction in the

---

[89] Kipling, R. (1893) The Deep-Sea Cables.

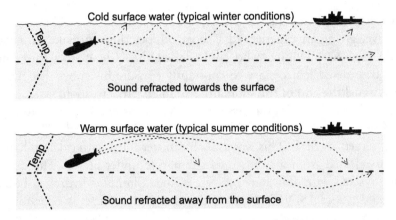

Fig. 5.9 Anti submarine warfare. The dotted lines show how the sounds from a submarine's propellers are refracted as they travel through water. The typical temperature profile beneath the sea's surface differs from summer to winter as shown by the dotted lines on the right of each figure

atmosphere, underwater refraction occurs in the vertical plane, which made it very difficult to get an accurate fix on the depth of a submarine using ASDIC. Underwater sound is also almost entirely reflected at the undersurface of the sea, or indeed any body of water be it a lake or a bathtub. The combination of reflection and refraction allows sound to travel considerable distances in the layer of the sea just below the surface. Refraction can also create sound shadows, i.e. regions where none of the sounds from a submarine reach the hydrophones of the attacking vessel. However, German submariners were seldom able to exploit these effects to their advantage because U-boats did not carry instruments that measured the temperature of the sea at different depths and so their crews were largely unaware of layers that might render their vessel inaudible to surface ships.[90] It was, however, well known by both sides that it was possible to confuse ASDIC in shallow water by descending to the bottom and remaining stationary. The operator would be unable to distinguish the echo from a submarine from that of a wreck (Fig. 5.9).

In fact, the most effective countermeasure to U-boats proved to be electromagnetic rather than acoustic. The greatest vulnerability of the submarines of that era was that they spent much of their mission on the surface, either sailing to and from their action stations or because they had to use their diesel engines to recharge their batteries. This enabled aircraft fitted with radar to locate and attack them. In the final year of the war the Germans

---

[90] The instrument, known as a bathythermograph, was fitted to WW2 US submarines, which operated against the Japanese in the Pacific Ocean. American submariners achieved what the Germans could only dream about: the almost complete destruction of the enemy's merchant fleet.

began to deploy a new design of U-boat that was much faster underwater and didn't have to surface to recharge its batteries because it could take in air through a tube known as a snorkel, allowing its diesel engines to be used while submerged. In addition it was coated in a rubber membrane intended to absorb sonar signals.[91] By then, however, the end was nigh and so the new phase in submarine warfare based on ultra quiet vessels that could remain submerged for days if not weeks was delayed until the Cold War was well under way.

Sonar is, if anything, more useful to post war navies than to those of first half of the twentieth century because technical developments have refined its capabilities. But the cat-and-mouse game between surface ships and submarines of WW2 is now played out between submarines. And for all the technical advances since WW2, tactical advantage often depends on a captain's skill and experience. Describing an encounter between the Royal Navy hunter-killer submarine HMS Warspite and a missile-carrying Soviet Union submarine in the late 1960's, Warspite's captain Sandy Woodward related that he realised that he could evade the Soviet submarine that was tracking him with its sonar by diving below the layer within which he was traveling.[92]

Sonar found other uses after WW2. Those occasional sonar encounters with whales during WW2 gave someone the idea that sonar could be of use to the fishing industry. There was a post war glut of relatively cheap sonar devices and it was found that not only can sonar locate a shoal of fish, the species can also be determined from the return echo. The strongest echoes are due to air-filled swim bladders and because these differ from one species to another, it is possible to distinguish between cod, pollock, mullet and sea bass. As with the bubbles of the Pillenwerfer, each of the multitude of swim bladders in a shoal of fish reflects sound due to the huge difference in acoustic impedance between seawater and the gas inside the bladder.

Passive sonar was also used during WW1 as a method of detecting enemy aircraft. Mobile sound locators were used by all armies on the Western Front to locate and track aircraft. They were in essence an array of large movable listening trumpets that made it possible to pick up and track the sound of the engine of an aircraft or airship. The British also constructed several stationary sound mirrors along the Northeast English coast to provide an early warning system to detect Zeppelins and Gotha bombers. These sound mirrors consisted of a huge vertical concrete slab with a shallow, circular

[91] This U boat version was designated as Type XXI.

[92] Hennessy, P., Jinks, J., 2015, The Silent Deep, The Royal Navy Submarine Service since 1945. Allen Lane, p 320.

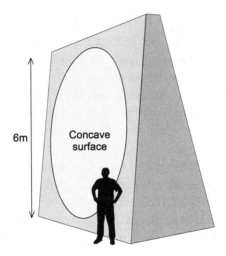

Fig. 5.10 Typical military sound mirror built by the British to pick up the noise of enemy aircraft approaching the coast

concave depression with a diameter of four or five metres on one face. The concave surface focussed the engine sounds of approaching enemy aircraft on the mouth of a horn which was connected to a stethoscope plugged into an observer's ears. The mirrors were not very effective; at best they gave only a few minutes warning. In any case, observers found it difficult to hear the low frequency engine sounds, so reliable detection of enemy aircraft was something of a hit and miss affair. Added to this, British air defences at the time were pitifully inadequate because few fighter planes or anti aircraft guns were available to attack and shoot down the enemy (Fig. 5.10).

The problem with a sound mirror is that is has to be very large to pick up the faint, low frequency sounds of a distant aircraft. Undeterred by the poor performance of these mirrors during the war, a determined band of scientists and engineers led by W.S. Tucker, a physicist who had worked on methods of locating enemy artillery using the sound of gunfire, somehow managed to persuade the War Office to subsidise research on improving their performance. And at Denge, near Dungeness on the Kent coast there are three surviving examples of these experimental military sound mirrors constructed between 1920 and 1930. The largest of these is a curved wall, 60 m long and 8 m high, with its concave side facing the English Channel. Under favourable atmospheric conditions it could detect an aircraft at a distance of 40 km, which, given the speed at which planes could fly in those days, provided at best a ten minute warning.[93] The engine sounds were picked up

---

[93] Scarth, R.N. (2017) Echoes from the Sky. Independent Books.

using an array of a new type of microphone capable of registering very low frequency sounds that Tucker had invented in 1916 to detect the sound of distant artillery.

The plan to build a chain of sound mirrors along the Kent Coast between Dover and Dungeness was abandoned soon after the large mirror at Denge was completed because the increased speed of aircraft made the early warning provided by the mirrors far too brief to be of use. Fortunately for Britain, radar, which was developed a decade later and employed high frequency radio waves, proved to be a far more effective method for locating enemy aircraft as, indeed, it proved to be for U-boats.

The War Office ordered that the mirrors at Denge be destroyed, though the order was never carried out. Instead, the mirrors fell into disrepair and might have collapsed into the water-filled gravel pit that now surrounds them. But in 2003 they were restored and are now open to the public a couple of times a year.[94] Incidentally, the wall of large sound mirror at Denge makes an excellent whispering wall, something that visitors quickly discover to their delight. Given Athanasius Kircher's interest in the unusual acoustics of buildings, he would have been delighted by these huge structures dedicated to hearing distant sounds.

Kircher, as we have seen, knew a great deal about the acoustic properties of curved surfaces. And in *Musurgia Universallis* he explained how two facing concave parabolic surfaces would enable two people to converse in a low voice with one another at a distance.[95] Possibly inspired by Kircher, in recent years some public parks and science museums have installed what have come to be called "whispering dishes" or "listening vessels".[96] These are pairs of large parabolic reflectors made of steel or concrete placed tens of metres apart with their concave surfaces facing each other. Their purpose is to enable two people to hold a conversation *sotto voce* well beyond normal earshot of one another. Sounds produced by someone at the focal point of one of the dishes is gathered by its concave surface and reflected towards the other dish. The parabolic surface of the receiving dish concentrates these sounds at its focal point. To hear them distinctly, you must place your ear at or very near that point. Some whispering dishes include a metal frame against which you place either your mouth or ear so you don't have to find the focal point by trial and error. Some concrete dishes have a bench built into the parabola so

---

[94] For a history of military sound mirrors and details of where and when to visit them (as well as non military sound mirrors) see http://www.andrewgrantham.co.uk/soundmirrors/ (accessed 27/07/2021).
[95] Kircher, A. (1650) Musurgia universalis sive Ars Magna Consoni et Dissoni, Vol II, Book IX. Rome, p 298.
[96] Online search for "whispering dishes" or "whisper dishes" will come up with several examples.

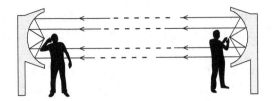

Fig. 5.11 Whispering dishes (not to scale)

that a person can sit within the concave hollow and listen to the soundscape. A solitary sound mirror faces the sea on the beach at Wijk aan Zee, a few kilometres west of Amsterdam. It is intended as an artwork and the idea is sit on a seat in front of it and listen to the surf and wind (Fig. 5.11).[97]

Reverberation

We all notice how different an empty room sounds compared to one that is furnished. That distinctive ring of emptiness is due to sounds that are reflected over and over again from its ceiling, floor and walls. But in an enclosed space it is usually not possible to distinguish the original sound from its reflection because the interval between them is too small for the ear to hear them separately. Recall that in most circumstances a distinct echo requires that the interval between the original sound and its reflection to be at least 1/10 s, which is possible only if the reflecting surface is more than 17 m from the sound source.

Distinct echoes are very unwelcome in enclosed spaces intended for human use. A notorious example of just how undesirable they are became evident during the opening ceremony of the Royal Albert Hall in London in 1871. Someone who attended the event wrote: "When the Prince of Wales read his address I heard every word repeated with perfect distinctness, the echo was pure and single, the two voices appeared like those of a prompter and a faithfully repeating speaker. The echo was remarkably well defined, and nearly as loud as the voice of the prince…[and] Every note of Madame Sherrington's solo was most vexatiously mocked."[98] It was jokingly said to be the only venue where a composer could be sure of hearing his work twice. It took a century to reduce the problem to an acceptable level, though it is still not

[97] Abri at Wijk aan Zee: https://www.artatsite.com/Nederland/details/Hooykaas-Madelon-Elsa-Stansfield-Abri-Aertszweg-Wijk-aan-Zee-ArtAtSite.html (accessed 27/07/2021).
[98] Williams, W.M. (1871) Nature, 3, p 469.

satisfactorily corrected, and probably never can be, given the dimensions and design of the auditorium.[99]

In an enclosed space where the delay between a sound and its reflection is significantly less than 1/10 s, multiple reflections will amplify and prolong the sound to a greater or lesser degree depending on its loudness and the degree to which it is absorbed by the surfaces it encounters. And the more often it is reflected, the longer a sound persists, albeit with diminishing loudness. There is, incidentally, an instructive optical parallel: a hall of mirrors in which we see endless reflections receding into the distance. The resulting visual confusion offers an insight into the auditory chaos wrought by multiple reflections of a sound.

The aggregate effect of multiple reflections of a sound separated by very short intervals is known as reverberation. Depending on the dimensions of an enclosed space, a reverberation that lasts more than a fraction of a second can make talking or playing music all but impossible because every syllable and note is drawn out, masking successive sounds. However, reverberation is not altogether undesirable. Indeed, as you will immediately notice within an anechoic chamber or in the middle of an empty field far from buildings and vegetation, the absence of reflections robs speech and music of depth and fullness. The secret of good room acoustics is to strike the right balance between reflection and absorption.

Sounds can't be reflected indefinitely because some of their energy is absorbed during each reflection, so sooner or later the reflections become inaudible. The degree of absorption of sound varies from one material to another, just as a room with black walls is far darker than one with white walls even when both are illuminated by light of the same colour and intensity. Hard surfaces such as stone brick or concrete are poor acoustic absorbers and will reflect a sound over and over again. Soft, porous surfaces like curtains, carpets, cushions and people are excellent absorbers, which is why there is little reverberation in a furnished room.

The acoustics of a room or hall is sometimes described as "dry" or "wet". In this context dry means that the sound is absorbed very quickly and wet that it persists (i.e. reverberates) for a while. An anechoic chamber is extremely dry whereas a large room with walls of stone or concrete is likely to be extremely wet. Sound absorption, however, doesn't just depend on the nature of the reflecting material, it also depends on frequency. Stone, brick and concrete are poor absorbers—i.e. excellent reflectors—of all frequencies, but carpets,

---

[99] The troublesome echo was finally brought under control in 1969 by suspending a large number of curved fibreglass discs from the ceiling. These intercept and scatter sound in all directions before they reach the top of the dome.

curtains and cushions absorb high frequencies much more effectively than low ones. This differential absorption is easily noticed: if high frequencies are absorbed, a room will sound hollow or booming; if low frequencies are absorbed it will sound piercing, sharp, or shrill.

Reverberation has another consequence: it increases the loudness of the original sound. If there are no reflections, then the loudness is simply that of the direct sound, which is less than that of the direct sound plus its reflections. This is why sounds must be louder if they are to be heard as distinctly in the open or in an anechoic chamber as they would within an enclosed space such as a room. In a room, sounds reach the ear both directly and by reflection from the walls, floor and ceiling. The overall loudness of a sound in that situation is the sum of sound reaching the ear directly and that reaching it after being reflected.

In a very reverberant space, those reflections reach you from every direction preventing you from being able to localise the source. In such an environment you feel completely immersed in sound in a way that can't be convincingly reproduced with loudspeakers. Should you find yourself alone in a large hall or church with walls, floors and ceiling of stone shout loudly or clap to experience what can only be described as a sound bath. Close your eyes and ignore the gradual reduction in loudness and concentrate on how you are enveloped in sound. You could say that you are immersed in the acoustic equivalent of a fog. Just as an optical fog prevents you from seeing anything beyond arms length an acoustic fog stops you hearing sounds clearly unless you are very close to their source.[100]

Reverberation is of particular interest to architects because it has such a marked effect on the acoustics of enclosed spaces. The first person to tackle the subject was the Roman architect, Vitruvius, who pointed out that reflection can interfere with the acoustics of a theatre. If a speaker is to be clearly understood, he said, his voice must reach the ears of the audience without reflections, otherwise "the case-endings are not heard, and it dies away there in sounds of indistinct meaning. The resonant are those in which it comes into contact with some solid substance and recoils, thus producing an echo, and making the terminations of cases sound double."[101] This is especially problematic for inflected languages such as Greek and Latin in which the hearer depends on case endings of words to make their meaning clear. (e.g. amo, amas, amat etc.).

---

[100] Blesser, B., Salter, L-R (2007) Spaces Speak, Are You Listening. MIT Press, p 144–49.
[101] Vitruvius (1914) The Ten Books On Architecture (trans Morgan, M.H.). Harvard University Press. Book 5, p 153.

But despite its very audible effect on the acoustics of halls, it wasn't until the very end of the nineteenth century that a systematic investigation into reverberation was first made. Wallace Clement Sabine, at the time a junior member of the physics department of Harvard University, was asked to look into the cause of the very poor acoustics of a new lecture hall and to come up with ideas on how this might be improved. The problem was that whenever someone spoke in the hall, their voice reverberated for several seconds, making even normal conversation all but impossible. Sabine had only an elementary knowledge of acoustics and had to devise his own approach to the problem. One of these was to use hundreds of cushions placed on the seats to simulate the effect of an audience on the acoustics of the hall.

As a result of his investigation he came up with the concept of *reverberation time*, which he defined as the time for the loudness of a sound to drop by 60 dB (equivalent to a million-fold reduction in intensity). He discovered that everything else being equal, reverberation time is principally determined by the total volume of a hall rather than its dimensions. Moreover, by trial and error, he found that the ideal reverberation time depends on whether a hall is used for music (i.e. a concert hall) or for speech (i.e. a lecture hall). Broadly speaking, the ideal reverberation time for speech is about one second and for music somewhat longer, depending on the genre. For chamber music it is about 1.5 s and for symphonic music it is about 2 s.

Given this, it is not surprising that reverberation has played an important part in the development of musical styles. The lengthy reverberation times of stone-walled Romanesque churches and the vast spaces of Gothic cathedrals suited Gregorian chant. Small concert halls favoured the faster pace of the chamber music of the Baroque age, and the larger concert halls constructed in the nineteenth and twentieth century are ideal for the Romantic music of the nineteenth century.

Although long reverberation times make speech and music unintelligible, reverberation can be exploited to dramatic effect. A well-known example of this is the echo in the Baptistery of St John in Pisa. It was visited by Mark Twain during his tour of Italy: "This Baptistery is endowed with the most pleasing echo of all the echoes we have read of. The guide sounded two sonorous notes, about half an octave apart; the echo answered with the most enchanting, the most melodious, the richest blending of sweet sounds that one can imagine. It was like a long-drawn chord of a church organ, infinitely softened by distance. I may be extravagant in this matter, but if this be the case my ear is to blame—not my pen. I am describing a memory—and one that will remain long with me."[102] The tradition of singing two or three notes in the Baptistery continues to this day and never fails to enchant its visitors.

Another famously reverberant space is the Hamilton Mausoleum in Hamilton, South Lanarkshire, Scotland. The chapel, completed in 1858, has a reverberation period of about 15 s, the longest of any building in Europe. In addition, the building also has four 'whispering alcoves' from any of which you can engage in a whispered conversation with a companion in the diagonally opposite alcove.[103] As for cathedrals, Cologne Cathedral in Germany has a reverberation time of 13 s, the longest of any gothic cathedral. And recently, Trevor Cox, an acoustic engineer at Salford University, established that the world's most reverberant enclosed space is a vast disused oil tank at Inchindown, near Invergorden in Scotland that was excavated deep within a mountain to store fuel for the Royal Navy at the start of WWII. It's dimensions are stupendous: 237 m long, 9 m wide and 13.5 m high and its measured reverberation time of 75 s is correspondingly huge.[104]

Specially built rooms with an exceptionally long reverberation time are used for research purposes and are often found alongside anechoic chambers. All the surfaces in these chambers are acoustically hard and none is parallel to any other to ensure that there is no resonance, which would emphasise some frequencies and interfere with the gradual, smooth decay of the reverberation. The reverberation chamber at the National Physical Laboratory in Teddington, London, has the longest reverberation time of any such room in Europe: 30 s for the lowest frequencies. Conversation within this chamber is impossible unless one is standing next to whoever one is speaking with. Musicianshowever, aping what is done in Baptistery of St John in Pisa, have exploited its prolonged reverberation to create haunting harmonic sounds. One hesitates to call the result music because it is not possible to sustain a melody when reverberation lasts more than a couple of seconds. Although the reverberation time of this room is 30 s, this can only be detected using sensitive instruments; to the unaided ear the reverberation lasts much less than 30 s because the intensity of later reflections falls below the threshold of hearing.[105] Indeed, this is true of all reverberant spaces: perceived reverberation time is always significantly less than the theoretical maximum based on Sabine's criterion of room volume.

---

[102] Twain, M. (1869) Innocents Abroad. Collins, p 162.

[103] These whispering alcoves are, in fact, whispering dishes.

[104] Cox, T. (2014) Sonic Wonderland, A Scientific Odyssey Of Sound. Bodley Head, p 53–7.

[105] When I visited the NPL reverberation chamber, I estimated that the sound of a clap lasted between 6 and 8 s until I could no longer hear it.

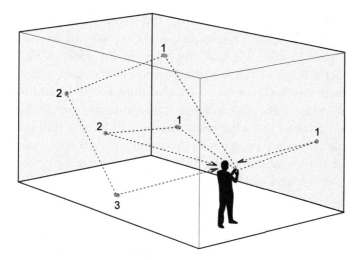

Fig. 5.12 Reverberation. Three possible paths by which the sound of a clap is heard within a room with hard surfaces due to one, two and three reflections. There are, of course, scores more routes, each involving multiple reflections

Reverberant sounds can travel huge distances. Taking the speed of sound in air as 340 m/s, the last audible remnants of sounds reflected in a room with a reverberation time of 2 s have travelled almost 700 m. Moreover, depending of the size of the room, it will have been reflected dozens of times. In the NPL reverberation chamber, the final detectable sound has travelled almost 10 km (30 s × 340 m/s). The corresponding distance at Inchindown is almost 26 km (Fig. 5.12).

Paleolithic Cathedrals

It has been suggested that the Paleolithic hunter-gatherers who inhabited Southern France and Northern Spain between 10,000 and 35,000 years ago, and who covered the walls of caves with depictions of animals, made use of reverberation and, more particularly, of resonance when choosing where to place the images.[106] Depictions of bison and deer, i.e. prey animals, are often found on the walls of chambers that are highly resonant, whereas lions, i.e. predators, are often found in non-resonant spaces. Particularly resonant spaces are also often marked with several dots of red pigment rather than with animals.

---

[106] Waller, S.J. (1993) Sound Reflection As An Explanation For The Content And Context Of Rock Art. Rock Art Research, 10.

The possibility of a link between image and sound was originally proposed in 1983 by Iegor Reznikoff, a specialist in ancient art and music, following a visit to a painted cave. He had long been in the habit of humming to himself to get a sense of the acoustic qualities of churches and cathedrals, and to discover unusually resonant spaces within them. He instinctively did the same when inside the cave and was surprised to discover several particularly resonant alcoves. But what really astonished him was that many of the less accessible resonant passages were decorated with red spots. He has since concluded that the caves were the Paleolithic equivalent of cathedrals, and that the resonant spaces had been identified by chanting shamans who would alter the pitch of their chants until they were reinforced by resonance within the alcove. Moreover, he found that reverberation in larger chambers within caves is reminiscent of that within Romanesque churches.

Resnikoff made another suggestion: the dimness of a sputtering flame of a tallow candle would limit visibility within pitch-dark caves and so these ancient people may have made use of resonance and reverberation to find their way though the maze of subterranean passages. By calling out and listening attentively they could supplement the meagre light of their lamps to get a sense of what lay ahead. Moreover, special notice would have been taken wherever their calls were reinforced by resonance, and these places were marked for special use. Resnikoff also noted that a male voice is more suited to this task than a female one, both because of it's greater power and it's lower pitch, which is not absorbed as rapidly by the walls of caves as that of a woman.[107]

But claims of a link between the acoustics of caves and the positioning of images within them have recently been questioned.[108] In the first place there is a problem with semantics. In these studies the word "echo" seems to be used to refer to just about any unusual acoustic effect, including distinct echoes, prolonged reverberations or strong resonances. This makes it difficult to establish with any certainty which of these effects might have determined the choice of where to place an image. There is also a problem with the methods used by proponents of the link between acoustics of a cave and the location of the paintings. They have usually relied on rough and ready methods using very simple equipment and the human voice.

---

[107] Reznikoff, I., Davois, M. (1988) La Dimension Sonore des Grottes Ornées. Bulletin de la Societe Prehistorique Francaise, Vol 85, p 238–46.

[108] Fazenda, B. et al. (2017) Cave Acoustics In Prehistory: Exploring The Association Of Palaeolithic Visual Motifs And Acoustic Response. The Journal of the Acoustical Society of America, 142, p 1332.

So while a correlation between acoustics and images can't be ruled out, there are alternative hypotheses. It could be that the choice of small spaces within caves in which to paint images may have been due to ease of access to the walls rather than their resonance, or because they are more inaccessible and so proclaim the painters prowess. After all, illicit graffiti that brazenly anounce "I was here" is often found on rooftops and other seemingly inaccessible locations. On balance, the case for sonorous Paleolithic cathedrals has yet to be made.

6

Making Noise

Abstract Accounts of interesting and unusual sounds and explanations of how they are produced, arranged according to the cause they have in common i.e. those due to impact, resonance, friction, bubbles, wind and shock fronts. Many of these explanations are of sounds mentioned in previous chapters.

Impacts

The earliest explanation we have for sound claimed that "there cannot be sound without the striking of bodies against one another".[1] And, on the face of it, it is tempting to assume that collisions are *the* archetypal source of loud, transient sounds. Indeed, until the seventeenth century thunder was ascribed one way or another to collisions between clouds. According to Aristotle "The windy exhalation in the clouds produces thunder when it strikes a dense cloud formation."[2] Lucretius claimed that thunder is due to wind-blown clouds crashing into one another.[3] Remarkably, much the same explanation was offered by Descartes almost 2000 years later "But for the storms that are accompanied by thunder...I have no doubt that they are caused by the fact that there are many clouds on top of one another, so that it sometimes

---

[1] Archytas, Fragment 1, in *A Source Book in Greek Science* ed M.R.Cohen & I.E.Drabkin Harvard University Press,1948, p 286.

[2] Aristotle, 350 BC, Meteorologica, Book 2, section 9. http://classics.mit.edu/Aristotle/meteorology.2.ii.html (accessed 18/03/2020).

[3] Lucretius, On the Nature of Things, Book 6, line 107.

J. Naylor, *Now Hear This*, https://doi.org/10.1007/978-3-030-89877-9_6

happens that the highest come down all at once upon the lowest...which will thus soon fall all together on the lower cloud with a great noise."[4]

The association of sounds with collisions may go some way to explain why scholars and musicians were prepared to take on trust the story of Pythagoras and the blacksmiths for close on 2000 years. How ironic that the science of acoustics was said to have been born in a noisy forge, a place where men unwittingly sacrificed their hearing as they pounded hot lumps of metal into all manner of useful artefacts.

But what an unconvincing tale it is. It's a pity that Pythagoras didn't pay more attention to what he is supposed to have seen and heard when he entered the smithy because if he had he would have realised that the sounds he heard came from the anvils and not the hammers and that the tones of those sounds were due to the size of the anvils rather than the weight of the hammers. He would also have realised that effect of hammering harder is to increase the loudness of the sound, it does not alter its fundamental tone. Wheeltappers heard and listened to the ring of the train's wheel, not that of the hammer. And the sonorous peal of a bell is due to the vibrations of the bell itself, not those of the clapper.

What if the smiths had been carpenters or shipwrights? Unlike an iron anvil, a balk of timber does not ring when struck a blow. The sound of hammers on wood is dull and hollow—closer to a thud than a clang. But a full explanation for that difference requires knowledge of the properties of matter and acoustics that wasn't available until the nineteenth century. What was known is that in every case the sound we hear when an object is struck or tapped is the result of vibration. Nor are these vibrations haphazard, they are determined among other things by the size, shape and composition of the object. All else being equal, large objects vibrate at lower frequencies than small ones—in the jargon, they are said to have a lower natural frequency.

As it vibrates, an object loses energy, partly to the surrounding medium (the vibrations of which we experience as sound) and the remainder to friction within the object itself. If the material from which the object is made is *stiff* (it's difficult to alter its shape), *elastic* (it returns rapidly to its original shape once the force that has altered its shape it is removed) and *homogeneous* (has a uniform structure devoid of internal cracks or discontinuities) then the vibrations die away much less rapidly than if the object is made from a material that lacks those characteristics. This is why in almost every instance

---

[4] Descartes, R., *Discourse on Method, Optics, Geometry, and Meteorology*, 1637. Translated with an introduction by Paul J. Olscamp, 2001, 324–5.

metallic objects ring when struck and wooden ones do not.[5] Not only is wood less stiff and elastic than metal, its fibrous structure rapidly converts vibrations into heat. And when vibrations die away rapidly, the resulting sound is equally short lived.

As we saw in chapter one, tapping is widely used as a test for the internal state of an object. A crack within a solid object made of a material that is stiff such as metal will prevent it ringing because the object can't vibrate as a whole: in place of a ring we hear a clunk. As for a watermelon, as it ripens its flesh becomes less firm and so the frequency of the sound when it is tapped is noticeably lower that when it is unripe. But identifying the precise pitch that indicates that it is ready to eat must take some practice because pitch will continue to drop as a melon over ripens. A similar explanation applies to a cooked loaf: when it is cooked the gluten and starch harden, trapping tiny pockets of gas. Hence the hollow sound when the stiff crust is tapped.

But however stiff, elastic and homogeneous, vibrations will die away rapidly if the sounding object is constrained in some way. A percussionist grasps the rim of a cymbal to cut short its clang. Indeed we all do this instinctively when we press our fingers against the surface of an object that is vibrating noisily. Doing so damps—i.e. supresses—its vibrations.

Most objects can vibrate at several frequencies at the same time, though it is usually the lowest of these that is the most audible. Increasing the force of the impact not only produces a louder sound, it also excites higher frequency vibrations, which make the resulting sound "brighter".

If you are near a table with a wooden top, tap it, first with a fingernail and then with a knuckle, and note the difference. A table top will vibrate at frequencies that depend on the wood it is made from, its dimensions (length, breadth and thickness) and how it is fastened to its legs. When struck it will vibrate most readily at its lowest or fundamental frequency. But fingernails are more rigid than flesh-covered knuckles and their impact causes the table top to vibrate strongly at frequencies that are much higher than those due to knuckles, and the addition of those higher frequencies to the fundamental frequency alters the timbre of the sound noticeably.[6]

There are, however, many situations where we hear the striker rather than what it strikes. The distinctive clip-clop as a horse walks slowly on a hard, rigid surface such as tarmac or concrete is due to vibrations of the hoof, not

---

[5] There are exceptions: lead (low stiffness, high density) emits a dull thud, some hardwoods (high stiffness, low density) emit a brighter tone (and are used in a xylophone).

[6] An audio spectrometer will clearly show the higher frequencies excited when tapping with a fingernail. A single tap is too brief for the spectrometer to register a clear trace, so it is necessary to tap several times in quick succession to reveal which frequencies are present in the sound you hear.

the tarmac. The same is true of the sound of noisy footsteps on paving: it's the heel we hear, not the pavement. Unlike a table top or a wooden floor, heavy, rigid surfaces such as a paved road or concrete pavement hardly move when struck by a small, light object. So what we hear must come from the latter.

You can confirm this with a simple experiment. Cut a long, thin wooden rod into sections of different lengths, say 20, 30, 40 and 50 cm. Drop them one by one from the same height onto a hard, rigid surface such as stone or concrete paving and listen to the sound each makes on impact. Now drop the rods onto a wooden floor. In the first case the sound differs from one rod to the next, but with the wooden floor they all make a similar sound.

With paving it is the vibrations of the rods that we hear, not those of the paving. Those vibrations depend, among other things, on the length and weight of the rod. In the absence of audible sounds from the paving you are able to hear the distinctive sound of each rod. A wooden floor is less rigid and less massive than paving and when struck its vibrations are vigorous enough that the resulting sound masks that of the rods. Moreover, the rate at which the floor vibrates is determined by the properties of the wood and how it is suspended and not by those of the rods falling on it. Hence the impact of each rod against the floor creates a sound that differs principally in loudness rather than timbre.

If you listen to the footsteps of people walking on paved surfaces while looking at their footware, you will realise that the size and composition of the heel determines the timbre of their sound. On a flexible surface, such as a wooden floor, all heels sound more or less the same.

Musical Stones

If you were asked to imagine the sound you would hear when a large rock is hit with a steel hammer, a discordant clunk would probably come to mind. But in several locations around the world there are rocks that ring with metallic tones when struck with a hammer. They are known variously as musical stones, bell stones or ringing rocks.[7] The sound of these stones is due to destructive interference between the many (inaudible) infrasonic vibrations set up by the impact of a hammer.[8]

---

[7] Ringing Rocks County Park: https://www.youtube.com/watch?v=fBiVt1pKnAQ (accessed 12/07/2021).

[8] To hear examples of the sounds of these rocks search online for "ringing rocks": https://www.youtube.com/watch?v=472PEHLpwTQ (accessed 19/07/2021).

There are several well know examples of euphonious stones in huge boulder fields, some as large as 60 hectares, dotted across eastern Pennsylvania and western New Jersey. One of these has been studied on and off for the best part of a century.[9] Not that their musical qualities are a recent discovery. It seems that the early European settlers assumed that these boulder fields were the work of Native Americans who had gathered the stones together for ceremonial purposes. Less credibly, it has been suggested that the boulder fields are the ruins of an ancient civilization or the landing sites for alien spacecraft. The idea that nature alone might be responsible seems not to have occurred to anyone until the turn of the twentieth century. But real progress in finding the answer took another 60 years.

It's now known that these boulder fields are due to glaciers that retreated at the end of the last ice age, 12,000 years ago. During the ice age, severe frost action shattered the rocks that lay under the glaciers. But why do the stones in some boulder fields ring, while in others they do not? The most likely explanation is that stones that ring are under a huge internal tension due to the expansion of the outer surface as it slowly changes its chemical composition as it weathers. The expansion puts the core under tension. In fact, this process occurs in most exposed rocks, but this usually causes the outer layer to break off in sections, a process known as *exfoliation*. Ringing rocks, however, are composed of a very hard type of volcanic rock known as diabase, which preserves the outer layer intact until it is hit with a hammer. The impact causes chips to fly off the surface (a process known as *spalling*) with considerable energy. Indeed, repeated heavy blows can cause a stone to stop ringing altogether because the loss of the external layer allows the internal core to contract, dissipating the tension.

A rockery made of ringing rocks may be appealing but is not possible in practice because it is difficult to replicate the climatic conditions that prevail in the boulder fields from which they would have taken. The most important of these is that the boulders are exposed to the sun and air allowing moisture to evaporate rapidly from their surface. At the same time, the ringing rocks rest on other boulders and so are not in contact with the ground. Experience has shown that a rock removed from its natural setting very quickly looses its ability to ring unless precautions are taken to keep it dry.

---

[9] Gibbons, J., Schlossman, S., (1970) "Rock Music", Natural History: Journal of the American Museum of Natural History, Vol LXXIX, No 10, p 36–41.

Acoustic Whistlers

While waiting on a platform for a train you may occasionally have noticed a decidedly otherworldly sound from the rails, either moments before the train arrives or shortly after it departs. But, as with many unusual sounds that are not loud, you probably haven't given this one a second thought.[10] In fact, this particular sound is an example an acoustic whistler. These are sounds that are due to a single, brief impact but are heard as a descending glissando, i.e. a drawn out sound that descends in pitch. We have already met examples of such sounds in chapter five: musical echoes, echo tubes and culvert whistlers. These are all due to multiple reflections of an initial brief, broadband sound.

But reflection plays no part in the glissando from railway tracks. The source of that sound is an impact between the train's wheels and the rails. The resulting vibrations travel as transverse waves within the rails at speeds that depend on their frequency, which is why they don't all arrive together.[11]

In fact there are many situations in which you can hear this type of acoustic whistler. Nor do you have to leave the comfort of home to do so because if you have a Slinky to hand you have all you need to create an acoustic whistler. Suspend the Slinky from your hand held at waist height so that the far end rests on the floor and, with a slight vertical jerk of your hand, send a pulse through the spring. When it reaches the floor, the coils at that end collide with one another because the floor stops the pulse dead in its tracks. If you hold your ear against the end of the coil in your hand you will hear a faint, drawn out metallic sound, just like the one you hear from the rails. This is the acoustic whistler. It will repeat several times as the pulse travels up and down within the wire of the spring several times—something that does not happen in a railway track.

The reason the sound is faint is that the surface area of the wire of the coil is very small and so very little of its vibrational energy is transferred to the surrounding air. To improve the transfer of the vibrations to the air, increase the area of contact by inserting a small postcard-sized sheet of stiff card between the last two coils at the top end. Alternatively, clamp a few turns of one end of the coil between your teeth with the other end resting on the ground and jerk your chin to start the pulse. If nothing else, this will convince you of the efficiency of bone conduction and provide compelling evidence of the role that the jaw played in the evolution of the ear in vertebrates.

---

[10] The sound is sometimes used in film sound tracks as a train is seen travelling away from a station.
[11] Speed of transverse waves in a solid is proportional to square root of their frequency—i.e. high frequencies travel faster than low frequencies.

The sound you will hear has been described as the discharge of a "Star Wars blaster". Indeed, it *is* the source of that sound because the distinctive sound was actually a recording of an acoustic whistler produced by hitting a long guy wire of a tall antenna tower with a hammer.

You may recall reading in chapter four that the speed of sound within any particular medium (gas, liquid or solid) is the same for all frequencies. Were that not so, individual tones in a piece of music would become increasingly out of step as they travel through air; and who knows what that would sound like—perhaps something like the distorted whale songs that are heard in the Deep Sound Channel mentioned in chapter four? But what you hear from the Slinky is not the sound of the collision itself but of sounds due to transverse vibrations travelling through the steel of the coils. And the speed of these does depend on frequency: high frequency transverse vibrations travel through the steel of the coils faster than low frequency vibrations. You can't hear these transverse vibrations directly, but they set up longitudinal waves in the surrounding air that you can hear (or within your jaw, if that is how you chose to hear the sound). The sounds due to high frequency transverse waves reach your ears a fraction of a second before those of lower frequency transverse waves. Hence the eerie, descending glissando.

Acoustic whistlers can also be heard from the surface of a frozen lake. Hurl a stone as far as you can onto the ice and as it skips across it you hear a series of whistlers because successive impacts of the stone causes the ice sheet to vibrate transversely. Those vibrations spread out from the point of impact and radiate as sound into the air above the ice so that the higher frequencies reach your ears before the lower frequencies.[12] The ice sheet floats on water, which allows it to vibrate up and down, something that can't happen with ice in an skating rink, which rests on a solid surface, or if the ice on a lake is very thick (Fig. 6.1).

Ralph Waldo Emerson, the American philosopher and poet, visited Thoreau in the winter of 1836 and recorded that "A thin coat of ice covered a part of the [Walden] pond but melted around the edge of the shore. I threw a stone upon the ice which responded with a shrill sound, and falling again and again, repeating the note with pleasing modulation. I thought at first it was the 'peep' 'peep' of a bird I had scared. I was so taken with the music that I threw down my stick and spent 20 min in throwing stones single or in handfuls on this crystal drum."[13]

---

[12] The speed of transverse waves in ice is more than ten times greater than the speed of sound in air and depends on the frequency of the wave.

[13] Emerson, R.W., Journal 10 December, 1836.

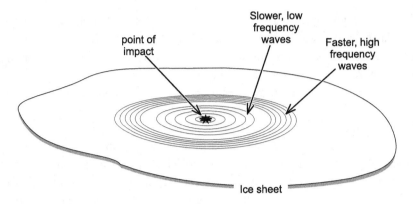

Fig. 6.1 Acoustic whistler from an ice sheet covering a lake. High frequency vibrations travel faster through ice than low frequency ones so we hear those first

Thoreau noted later "My friend tells me he has discovered a new note in nature, which he calls the Ice-Harp".[14]

Acoustic whistlers are also sometimes heard when people skate on a frozen lake.[15] Moreover, the pitch of the sound is related to the thickness of the ice sheet: the higher the frequency, the thinner the ice. In fact, in principle it is feasible to gauge whether an ice is thick enough to skate on safely because it should be possible to estimate ice thickness simply from the tones of the acoustic whistler. Ice thinner than 5 cm emits a sound of 660 Hz and is *not* safe to skate on. But, as we have noted, an acoustic whistler can be heard only by someone who is several metres from its source. Skaters can't themselves hear the whistler of which they the source so they would have to rely for a warning on someone who is several metres away—someone blessed with perfect pitch or with an audio spectrometer to hand. Gunnar Lundmark, a Swedish acoustician, has investigated acoustic whistlers produced by skaters and offers the following advice: "it is a good idea never to skate first. It is safer 20 m behind the leader—then you can hear the tone."[16]

Acoustic whistlers are also produced when an ice sheet expands or contracts during the day or night. Thoreau described the phenomenon in "Walden or Life in the Woods". He was at Flint's Pond on 24 February 1850 when "The pond began to boom about an hour after sunrise, when it felt the influence

14 Thoreau's Journals, 5 December, 1837.

15 Search online for "The sound of ice" video: https://www.youtube.com/watch?v=66a3_MGTDoA (accessed 20/04/2021).

16 Lundmark, G., Skating on thin ice—And the acoustics of infinite plates, Paper Presented At The 2001 International Congress And Exhibition On Noise Control Engineering, The Hague, Netherlands, 27–30 August 2001. https://silentlistening.files.wordpress.com/2010/01/skating-on-ice.pdf (accessed 20/08/2019).

of the sun's rays upon it from over the hills; it stretched itself and yawned like a waking man with a gradually increasing tumult, which was kept up 3 or 4 h. It took a short siesta at noon, and boomed once more toward night, as the sun was withdrawing his influence... The pond does not thunder every evening, and I cannot tell surely when to expect its thundering; but though I may perceive no difference in the weather, it does. Who would have suspected so large and cold and thick-skinned a thing to be so sensitive? Yet it has its law to which it thunders obedience when it should as surely as the buds expand in the spring."[17]

The sound described by Thoreau is sometimes referred to as "ice yowling"[18]

Good Vibrations

This chapter began by revisiting Pythagoras and his hammers. Setting aside a well-founded scepticism concerning specious claims about the harmonious sounds of hammers striking anvils, we noted that hitting an object will make it vibrate and that how it vibrates depends on its shape, size and composition among other things. So much so, that most objects can vibrate readily only at very particular frequencies, the lowest of which is known as the object's natural frequency. Moreover, those vibrations are communicated to the surrounding air and, if they fall within the range of frequencies audible to our ears, are heard as the sound that we learn to associate with the object that is its source. As you will know, we can often identify an object and the substance from which it is made merely from the sound it makes when it is lightly tapped.

But, as we saw in chapter two, there is another way to make an object vibrate, and that is to subject it to vibrations that are very close to or which match its natural frequency. We come across examples of this every day. Various bits of a stationary bus, often including its passengers, will shake while its motor is running, as will a car travelling rapidly over rumble strips on a road surface. A washing machine vibrates noisily during its spin cycle. During an earthquake, tall structures such as buildings, trees and lamp posts will sway if their natural frequency matches the rate at which the ground is vibrating due to the seismic waves that reach the surface upon which

---

[17] Thoreau, H.D., 1854, "Walden or Life in the Woods", p 322–23.

[18] Online video "Ice cracking on the pond. Sounds like Star Wars!": https://www.youtube.com/watch?v=Lr9_DvsJFXE OR "Dispersion of Sound Waves in Ice Sheets" (a Sound file) https://silentlistening.wordpress.com/2008/05/09/dispersion-of-sound-waves-in-ice-sheets/ (both accessed 12/07/2021).

they stand. And there are numerous examples of bridges, piers and buildings swinging about, sometimes wildly, due to crowds inadvertently moving in step with the natural frequency of these structures—though in these situations, the movement is inaudible because the vibrations are well below audible frequencies.[19] Even something as insubstantial as air will resonate if it is confined within an enclosed space.

In fact, resonance of the air within a partially closed space is a major source of sounds. Sometimes the sounds are unexpected and interesting, at others they are unwanted and troublesome. But without resonance, we wouldn't be able to speak and musical instruments would be silent, as we shall now see.

Listening to a Seashell

You probably know how to make a pen top hoot by putting the open end to your lips and blowing over it. In the process you may also have discovered that the pitch of the sound it makes depends on the length of the top's cavity. As you blow, your breath creates eddies across the open end causing the pressure at the opening to rise and fall repeatedly, which makes the air within the pen top vibrate. Depending on how hard you blow there is a point at which the vibration and the changes in pressure due to eddies at the open end are in step with one another. And when this happens, as long as you keep blowing, you hear a loud, continuous tone due a standing wave in the air within the cavity.

A standing wave occurs when a wave and its reflection combine so that at some points they are in step with one another, causing the medium through which they are travelling to vibrate strongly, while at others they are completely out of step and the medium remains motionless. The places where this movement is greatest are known as antinodes, and those where there is no motion are known as nodes. In the case of the pen top, the standing wave is due to changes in air pressure: it rises and falls repeatedly at an antinode

---

[19] A notorious example of a swaying pedestrian bridge is London's Millenium Bridge that spans the Thames. On the day it was opened in June 2000, people reported that it swayed from side to side. The natural sway as a person walks multiplied by several hundred people crossing the bridge at the same time caused it to begin to swing from side to side. This caused walkers to synchronise their sway with that of the bridge, which increased the amplitude of the bridge's sideways oscillations. The bridge was closed down two days after it opening and it took a year and a half to identify the cause of the problem and correct it. That done, the bridge reopened in January 2002. For further examples of resonance: *Buildings*: Dotti, N. R., 2011, A building moves with the music, Physics Today, 64, 1, 8 // *Bridges*: Fall of the Broughton Suspension Bridge, Near Manchester, Philosophical Magazine, Vol IX, January-June, 1831, 384–9 // *Pleasure Piers*: Capstick, J.W., Sound, An Elementary Textbook For Schools And Colleges, 1913, p 130–31.

and remains unchanged at a node. Moreover, the pitch of the sound you hear is the lowest possible frequency of the standing wave, i.e. its fundamental frequency (or its natural frequency, though these terms are interchangeable).

Because a standing wave is a combination of two waves travelling in opposite directions, its maximum amplitude is twice as great as that of the individual waves. This is why the fundamental frequency of a musical instrument is particularly loud and dominates the sound it makes. Resonance of the ear canal, which is in effect a short cylinder some 3 cm long, lies between 3000 and 4000 Hz, which may explain why many people who suffer noise induced hearing loss do so between those frequencies.

Panpipes are, in effect, a series of pen tops of different lengths. The length of each of the tubes of a panpipe is chosen so that its pitch fits into a musical scale. Organ pipes can be thought of as gigantic pen tops. Indeed, organ makers tend to refer to the length of an organ pipe rather than its pitch. Some of the very largest organs have a pipe that is 10 m long, the pitch of which is just below the threshold of hearing and so is felt rather than heard.[20]

In fact, any tube that is narrow compared to its length will emit a sound when the air within it is made to resonate, even if both ends are open. The only difference is that the resonant frequency of the column of air within a tube that is closed at one end (e.g. a pen top) is twice that of the same tube with both ends open. And this is exploited by musical wind instruments. Wind instruments such as flutes, clarinets and trumpets all depend on the resonance of the air within a long, narrow tube to create particular notes. Each note is determined by the length of the air column, a length that can be altered by uncovering holes in the side of the tube (flutes and clarinets) or increasing its length by pressing valves (trumpets).

On a windy day, hollow tubes that have one or more openings may emit sounds depending on the speed of the wind and the length of the tube. Hence the low-pitched moan of wind from a chimney. All around the world there are sound sculptures that rely on an array of tubes of different lengths to produce a low moan, hum or howl when there is a wind.[21] The hollow tubes of a roof rack sometimes surprise the occupants of moving car by emitting a whistle, which will come and go depending on the speed of the car.

Another sound you may have been dimly aware of when travelling in a car when one of its windows is open is a very low frequency throb within the cabin. It's known as "side window buffeting" and is particularly noticeable when one of the rear windows is open. The flow of air over the surface of the moving car creates eddies at the open window, causing a periodic rise and fall

[20] Johnston, I., 1989, Measured Tones, The Interplay Of Physics And Music. Adam Hilger p 214.
[21] Online search for examples of "sound sculptures using wind" (accessed 19/07/2021).

in air pressure within the cabin. Given the large internal volume of the car, the air within the cabin expands and contracts slowly, hence you hear a throb (described as "wubwubwubwub") rather than the high pitched whistle of the pen top.

The sound from an empty bottle when you blow across its open end is also due to a vibrating column of air, though the pitch of the sound depends on the interaction between the air in the neck of the bottle and the air within the body of the bottle rather than the air as a whole. Eddies created as air flows across the open end cause the air within the neck to vibrate, just as it does in a pen top. But in this case that motion is modulated by the mass of the air in the body of the bottle. The process was explained by Herman Helmholtz in 1862, hence any cavity that consists of an enclosed space connected to the outside by short, narrow neck is known as a Helmholtz resonator.

The pop you hear when a cork is pulled rapidly from an unopened bottle of wine depends on the length of the column of air between the bottom of the cork and the wine, as does the pop when a cork is propelled from a bottle of champagne. Both are examples of Helmholtz resonance. As a cork moves out from the neck, the volume of the space between the wine and cork increases and the pressure of the air trapped therein decreases. The moment the cork is free of the bottle, the air outside rushes into the neck, briefly increasing the pressure of the air there above atmospheric pressure. That in turn forces some air out of the neck and the resulting drop in pressure allows air to rush back into the neck. This back and forth motion continues for a few thousands of a second until the all the energy of the vibration is all lost as heat.[22] That vibrating column of air is the source of the sound you hear.

The initial gurgle that you hear when you begin pouring from a newly opened bottle of wine (or water) is also due to resonance. Air must enter the bottle to allow liquid to leave it. Initially the neck of the bottle is full of liquid that prevents air entering the bottle. As you tilt the bottle, a growing pocket of air forms in stages just beyond the neck as liquid pours out. Each time the air breaks through the liquid and enters the pocket it causes the air within it to vibrate. If you pour carefully and very slowly to control the entry of air you will notice that the pitch of the sequence of glugs diminishes slightly each time the volume of air in the pocket increases. This is to be expected because, as Helmholtz discovered, the resonant frequency of an air filled space depends, among other things, on its volume: the greater the volume, the lower the resonant frequency. The shape of the bottle pays a part.

---

[22] For a video demonstration of the pop of a champagne bottle search online for "The Actual Mathematics of Popping Champagne Corks": https://www.youtube.com/watch?v=xTcvl-kw9fU (accessed 12/07/2021).

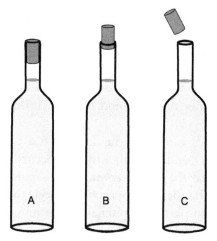

Fig. 6.2 Pop of a cork. As the cork is drawn from the neck of the bottle (A to B) the volume of the space above the wine increases and the pressure of the air within drops slightly. At the moment the cork leaves the neck (C) atmospheric pressure compresses air within and sets up a brief oscillation in air pressure that is the source of the pop that you hear

Gurgling is affected by the shape of the bottle. It is far more obvious with a bottle that has a wide shoulder at the base of the neck (known as a Bordeaux bottle) which traps the pocket of air more effectively than with pear shaped bottles (known as Burgundy bottles) (Fig. 6.2).[23]

Cavities can also resonate in the presence of sustained ambient sounds, which are, of course, themselves vibrations of air. The faint murmur of the sea that people imagine hearing when they put a large conch seashell to their ear is not an illusion, though it has nothing to do with waves breaking on a seashore. What you hear is due to the air within the cavity of the shell resonating to ambient sounds. Perhaps the association between shells and the sea is the reason why those resonant sounds are heard as a marine murmur. In fact, as long as there are ambient sounds composed of a broad range of frequencies, you can hear the murmur of the sea simply by holding one end of a cardboard tube to your ear.[24] The length of the tube determines the frequency of the sound you hear: longer tubes resonate to lower-pitched ambient sounds. Covering the open end of the tube with your hand will raise its resonant frequency by an octave, a change that may make it easier to notice

---

[23] Should you wish to experiment with gurgles, I recommend you use an empty bottle and fill it with water because it takes practice to produce a series of air pockets and hear distinctly the drop in pitch that accompanies them.

[24] The sizzle as water is heated to boiling in an electric kettle, the thrum of a fan oven and traffic are all good sources of continuous sound that will cause the air within a cavity to resonate.

resonance. You will also find that cupping your hand around your ear when straining to hear a faint sustained sound can emphasise some frequencies because, as with a tube, the air enclosed by your cupped hand will resonate due to frequencies present in external sounds. The timbre of the sound will change noticeably as you slowly curl your fingers to make an open tube and change again if you cover the open end of your closed fist with your other hand.

That ambient sounds can cause the air within a cavity to resonate was known in the ancient world, though why this occurs was not understood. Vitruvius recommended the use of *echea*, bronze vases placed between the seats of a theatre.[25] The idea appears to have been that the air within the vases would resonate to particular pitches of the musical instruments used during performances and hence enhance their sound. No such vases have ever been found so it's a mute point whether they were ever used. Moreover, recent investigations have concluded that the effect of such vases would have been barely audible.

In fact, you are almost certainly aware of the link between the length of the column of air within a container and its resonant frequency because whenever you pour water into a jug or a glass the pitch of resulting sound as it fills up rises at an ever increasing rate. Indeed, the acoustic clue you often unconsciously rely on to judge when it is full is that the pitch stops rising. Try this: with your eyes closed, pour water from a height into a tall, narrow container such as a glass or a vase, ensuring that it creates a mass of noisy bubbles, and stop when you judge that it is full. Open your eyes to check whether you have judged it right. The pitch rises because as the length of an air column gets shorter, its resonant frequency increases: the resonant frequency doubles if the length of the column of air is halved. If water is poured at a steady rate, the time taken to successively halve the column of air will decrease more rapidly. Hence the pitch changes at an increasing rate as the container fills up and the column of air above the water becomes shorter. The pitch of the sound stops rising when the water has reached the brim, i.e. the length of the column of air is zero (Fig. 6.3).

Another way to make the air within a small hollow space such as a vase or a tube resonate is to tap the open end sharply with your fingers or palm. You may have to experiment to find the best way to do this. The sound you hear is determined by the size of the container and is the basis for a number of musical instruments, the best known of which is the udu. The udu is essentially a large vase with two openings, one being the neck of the vase and the

---

25 Vitruvius (1914) The Ten Books On Architecture (trans Morgan, M.H.). Harvard University Press. Books 5, p 143.

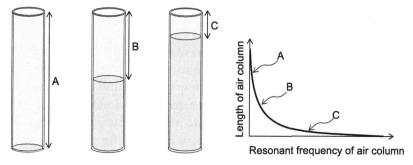

Fig. 6.3 Resonance of a column of air within a narrow cylinder. As the cylinder fills with water the sound you hear rises in pitch. The increase in pitch is more rapid and therefore more noticeable as the water level rises and the air column becomes shorter. Here the length of the air column B is half that of A. But the change in frequency of the air column from A to B is several times less than that from B to C. The frequency changes much more rapidly from B to C than from A to B

other a circular hole in its side. The player strikes the hole with the fingers or palm, which makes the air within the vase resonate briefly. The udu is a Helmholtz resonator, hence the frequency of the resonant sound depends on the volume of the vase and whether the player keeps his fingers/palm on the hole as the air within it resonates (i.e. the hole is closed resulting in a low frequency) or lifts his fingers/palm immediately after hitting the hole (i.e. hole open, which doubles the frequency compared to keeping the hole closed).[26]

Another instrument based on this principle is the tubulum. It consist of a number of plastic drain pipes of different lengths on a rigid frame and in its simplest form resembles a huge pan pipe.[27] The player strikes the open end of the tube with a rubber paddle or the sole of a flip-flop. In the right hands it makes a passable, if limited, musical instrument.[28]

Incidentally, with an empty bottle you can confirm that the frequency of the sound produced when a hollow vessel with a narrow neck is tapped (hold your index and middle finger together and tap the open end sharply) is the same as that produced when you blow across the open end. The only difference is that the latter sound is louder than the former.

---

[26] How to play an udu: https://www.youtube.com/watch?v=Mru1gCuQu8Q (accessed 19/07/2021).

[27] How to make a tubulum: https://drumsoul.files.wordpress.com/2016/12/how-to-build-a-blue-man-tubulum.pdf (accessed 22/07/2021).

[28] Snubby   J: https://www.youtube.com/channel/UCpbzH4Vyer8K3z5omjXGMlQ   (accessed 19/07/2021).

The Sound of Froth

At the chime of midnight, as the old year makes way for the new, we charge our glasses with champagne and wish one another good health and good fortune in the coming year. But the accompanying sound of clinking glasses is not a cheery tinkle but a dispiriting, leaden clunk. Indeed, for a moment you might wonder whether the glasses are made of plastic rather than glass. But replace champagne with still wine or water, and the glasses regain their merry tinkle.

The dull clunk of champagne was remarked upon by John Herschel.

A pleasing example of the stifling and obstruction of the pulses propagated through a medium, from the effect of its non-homogeneity, may be seen by filling a tall glass (a Champagne glass, for instance) half full of that sparkling liquid. As long as its effervescence lasts, and the wine is full of air bubbles, the glass cannot be made to *ring* by a stroke on its edge, but gives a dead, puffy, disagreeable Sound. As the effervescence subsides the tone becomes clearer, and when the liquid is perfectly tranquil the glass rings as usual; but on reex-citing the bubbles by agitation, the musical tone again disappears ... This neat experiment seems to have been made originally by Chladni (Acoustique, §214)[29]

Herschel was right that bubbles are the cause of that "dead, puffy sound," but wrong about how they bring that about. The bubbles don't obstruct the passage of sound, rather their presence makes it easier to compress the liquid without appreciably changing its density. As we noted in chapter two, the speed of sound in a liquid is governed by a combination of its compressibility and its density. That speed drops dramatically when a vast number of tiny bubbles are present because they make the liquid more compressible, just as a sponge is easy to squeeze because it is full of holes. And the change in speed in turn affects the resonant frequency of the liquid within the glass and consequently the pitch of the sound you hear.

There is another, cheaper, way in which you can hear the effect of altering the compressibility of water on the speed of sound. If you drink instant coffee or hot chocolate you may have noticed that as you stir to dissolve the powder in hot water or milk the pitch of the pitch of the spoon as it strikes against the side of the mug rises. The rise is even more obvious if you tap gently

---

[29] Herschel, J. (1830) Sound. In: Smedly, E. (ed) (1830) Encyclopaedia Metropolitana, Vol II. Baldwin and Cradock, London, p 771.

against the bottom of the mug with a soft mallet such as the tip of the handle of a wooden spoon because this does not make the sides of the mug ring.

Instead, what you hear is due to resonant sounds within the liquid. The frequency of these sounds depends on the depth of the liquid in the mug and the speed of sound in that liquid. All else being equal, the greater the depth of the water, the lower the resonant frequency of the water column. But when a powder such as instant coffee granules is dissolved in water, it introduces a huge number of tiny bubbles of air. And, just as with champagne, this reduces the speed of sound within the bubbly mixture, which in turn lowers the resonant frequency of the column of liquid in the mug.

At first you hear a low pitched thud that corresponds to the maximum number of bubbles within the liquid. But as they rise to the surface and escape into the air, the speed of sound in the liquid increases, which raises the resonant frequency of the space occupied by the liquid. Hence the rise in pitch if you continue to tap as the bubbles rise to the surface. It has come to be known as the "hot chocolate effect" after a paper written in 1982 by Frank Crawford.[30] The scientific name for Crawford's hot chocolate effect is the allasonic effect.

Herschel's champagne and Crawford's hot chocolate are not the only examples of the consequences of altering the speed of sound has on resonant sounds because inhaling helium alters the timbre of one's voice (Fig. 6.4).

The Human Voice

The human voice is arguably among the most unexpected and counter intuitive examples of resonance within a cavity. Contrary to expectation, the vibration you experience in your throat when you speak is not the sound that emerges from your mouth. Those vibrations are due to air from the lungs flowing through the vocal folds, which makes them open and close at a rate that depends on the speed at which air flows through them and the size and mass of the flaps. Male vocal folds are larger and heavier than female ones and so open and close at a lower rate, typically 120 times a second compared to 220 times a second. These are the fundamental frequencies of the sounds produced by the vocal folds. However, the sound produced in this manner is just a monotonous buzz. Nor do we hear it directly because what emerges from the mouth is an altogether different sound.

---

[30] Crawford, F. S. (1982) The Hot Chocolate Effect. American Journal of Physics, 50, 398–404.

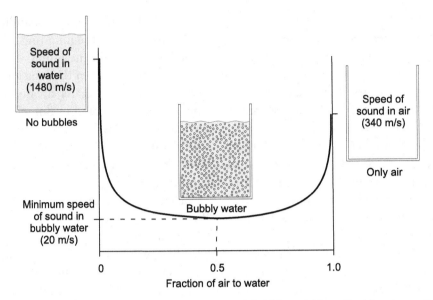

Fig. 6.4 Hot chocolate effect. The presence of bubbles within a liquid reduces the speed of sound, which in turn lowers the fundamental frequency of the mixture. Note that to accommodate the range of velocities from liquid to gas, the vertical axis of the graph has to be logarithmic, not linear. The speed of sound in bubbly water is least when the water to air bubble ratio is 50:50 as shown on the graph

The vocal buzz, which consists of the fundamental frequency at which the vocal folds open and close, plus a large number of higher harmonics, is transformed into the phonemes of speech by resonance of the air within the vocal tract. This consists of the section of the throat above the vocal chords (known as the pharynx) and the cavities of the mouth, the opening of the mouth and the nasal passages. By altering the shape and volume of the pharynx and mouth cavity, the resonant frequency of the vocal tract is altered so that some frequencies present in the vocal buzz are emphasised and others are suppressed. The vocal tract resembles a pen top in that it is closed at one end—the vocal folds—and open at the other—the mouth.

The shape and size of the vocal tract is altered primarily by the tongue. In effect the tongue creates two independent spaces: the first of these lies between the vocal folds and the back of the tongue. The second is the space between the front of the tongue and the lips. That space can be further altered by opening and closing the lips of the mouth. Each of these spaces has it own particular resonant frequency, known as a formant, which depends on the position and shape of the tongue and of the lips. The sound that emerges from the mouth is a combination of these formants.[31]

Helium Voice

The shape of the vocal tract is not the only thing that affects the frequency of the various formants that give your voice its particular timbre, as you will have noticed if you have ever spoken after inhaling some helium. Doing so appears to raise the pitch of your voice and make you sound like Donald Duck. Helium has no effect on the rate at which the vocal folds open and close, but because the speed of sound in helium is some two and half times greater than the speed of sound in air, the increase in the speed of sound means that the gas within the various sections within the vocal tract now resonate at frequencies that are two and a half times greater than in air. At the same time the relative loudness of these formants also change. As a result your voice is now composed of higher frequencies.

Deep sea divers breath a helium-oxygen mixture containing 68% helium by volume, so the overall increase the pitch of the voice is only one and a half octaves. However, at the very high pressures of the helium-oxygen atmospheres breathed by saturation divers working at great depths, speech becomes unintelligible, and poses a very real impediment to communication. The problem is slightly mitigated by electronic audio processors known as "Helium speech unscramblers".

Helium voice is widely considered to be a harmless party trick. But be warned, it is not without danger. Inhaling helium is potentially lethal because by filling your lungs with this inert gas you displace air and run the risk of asphyxiation. Hence it is vital to breath deeply for a while *after* having inhaled the gas to expel the remaining helium from your lungs. What ever you do, don't inhale more helium immediately after the first inhalation: breathe deeply to purge your lungs thoroughly with air before inhaling more helium. And always inhale helium from a balloon and never directly from a pressurized tank of helium. The high pressure of the gas will damage your lungs.

Helium was first detected in the Sun's spectrum in 1868 but wasn't found on Earth until 1895 when it was discovered as a gas emanating from a sample

---

[31] For an interactive simulation of the effect of tongue position within the mouth cavity on speech sounds, see "pinktrombone": https://dood.al/pinktrombone/ (accessed 12/07/2021).

of Uranium.[32] But "helium voice" was almost certainly known, though not understood, a century earlier.

In 1799, a Swiss surgeon, J.P. Maunoir, had been amusing himself by inhaling hydrogen and was "astonishingly surprised at the sound of his voice, which was become soft, shrill, and even squeaking, so as to alarm him."[33] The dangers of breathing hydrogen were already known: not only can it result in asphyxiation, there is also the possibility of explosions. Several years before Maunoir's discovery, inhaling hydrogen had been employed as a party trick by a pioneering French balloonist, Pilatre de Rosier. He would inhale a mixture of hydrogen and air and breathe it out through a long tube, and to prove that hydrogen was present he would ignite the mixture as emerged from the tube. On one occasion the mixture exploded in his mouth "…and almost stunned him. At first he thought that the whole of his teeth had been driven out, but fortunately he received no injury whatever."[34] Rosier was killed in an attempt to cross the English Channel by balloon. The upper part was filled with hydrogen, the lower with hot air from a brazier slung underneath. Shortly after take-off, the hydrogen ignited and he and his companion fell 500 m to their deaths. Nevertheless, despite the dangers, inhaling hydrogen for amusement became something of a pastime during the early decades of the nineteenth century.

Inhaling a gas in which the speed of sound is much less than that in air will have the opposite effect: the dominant formants of the vocal tract will be much lower than in air, making the voice sound deeper. To demonstrate this, sulphur hexafluoride, a colourless, odourless, non-toxic gas is often used. The speed of sound in sulphur hexafluoride is two and a half times less that it is in air which results in formant frequencies that are correspondingly lower than they would be in air. The result is sometimes described as a Darth Vader voice.[35]

Although sulphur hexafluoride is not toxic, the same caveats about inhaling helium apply: it can asphyxiate you. There is an additional danger with sulphur hexafluoride because it is much denser than air and so is more difficult to rid one's lungs of the gas after inhaling it. To hasten exhalation

---

[32] When Uranium decays it emits alpha particles, which become the nucleus of a helium atom by capturing electrons from surrounding atoms. Helium cannot be manufactured, hence supplies of helium are limited and may run out in the not too distant future if people continue to use the gas in party balloons.

[33] Maunoir, J.P (1799) Des Effets du gaz hydrogèn sur la voix. Journal de Physique, de chimie, d'histoire naturelle et des arts, Vol 48, p 459.

[34] Dollfus, A. (1993) Pilâtre de Rozier, premier navigateur aérien, première victime de l'air. Association française pour l'avancement des sciences.

[35] Video of "sulphur hexafluoride voice" https://www.youtube.com/watch?v=u19QfJWI1oQ (accessed 19/07/2021).

one is advised to bend forward to lower one's head and breath deeply with one's mouth wide open.

Singing in a Bathroom

Should you be one of those people who like to sing while taking a bath or a shower, you may have wondered why your voice can sound so much richer and louder than it normally does. There are at least two reasons. The tiled surfaces and smooth, bare plaster walls of the bathroom are poor absorbers so sounds don't die down as quickly as they do in larger rooms full of furniture, carpets and curtains. At the same time the dimensions of the bathroom or shower cubicle will emphasise some frequencies present in your voice as a result of resonance.

Those reflections are the source of reverberation that lengthen each note allowing it to blend with those that follow, supressing unwanted changes in pitch or loudness as you sing. Reverberation also increases the loudness of your voice because the reflected sound is added to the sound issuing from your mouth.

At the same time, as you sing (or speak) you set the air within the bathroom vibrating at several frequencies that correspond to some of the formants of your voice. Those formants whose wavelengths match the distance between opposite walls or the floor and ceiling will be emphasised because they create standing waves between opposite surfaces within the bathroom, say between ceiling and floor or between walls (Fig. 6.5).

The formant frequencies that are most likely to set up audible standing waves are at the lower register of your voice because their wavelengths are going to be closest to the dimensions of a small room. Moreover, the resonant sounds are loudest at the antinodes of a standing wave, something you can discover for yourself by raising or lowering your head until your ears are at an antinode. Indeed this is how you can recognise a resonant sound: it is always noticeably louder than the original sound.

The so called Oracle Room within the 5000 yearold Hal Saflieni Hypogeum underground temple complex in Malta is known to resonate strongly at around 70 and 110 Hz, frequencies that are close to the fundamental frequency of the male voice. Researchers have speculated that the priests may well have harnessed these resonances in their rituals.[36] And

---

[36] Debertolis, P., Prof.agg, Coimbra, F., Dr., Eneix, L. (2015) Archaeoacoustic Analysis of the Hal Saflieni Hypogeum in Malta. Journal of Anthropology and Archaeology, June Vol. 3, No. 1, pp. 59–79.

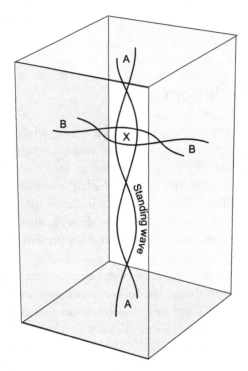

Fig. 6.5 Two of the many possible standing waves within a shower cubicle. Standing waves are set up between floor and ceiling (AA) and between walls (BB). Resonance will be most noticeable when your ear is at an antinode of both standing waves (X)

what of the claim that our forebears used both resonance and reverberation when selecting where to paint the images of animals within caves? Does the experience of singing in a bathroom shed any light on the matter?

The Sounds of Friction

Friction is responsible for some of sweetest sounds (a violin in the hands of a virtuoso) as well as some of the worst (scraping glass with a metal blade). The variety of friction sounds is huge, as is the number of distinct words with which to describe them: squeals, squeaks, screeches, groans, chirps, creaks, chatters and scrapes. And, as you will know, they are usually unwanted and frequently annoying. But there are many situations in which we should listen to them because they are often the first indication we have of a fault.

In every case these sounds are due to the same cause. When two smooth surfaces in close contact slide past one another their motion often proceeds in a series of tiny jerks. The motion is jerky because the surfaces keep on sticking together and stretching slightly before breaking free. When they break free, the surfaces spring back and set up vibrations within the objects that are sliding over each other, vibrations that can be ear-splittingly loud if they happen to match the natural frequency of one or other of those objects, particularly if the surface area of one of them is large.

The jerky movement is described as "stick-slip", with "stick" being the result of adhesion and "slip" the source of vibration. In both phases, friction is in play because surfaces in close contact adhere to one another. That adhesion is the source of friction. Moreover, adhesion is always much greater when the surfaces are both at rest than when they slide over one another. Friction between surfaces at rest is called "static friction" and "sliding friction" when they are in motion. You will have often noticed that you have to push much harder to get an object to begin to slide than you do to keep it sliding. A lubricating layer of oil or grease that keeps the surfaces apart will prevent them sticking and slipping.

A squeal occurs when the stick-slip phase keeps repeating rapidly, thus maintaining a continuous note. But if static friction between surfaces is large the moving surfaces are unable to vibrate freely when they break free of one another and the result is a series of brief, discrete broadband sounds that we hear as creaks or groans.

Here, then, is the source of the *squeal* of rubber soles on polished wooden floors, the squeal of car tyres when braking hard or turning into a corner at speed, *screeching* train wheels, *squeaky* hinges, *squealing* brakes, *scraping* chair legs and *creaking* floorboards. Stick-slip is the source of the *squeak* when you rub an inflated balloon with a finger or wipe a large sheet of glass such as a window pane with a damp rag, a squeegee or styrofoam, the *chatter* of wipers on a dry windscreen, the *crunch* when you walk on cold snow or on a pebbly beach, the *creak* as central heating pipes warm up or cool down, causing them to rub against the surfaces they are in contact with as they expand or contract, and even the alarming *rumble* of an earthquake. You may also have noticed that glassware from a dishwasher or newly washed wet hair both squeak when rubbed with clean, slightly damp fingers—hence, perhaps, the cliché for honesty: squeaky clean. Even bread and cheese can get in on the act: you may sometimes hear faint squeaks as you slice through a dense loaf of sourdough bread or chew a piece of halloumi cheese.[37]

It is often supposed that the archetype of sick-slip sounds is the squealing chalkboard, though in these days of digital smart boards the number of people who are familiar with that sound can't be very large. The sound is due

[37] After much trial and error, Trevor Cox managed to play a few bars of Ode to Joy by rubbing wine glasses with halloumi. http://trevorcox.me/halloumi-glass-harmonica (accessed 10/06/2021).

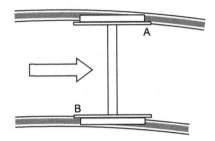

Fig. 6.6 Wheel Squeal. The outer wheel presses against the outer rail at A, the inner one at B. Most of the squeal comes from the inner wheel

to the vibration of the chalk itself as it sticks and slips. When sticking, the chalk momentarily bends and then springs back when it breaks free and slips and vibrates rapidly for a brief instant. The squeal is amplified if the board itself also vibrates. In fact, many stick-slip sounds would be all but inaudible were it not that their vibrations are coupled with a large surface that sets up vibrations in the surrounding air. Such a surface is known as a sound board.

The ear-splitting noise you hear when a train is travelling around a bend is known as wheel squeal. Unlike the wheels of a car, those of a train are fixed and so can't turn to follow the curving track. Hence its wheels slide sideways and scrape against the track, constantly sticking and slipping, which in turn can set them vibrating at their natural frequency. Most of the squeal comes from the leading inner wheel because the outer wheel presses more closely against the rail, which prevents it from vibrating freely. Wheel squeal is predominantly an urban phenomenon because in towns and cities railway lines are more likely to have more curves to fit in with the local layout (Fig. 6.6).

To the potentially endless list of stick-slip sounds, we might add the crunch that you sometimes hear when walking on snow. When snow has been lying for a day or two in cold conditions individual flakes start to stick together. When you step on the snow your weight breaks the bonds between the flakes and this sets up vibrations in the crystals that are heard as crunch.

Singing Wine Glass

In his *Dialogues Concerning Two New Sciences*, Galileo mentions in passing that "a glass of water may be made to emit a tone merely by the friction of

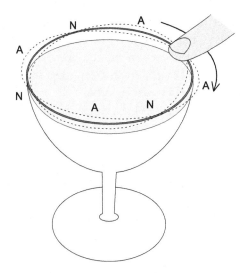

Fig. 6.7 A singing wineglass. The four antinodes (marked A) are the maximum movement of the rim. The rim is momentarily stationary at the four nodes (marked N). The point of contact between the finger and the rim is always a node

the finger tip upon the rim of a glass"[38] In fact, you need to ensure your finger is both free of grease and slightly moistened in order to get the glass to ring. As it slides around the rim your finger is continuously sticking and slipping. The wine glass rings when the rate at which your finger sticks and slips matches the natural frequency of the wine glass (which you can find by gently taping the side of the bowl of the glass, say with the handle of wooden spoon). The natural frequency of the glass determines the fundamental tone, but the sound you hear will be composed of the fundamental tone and its harmonics (Fig. 6.7).

When your finger and the glass stick together the rim is momentarily pulled in the direction in which you are moving your finger. When they break free from one another, the rim springs back causing the glass as a whole to vibrate at its natural frequency, a frequency that depends on the thickness of the rim and the diameter of the open end of the glass. Increasing the thickness of either the rim or the diameter of the open end both lower the natural frequency of the wine glass. Filling the glass with liquid also reduces its natural frequency because the mass of the liquid decreases the rate at which the glass can vibrate freely.

---

[38] Galileo Galilei (1914) Dialogues Concerning Two New Sciences, translated from the Italian and Latin by Henry Crew and Alfonso de Salvio. The Macmillan Company, p 99.

As we saw in chapter three, Benjamin Franklin's glass armonica exploited this phenomenon. The "Cristal Baschet", invented by François and Bernard Baschet in 1952, is a modern version on the glass armonica. It consists of a large number of glass rods of different lengths, the ends of which are embedded in a metal block. The player strokes the rods with wet fingers or hands creating sustained, ethereal tones which are far warmer than those of the armonica.[39]

Shattering a Wine Glass with the Human Voice

The possibility that a wine glass can be shattered by the unaided human voice has long been known, though how often, if at all, it has been actually been achieved is open to doubt.

In November 1670, Daniel Morhof, an overseas member of the Society, wrote a letter to Henry Oldenburg, the secretary of the Royal Society, describing how he had seen a wine glass known as a *rummer* shattered by a young man singing a prolonged note that made the glass vibrate vigorously while he was in Amsterdam.[40] Morhof claimed that after witnessing the demonstration "I tried a few times to see if I could imitate him, but the attempt was vain for my voice was thick and unsteady with the strain, until in the end I succeeded with a thinner glass." A few years later, John Wallis, an English clergyman and mathematician, noted in passing that "I have heard of a thin fine Venice glass cracked with the strong and lasting sound of a trumpet or cornet near it, sounding a unison or consonant note to that of the glass. And I do not judge the thing very unlikely"[41] Hooke said that he had himself tried a similar experiment, but had only ever got the glass to vibrate rather than to break.

The sound emitted by the wine glass is, as we have seen, due to vibrations of the rim. The larger the amplitude of the vibrations, the louder the sound. And if there are defects or microscopic cracks in the glass, large distortions of the rim may cause the glass to shatter. But it's not possible to shatter a wine glass by rubbing the rim because the finger pressing on the rim supresses large vibrations. But directing a very loud sound at the glass that matches its natural frequency can cause vibrations that are large enough to shatter the

[39] "Eric Satie Gnossienne 1 played on a Cristal Baschet": https://www.youtube.com/watch?v=8cmzjI kx6b8 (accessed 19/07/2021).

[40] Oldenburg, H., *Correspondence*, ed. A.R. Hall and M.B. Hall, 13 Volumes, 1965–86. University of Wisconsin Press, vol. 7, p 229–34.

[41] Wallis, J., (1677) On the trembling of consonant strings. Philosophical Transactions, vol 12, p 842.

glass. This can be achieved if the sound is electronically amplified and fed to a loudspeaker placed very close to the glass. There are several YouTube videos showing this. There are far fewer showing that it can also be achieved *without* electronic amplification of the voice.[42]

Squeals of Delight

By now you may have concluded that no good can come from friction's sounds. But stick-slip is also responsible for what is arguably one of the sweetest sounds that humans have managed to create. Is there a more sublime sound than that from a violin in the hands of a virtuoso?

The source of that sound is the stick-slip interaction between the strings of the bow and those of the violin. The strings of a bow are made of several strands of horsehair, a material which has tiny scales that all face in one direction. The horsehair strands are rubbed with rosin to increase the friction between them and the violin strings. As the bow moves across the string of the violin's these scales momentarily catch and drag the string sideways, i.e. they stick. But the tension in the stretched string soon becomes large enough so that it breaks free and snaps back, sliding across the horsehairs until it comes to rest, at which point it once again sticks to the bow strings. The to and fro motion of the strings repeats for as long as the strings of the bow slide across them in the same direction.

The back and forth motion of the violin strings are almost inaudible, however. Being so narrow they can't set up strong vibrations in the surrounding air. But their motion is communicated to the body of the violin through the bridge over which they pass, so it is the vibration of the upper surface of the instrument, known as the belly, that is the principle source of the sounds we hear from the instrument.

Singing Sands

Among the most puzzling and arresting examples of friction-generated sounds in nature are those of so-called singing sands that are found on beaches the world over and occasionally from desert dunes when its surface is disturbed when one walks over it.

---

[42] Shattering a wine glass with the unassisted human voice: http://www.metacafe.com/watch/3779902/how_to_shatter_glass_with_your_voice/ (accessed 28/07/2021).

Singing sands are not uncommon. They are usually found on seacoasts but can also be found on lakeshores and river banks.

> The so-called musical sand may be found at a number of places around the coast, where the quartz grains which make up the bulk of natural sands are dry and of fairly uniform size. On a small scale, the sand may be made to sing by walking over it or by thrusting a stick into it. At each footfall or impulse from the stick a momentary whistling sound is set up.[43]

In my limited experience of singing sands, the sound is similar to the squeak that is produced when a rubber sole slides on a smooth wooden floor. The sound of singing sand ranges in frequency between 800 and 1,200 Hz. It is most readily produced by sand which has dried out following rain since it is then at it's cleanest. Impurities or moisture between the grains can prevent the effect from occurring. Sites of singing sands are frequently given in geological guidebooks. In the British Isles there is a well-known site on the north-western coast of the Isle of Eigg. Indeed, there are several beaches on the Northwest coast of Scotland that have singing sands. Other well-known sites are on the Hawaiian island of Kauai and McMasters beach just north of Sydney, Australia. The sand of Bondi beach can sometimes sing when walked upon.

The cause of the sound is as yet something of a mystery, though this much has been established: the grains of sand must be of quartz, of similar size to one another and must be smooth and round. The sound is probably due to friction between grains as they move past each other, in other words due to stick-slip.[44]

Booming Sands

The sound produced by desert sand is very different to that heard in coastal sand; it is also much rarer. The characteristic frequency of these desert sounds lies in the region of 80 Hz. It is variously known as desert thunder, booming sands, booming dunes and roaring sands. The English explorer and geologist, R.A. Bagnold, often heard these sounds during his travels in the Sahara Desert.

---

[43] Richardson, E.G. (1947) Sound and the Weather. Weather, p 205.

[44] For the sound of true singing sands (not to be confused with booming sands: https://www.you tube.com/watch?v=hc7Gt4yqHlg OR 2 min into: https://www.youtube.com/watch?v=Qy4cTqAMy7k (both accessed 19/07/2021).

I have heard it in south western Egypt 300 miles from the nearest habitation. On two occasions it happened on a still night, suddenly, a vibrant booming so loud that I had to shout to be heard by my companion. Soon, other sources, set going by the disturbance, joined their music to the first, with so close a note that a slow beat was clearly recognised. This weird chorus went on for more than five minutes continuously before silence returned and the ground ceased to tremble.[45]

As you might expect, there are booming sands in several deserts. Well known ones include La Mar de Dunas and El Cerro Bramador in the Atacama desert in Chile, the Kelso Dunes near California's Mojave Desert, the Desert Dunes of Badain Jarin, Inner Mongolia, and the dunes near Liwa in area known as the Empty Quarter in Arabia. In almost all cases the sounds are produced when the lee of a dune collapses and slides down to its base. This collapse can be induced simply by pushing some sand over the top of a dune with one's foot.

The exact cause of the sound in these booming dunes is not fully understood, though the following facts have been established. The most important requirement is that the grains of sand are all of a similar size. They also have to be smooth but need not be spherical. Since the only sorting agent available to grade the grains is the wind, booming dunes are most likely to occur at the downwind end of a field of dunes. It is absolutely essential that the sand is dry. Even sand that feels dry to touch may contain enough moisture to prevent booming.

The friction between individual grains that is the most likely cause of the high pitched squeak of beach sand cannot also be the cause of the much lower frequencies heard in booming dunes. The most likely explanation is that these low pitched sounds result from an entire layer of sand just below the surface being set into vibration by a large quantity of sand sliding down the side of the dune.[46]

Babbling Brooks and Hissing Kettles

"I can recommend any reader who is not afraid of being late for breakfast to keep a bag of marbles in his bathroom."[47] We owe this timely advice

---

[45] R.A. Bagnold (1941) The Physics of Blown Sand and Desert Dunes. Methuen & Co, London, p 250.

[46] Video of "booming sands" https://www.youtube.com/watch?v=0XF6kGDLcVE (accessed 19/07/2021).

[47] Worthington, A.M. (1908) A Study of Splashes. Longmans, Green and Co, London, p 121.

to Arthur Worthington, an English physicist and the author of *A Study of Splashes*. The book is about splashes rather than the sounds associated with them. But since splashes are seldom silent, marbles come in handy should you wish to listen rather than look.

The sounds of dripping taps, tinkling fountains, babbling brooks and breaking waves have a common cause, though it is not, as you might expect, due to the impact between a drop and the surface of the water with which it collides. Nor is it the result of turbulence created as flowing water swirls around obstacles, as it does in a fast flowing stream or river. Instead, the noise is due to rapid vibrations of tiny bubbles of air just beneath the surface of water that are created following the impact or by turbulent flow. In fact, one way or another, almost all sounds that we associate with water involve bubbles. These include the sizzle as water is heated to boiling point in an electric kettle and the sound of a ship's rotating propellers. Bubbles that rise from within a liquid and bust at the surface are, by comparison, almost silent: a fizzy, frothing beverage is all but inaudible.

In Praise of Dripping Taps

There are people for whom the plip-plop of a dripping tap is an unbearable torture; I am not one of them. It's not that I am indifferent to the sound, it's that I find it both melodic and fascinating. Let me explain.

Worthington was the first person to photograph splashes and what these photographs revealed is that when a drop hits the surface of water it throws up a jet and forms a small crater in the liquid. But it was several decades before anyone realised that the splash itself is usually all but inaudible. The "plink" you hear is due to a tiny bubble of air that breaks free from the bottom of the crater and vibrates furiously just below the surface for a fraction of a second. The bubble itself is too small for its vibrations to be directly audible, but they make the surface of the water vibrate, which transfers the vibrations of the bubbles to the surrounding air (Fig. 6.8).

Although it may not be immediately obvious, the sound of a vibrating air bubble has a well-defined pitch that depends only its size. The smaller the drop, the greater the frequency at which it vibrates. It is this that gives it its musical quality. But that is only apparent when you can hear individual drops. The secret of these bubbly sounds was eventually discovered by Marcel Minnaert, a Belgian astrophysicist in 1933.[48]

[48] Minnaert, M. (1933) On Musical Air-Bubbles And The Sounds Of Running Water. Philosophical Magazine, 16 (7), p 235–248.

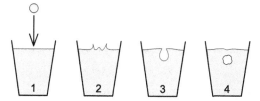

Fig. 6.8 Formation of an underwater bubble of air following the impact of a drop of water. The oscillations of the underwater air bubble in 4 is the source of the sound associated with a impact of drop of water

The average size of these bubbles depends among other things on the temperature of the water, something that you will notice by comparing the sound of pouring cold water from a height into a container such as a glass or a cup and with that of freshly boiled hot water.[49] Bubbles formed in cold water are on average smaller than those in very hot water because the size of the bubbles depends in part on the surface tension of water. Surface tension, which is due to the attraction between molecules of water, decreases as the temperature of water increases so that a given pressure of the air within the bubble will make it larger. And larger bubbles vibrate at lower frequencies than small ones.

Whenever you come across a noisy brook you will notice that the sounds you hear come from places where there are bubbles. These are produced where flowing water encounters solid objects or changes in height that cause the stream to break up and create bubbles. A foaming mass of bubbles appear white because light is reflected multiple times as it makes its way through them. In fact, white water is noisy, clear water is silent.

There are many situations where this is very obvious. The sound from the foaming mass within the pool at the foot of the waterfall is due to the bubbles created by the impact of the waterfall. Moreover, that foaming mass is made up of a huge number of bubbles of different sizes and so the resulting sound will be composed of a broad range of frequencies. Bubbles trapped within

---

[49] The difference was noticed over 2000 years ago. Aristotle, Problemata, Book XI, section 10: "Why does cold water poured out of a jug make a shriller sound than hot water poured from the same vessel? Is it because the cold water falls at a greater speed, being heavier, and the greater speed causes the sound to be shriller?" His claim that cold water falls faster than hot water is, of course mistaken. For Aristotle, the speed with which an object falls to the ground (which he termed natural motion) depends on the density of the material out of which it is made. Some 2000 years later Galileo proved that this is assumption is unfounded.

waves as they topple over as they reach the shore, are the source the characteristic sound of surf. Rain falling on water becomes noisy when the size and speed of the raindrops are large enough to create underwater bubbles.[50]

The Kettle's Song

Osborne Reynolds seems to have been the first person to have taken a serious scientific interest in that most mundane of domestic sounds, that of a boiling kettle.

> Among the many phenomena, the secrets of which have been preserved by the deadening influence of familiarity on curiosity, there is perhaps none more remarkable than that of the "singing of the kettle on the hob", which has many times been the subject of sentiment and verse but not, it would seem, hitherto a subject of physical study which like the study of the rainbow might afford evidence as to the conditions under which we exist.[51]

But his interest in the sounds produced as water is heated up was merely a preamble to a wider interest in a phenomenon now known as cavitation.

> That the cheering evidence of the readiness of the social gathering is not the only evidence to be obtained from the song of the kettle will in the first place be demonstrated in these experiments. Thus, having analyzed by experiment the physical causes of this sound and its variations, the purpose of the experiments is to demonstrate the relation which exists between sounds in the kettle and sounds produced by the motion of water, or any liquid, under certain common conditions. And, in the third place, to demonstrate the general fact that liquids flowing between fixed boundaries emit no sound as long as they continuously occupy the space between the boundaries, and thence to demonstrate that when such sound occurs it is evidence of the boiling of the water.[52]

You are probably already aware that when water is boiled in an electric kettle it is accompanied by a distinct sequence of sounds, beginning with a gentle hiss, which then rises to a crescendo of loud pops and crackles and

---

[50] It is only drops with a diameter greater than 1 mm that create the underwater bubbles necessary for sound.

[51] Reynolds, O. (1894) Experiments Showing The Boiling Of Water In An Open Tube At Ordinary Temperatures. In: Reynolds, O. (1901) Papers On Mechanical And Physical Subjects, Vol 2. Cambridge, p578-587, see p 578.

[52] Reynolds, O. (1894) ibid. p 578.

finally subsides to a quiet bubbling as the water reaches a steady boil. But the only one of this sequence of sounds to which you usually pay heed is the final one because it informs you that the water has reached its boiling point.

The initial hiss is due to bubbles of water vapour that form within the scratches in the base of the container as the water warms up. Initially, these collapse before they can break free, as you will see if you look at the base of saucepan in which you are heating water. Each collapse results in a sharp click. Later, when the water at the base of kettle is much hotter, the vapour bubbles are able break free and rise up into cooler water, where they collapse with a sharp click, though these are so numerous that they merge into a cacophony.[53] In the enclosed space of a kettle, these sounds are enhanced due to resonance within the water and the air space above it. As proof that resonance is involved, remove the lid of the kettle during this phase of boiling and note that the overall pitch drops. Eventually, the water as a whole becomes hot enough for the vapour bubbles to reach the surface without collapsing, and the only audible sound is of bubbles bursting at the surface.[54]

But, as Reynolds realised, it isn't necessary to heat a liquid such as water in order to produce bubbles of vapour within it. All that is required is a sudden drop in its pressure and bubbles will form within scratches and pits on the surface in contact with the liquid. This is the reason why the contents of a can or bottle of fizzy drink will sometimes spray violently out of its container when opened. Opening the container allows the gas in the space above the liquid to escape, which in turn causes a drop in pressure within the liquid. The dissolved gas then comes out of solution as a mass of bubbles, forcing the contents of the container to spray out.

A rapidly rotating propeller is another situation in which bubbles form and collapse without a liquid been heated. This is the source of the sounds created by ship's propeller. If the propeller is moving rapidly, the pressure of the water in contact with its surface will drop to the point where bubbles of water vapour form.[55] But when they break free and move from low pressure

---

[53] The formation of bubbles of water vapour prevents the water from superheating, which can occur when water is heated in a microwave. Microwaves are absorbed within the liquid and not at the surface of the container. Hence no vapour bubbles form until air bubbles are introduced into the superheated liquid, either by stirring it or adding sugar, which may result in an explosion of hot foam.

[54] Walker, J. (1982) What Happens When Water Boils Is a Lot More Complicated Than You Might Think. Scientific American, Dec.

[55] This is an example of the Bernouilli principle: when the speed of a fluid is increased, the pressure it exerts on a surface with which it is in contact decreases. Take an A5 sheet of paper and hold it with two hands up to your lips. When you blow over it, the far upper surface which initially droops downwards will rise up because the air flow of your breath reduces the pressure acting on the upper surface allowing the pressure on the lower surface to push the paper up. This is why air flowing over wings enable planes to stay up in the air.

at the surface of the propeller to the higher pressure of the surrounding water they collapse suddenly with a loud a click. Not only is that rapid collapse a source of sound, it releases enough energy to damage the surface of the propeller.

Surprisingly, cavitation is employed by a tiny crustacean known as a snapping shrimp to catch its prey. The creature has a huge claw that it can close so rapidly that it creates a bubble of water vapour due to cavitation. The resulting bubble collapses very rapidly, producing a shock wave powerful enough to stun its prey. The victim is grabbed by the smaller claw and gobbled up. These shrimps live in large colonies and, as we noted in chapter five, the collective sound of those collapsing bubbles once perplexed submariners who took the source to be due to enemy action.

Bees in Trees

For a week or two in April, when the cherry trees that line the local roads in my neighbourhood are in full blossom, they thrum with the sound of bees gathering nectar, a sound that waxes as I walk under each tree and wanes when I am between them. The trees gradually fall silent as they shed their blossom, their petals swirling about in the breeze and gathering in drifts on the pavement. And for those few days of early spring I am reminded of an unexpected connection between the hum of bees in flight and the swirl of petals in a breeze: they are both due to whirling eddies of air.

Eddies, also known as vortices, are ubiquitous in nature. They form as a result of friction whenever a fast flowing fluid such as water or air rubs up against a solid surface. Friction reduces the speed of the fluid close to the surface while that of main body of fluid continues unimpeded allowing the faster moving fluid to get ahead of the slower fluid. The result is a series of rotating eddies within the fluid close to the surface (Fig. 6.9).

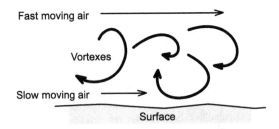

Fig. 6.9 Boundary layer eddies. Notice that the vortexes spin in the opposite direction to the movement of the air

Leonardo famously studied and sketched eddies in flowing water and many of his insights into their nature have stood the test of time.[56] But in air, eddies are invisible without lightweight matter such as smoke, clouds of condensed water vapour, dust, petals or leaves to reveal their presence. And in some circumstances a rapid succession of aerial eddies can be heard because each eddy exerts a tiny force on the surrounding air.

The rapid succession of tiny eddies created by the beating wings of an insect in flight is the source of the buzz or whine that announces their presence. An eddy forms with every stroke of the wing so the faster it flaps, the higher the pitch of its sound. In fact, the frequency of the sound you hear is the same as the rate at which the wings are beating, as Robert Hooke explained to Samuel Pepys during a chance meeting in a London street.[57]

Up, and with Reeves walk as far as the Temple, doing some business in my way at my bookseller's and elsewhere, and there parted, and I took coach, having first discoursed with Mr. Hooke a little, whom we met in the streete, about the nature of sounds, and he did make me understand the nature of musicall sounds made by strings, mighty prettily; and told me that having come to a certain number of vibrations proper to make any tone, he is able to tell how many strokes a fly makes with her wings (those flies that hum in their flying) by the note that it answers to in musique during their flying. That, I suppose, is a little too much refined; but his discourse in general of sound was mighty fine.[58]

Unsurprisingly, Leonardo da Vinci had noted that beating wings are the source of that buzz almost a couple of century earlier.

That the sound which flies make proceeds from their wings you will see by cutting them a little, or better still by smearing them a little with honey in such a way as not entirely to prevent them from flying, and you will see that the sound made by the movement of the wings will become hoarse and the note will change from high to deep to just the same degree as it has lost the free use of its wings.[59]

---

[56] Ball, P (2009) Nature's Patterns: A Tapestry In Three Parts: Flow. OUP, see chapter 1.

[57] You can measure the rate of wing beat with an audio spectrometer.

[58] Pepys, S. Diary entry Wednesday 8 August 1666.

[59] The Notebooks of Leonardo Da Vinci (1938) Arranged, rendered into English and Introduced by Edward MacCurdy, Volume 1, Jonathan Cape, p 288.

The Howl of the Wind

The hiss, whistle or moan that we hear on a windy day when air encounters narrow cylindrical objects such as pine needles, twigs, branches or wires go by the collective name of aeolian tones or sounds. In Japan the sough of a pine forest is called *maksukaze*, and in earlier times its sound was much appreciated.

The cause of these sounds is the succession of tiny eddies that develop alternatively on either side of an obstacle and spin away downwind. As with beating wings, as each eddy is formed it is accompanied by a tiny rise and fall in air pressure. If the rate at which this occurs lies within the limits of audibility we hear a sound. The necessary eddies can form only within a particular range of air speeds. At low speeds air flows silently over an obstacle because no eddies are formed. At intermediate speeds, depending on the dimensions of the obstruction, we hear aeolian tones with a characteristic pitch. But at very high speeds, airflow becomes turbulent and result is a loud, chaotic, broadband noise. Turbulence is responsible for characteristic roar of the exhaust gases as they exit the rear of the aircraft engine at high speed and mix with the surrounding air.

An investigation into the physics of aeolian tones by the Bohemian physicist, Vincent Strouhal, established that the dominant pitch of these sounds increases with the wind's velocity and that it decreases as the diameter of the obstacle is made larger.[60] Thus for a given wind speed, narrow obstacles such as pine needles produce high frequencies while thicker ones, such as twigs or branches, produce low frequencies.[61] And since the same wind blowing through a wood encounters obstacles of different widths such as the twigs and the trunks of trees, aeolian sounds in a wood will be made up of a wide range of frequencies. Indeed, this is the explanation for Hardy's observation that "To dwellers in a wood almost every species of tree has its voice as well as its feature."[62]

---

[60] Strouhal, V. (1878) Uber eine besondere Art der Tonerregung. Ann.Phys.Chem, 5, p 216.

[61] The mathematical relationship between wind speed and diameter of obstacle discovered by Strouhal is surprisingly simple: Frequency = 0.18 x (wind velocity ÷ obstacle diameter). The frequency of the sound produced when a gentle breeze (say Beaufort Number 3, which corresponds to a velocity of 4 m/s) blows through a pine forest in which the needles have an average diameter of 1.5 mm will be 480 Hz.

Using this formula, the dominant frequency of the sound produced by a gentle breeze, say Beaufort Number 3, which corresponds to a velocity of 4 m/s, blowing through a pine forest in which the needles have an average diameter of 1.5 mm is 480 Hz. Incidentally, Strouhal's relationship predicts that for a wind velocity of 4 m/s any object wider than 4 cm will not produce an audible frequency. Does this tally with your experience?

[62] Hardy, T. (1920) Under The Greenwood Tree. E.P Dutton and Company, p 3.

Aeolian sounds are not confined to pine needles, twigs or branches. On a windy day, telephone cables, wire fences, guy ropes and slated fences are all potential sources of such sounds. Occasionally, the frequency of an aeolian tone will match the natural frequency of the obstacle that is its cause. When this happens, the obstacle itself vibrates and will reinforce particular aeolian frequencies, making them louder than others.

"At the entrance to the Deep Cut, I heard the telegraph wire vibrating like the Aeolian harp", Thoreau noted in his diary.[63] A few days later he added "The telegraph harp sounds strongly to-day, in the midst of the rain. I put my ears to the trees and I hear it working terribly within... the sound seems to proceed from the wood. It is as if you had entered some world-famous cathedral, resounding to some vast organ...The wire vibrates with great power, as if it would strain and rend the wood".[64]

Telegraph wires were long ago replaced by telephone cables, but they too are often strung between poles. The frequency at which a wire strung between poles can vibrate depends, among other things, on the tension of the wire—something that as we saw in chapter two was discovered by Marin Mersenne almost 400 years ago. Hence the hum that is heard from a vibrating wire on a warm and windy day may become a whine on very cold days because the wire contracts as its temperature decreases, which in turn increases its tension. And just like the string of a guitar, increasing the tension of a wire will increase the frequency at which it vibrates. Follow Thoreau's advice and press your ear against one of the poles between which the vibrating wire is strung the better to hear its aeolian hum.

Aeolian sounds can also be produced by superstructures on buildings such as arrays of decorative slats or slatted sunscreens. A notorious example of the consequence of adding decorative slats to a building is what occurred when the finishing touches were added to Beetham Tower, in Manchester (U.K.). Without considering the acoustic consequences, the architect insisted on placing an array of horizontal slats on the roof of the building to increase its height and emphasise its slenderness. Unfortunately, on very windy days the slats produce loud aeolian tones variously described by Mancunians as a howl, a scream, a wail or a hum. The cause has been traced to eddies created as fast moving air flows over the slats and which cause the air in the gaps between the slats to resonate. The slats themselves will sometimes vibrate if the tone matches their resonant frequency, and this increases the loudness of

---

[63] Thoreau, H., Journals, entry for 12 September, 1851.
[64] Thoreau, H., Journals, entry for 23 September, 1851.

certain tones. Attempts to eliminate the sound by altering the profile of the slats has not been entirely successful.

In 2020, alterations to the profile of pedestrian railings of the walkway on the western side of Golden Gate Bridge in San Francisco has resulted in an aeolian moan on very windy days than can be heard over a large area of the city and surrounding countryside.[65] The cause of the sound is that the railings were made thinner to increase airflow and reduce the air resistance of the bridge as a whole.

The sound of a jet engine and of a blow torch also involve air in motion and occurs when a stream of very hot, fast moving gas meets stationary air. The resulting turbulence is the source of the loud, broadband, low frequency roar that can be loud enough to deafen anyone close by.

Aeolian tones differ from the resonant sounds heard when a current of air flows across the open end of a tube. Although similar mechanisms are at work in both situations—eddy currents and resonant vibrations—unlike Aeolian tones, which increase continuously in pitch as the velocity of the wind increases, sounds due to the resonance of air within cavities occur only at particular wind speeds.

Aeolian Harps

Beetham Tower's unwelcome hum is a classic example of sound out of place, of unwanted noise. But winds have long been deliberately harnessed to create sounds using all manner of devices. One such is an instrument known as an Aeolian harp. The very first one was designed and built by Athanasius Kircher in the seventeenth century.[66] It consisted of a hollow wooden box with openings (i.e. a sound box like that used in stringed instruments) across which were fixed several strings of catgut of different diameters all tuned to the same low fundamental pitch. He placed his harp on the sill of an open window to catch the wind and listened to its tuneless hum from within the room. Depending on the speed at which air flows across strings one or more of the strings will vibrate and the air within the sound box will resonate, thus increasing the loudness of certain tones. Because all the strings are tuned to the same low frequency, they all have identical higher harmonics and so when two or more strings are vibrating at the same time, the different tones are

---

[65] "Why Does The Golden Gate Bridge Sound Like It's Singing?": https://www.youtube.com/watch?v=WTTUluvvpls (accessed 19/07/2021).
[66] Kircher, A. (1650) Musurgia Universalis sive Ars Magna Consoni et Dissoni, Vol II, Book IX. p 352–3.

in harmony with one another. Aeolian harps are still being made both for private use and public display. In fact, as often as not, acoustic sculptures are variations on Kircher's Aeolian harp.

Aeolian sounds are associated with wind, but it is relative motion between the object and the air that is important. So if a stationary object causes an aeolian tone on a windy, shouldn't the reverse also be possible? Indeed it is. Slash at the air with a length of cane as quickly as you are able and it will swish or swoosh depending on its thickness and how rapidly it moves.

Whistle While You Work

You know how to whistle, don't you? Of course you do: just put your lips together and blow.[67] Actually, it's not quite as simple as that because whistling also involves your tongue, so you have to be taught how to whistle. Nevertheless, it's by no means improbable that humans have whistled for one reason or another since they first roamed the African savannah. It's something that the anatomy of our lips and vocal tract enable us to do with ease once we have learned how to coordinate our lips, tongue and breath. People whistle to attract attention, express surprise or admiration, issue instructions (shepherds to their sheep dogs), subvert authority (whistling 'Colonel Bogey' in David Lean's Bridge on the River Kwai), make music (Roger Whittaker, Geert Chatrou) and communicate (whistled languages).[68]

But just how puckering one's lips and blowing creates a high pitched, pure tone has only recently been fully understood. It has been known for at least a century that as a stream of air issues from a circular opening such as slightly parted, puckered lips, the air in contact with their circumference moves more slowly than the air in the centre of the flow resulting in succession of anti-clockwise vortices in the air just beyond the opening. But it has since been discovered that the whistle itself comes mainly from within the mouth. The pulses of pressure due to a rapid succession of vortices set up resonant vibrations within the cavity of the vocal tract. To change the note of a whistle you have to change the volume of your vocal tract by altering the shape of your tongue. Whistle a low note followed by a higher note and you will notice that as you change from the former to the latter you press the tip of your tongue against your bottom teeth and arch your tongue slightly, which reduces the volume of the vocal tract and raises its resonant frequency.[69]

---

[67] Lauren Bacall's oft-quoted line from the film *To Have and Have Not*.

[68] Lucas, J., Chatburn, A. (2013) A Brief History Of Whistling. Five Leaves Publications.

[69] Lip whistling is an example of an *orifice tone*.

But that's not the whole story: air pressure also plays a part. In 1972, several astronauts training for a mission in Skylab, the first American space station, spent 56 days together in a hyperbaric chamber. To replicate conditions on Skylab, the atmosphere within the chamber was mixture of 70% oxygen and 30% nitrogen at one-third normal atmospheric pressure. Among other things, the three-man crew discovered that this low pressure affected sounds within the chamber. "Their initial observation in the 5-psi atmosphere was that sound seemed to be further away and somewhat softer, so for a few days they become aware of shouting to each other and becoming hoarse. Another consequence of the reduced pressure was an inability to whistle and a sneeze was far more milder than expected."[70]

The likely reason these astronauts were unable to whistle within the chamber is that the creation of the necessary train of vortices responsible for a lip whistle depends among other things on the density of air. The density of a gas depends on its pressure, so a low pressure probably prevents vortices forming at a rate that will set up the necessary resonant vibrations in the vocal tract to create a whistle.

A later generation of astronauts seem not to have known about these unusual acoustic phenomena and were surprised to discover that it was impossible to whistle while wearing spacesuit, though they were able to do so inside the International Space Station. The reason is the atmosphere in the ISS has the same composition and pressure as that at sea level on Earth, whereas in a spacesuit astronauts breathe pure oxygen at a third of normal atmospheric pressure.

The effect of atmospheric pressure on whistling might also be noticeable on Earth. The pressure of air at the summit of Mount Everest is only marginally more than the air inside Skylab or a spacesuit, which suggests that mountaineers would be unable to whistle as they near the end of their climb. In fact, Dan Pettit, an American astronaut, once mused that whistling could be used by mountaineers as a simple test of whether they have reached the so-called 8000 m "death zone", an altitude at which atmospheric pressure is about one third that at sea level. Disappointingly, to date no mountaineer seems to have followed up Pettit's suggestion—they probably have other things on their mind as they near Everest's summit.[71]

Physical whistles come in different forms but all employ a combination of steady stream of vortices and a resonant cavity. The whistle in the spout of a kettle consists of a small, closed cylindrical chamber with a hole at either end.

[70] Shayler, D.J. (2001) Skylab, America's Space Station. Springer, p 155.
[71] Whstling on the sumit of Everest: https://pythom.com/PythomLabsNASA-climbing-challenge-Find-Mount-Everest-quotWhistle-Linequot-2016-02-29-27573 (accessed 28/07/2021).

Steam from the kettle entering the chamber through the first hole collides with the surface at the opposite end, causing the pressure within the tiny chamber to rise and fall. This leads to a steady stream of vortices in the steam issuing into the air from the second hole and which are the source of the whistle that is heard.[72]

A delightful example of a device that employs a stream of air to produce a whistle is the Peruvian whistling jar. These pottery jars consist of two spherical chambers, both open to the air and connected by a hollow tube. The double chamber is partially filled with water through a spout in the first chamber. When the jar is tipped, the water in the second chamber rises, forcing air through a whistle at its mouth. Depending on the design of the whistle, several distinct sounds can be produced, the range of which is quite astonishing.[73] The purpose of these jars is to make sounds and the second chamber has a zoomorphic shape reprenting the creature associated with the sound of the whistle. They are not water jugs, so the water remains within the jar as it is tipped back and forth to repeat the whistle.[74] Simpler forms of these jars have just one chamber that is filled with water and are really water jugs that make a sound when tipped. As water pours from a spout it draws in air through a whistle in the upper end of the hollow handle.

Shock Waves

During the 1870–71 Franco-Prussian War, the Prussians accused the French of using explosive bullets based on the gaping wounds inflicted on their troops by French rifle fire. A Belgian physicist, L.H.F. Melsens, who had made a study of gunshot wounds, believed he knew the reason: those gaping wounds were due to a cone of compressed air in front of a bullet being forced into the body at the moment of impact, not to an explosion of the bullet. Ernst Mach, the renowned German physicist and philosopher of science, was intrigued by Melsens' theory and set out to obtain photographic evidence of the effect of the passage of a high velocity bullet through the air.

The technique he employed was to photograph the shadow cast by the bullet as it flies through the air when it is illuminated by a brief, bright electric

---

[72] Henrywood, R. H., Agarwal A. (2013) The Aeroacoustics Of A Steam Kettle. Physics of Fluids, 25, October.

[73] Pan whistling jug: https://www.pierrecharrie.com/pan AND Chirpy Jug: https://www.youtube.com/watch?v=k_u7ot61tyA.

[74] Peruvian whistling jars: https://www.youtube.com/watch?v=ZzoiL7x56Eo.

spark.[75] He found that to create a cone of compressed air the bullet has to be travelling faster than the speed of sound. He called this a "head wave" and it later came to be known as a shockwave.

Mach concluded that the injuries sustained by the Prussians were not due to the "head wave" but rather to the impact of the bullet itself. He also realized that a head wave would accompany the bullet as long as it was travelling supersonically. This would explain accounts that on some occasions during a battle two sounds were heard when a shell was fired from a cannon and at others only one. Mach surmised that when two sounds were heard one of them is due to the head wave of the projectile while it travelled supersonically. The other was due to the ejection of gas from the muzzle of the cannon when it was fired, known as a blast wave. If a shell travels subsonically, only the blast wave from the cannon would be audible.[76]

Mach's "head wave" comes about because a body travelling faster than sound compresses the air in its path to such a degree that it creates a spike of very high pressure within the air that trails behind it and resembles the wake of a ship. The spike in pressure is experienced as brief, loud sound, a so-called sonic boom, as it sweeps past you. In fact, by the time it reaches the ground the pressure within the shock wave due to a large supersonic aircraft such as Concorde flying at altitude is only about one thousandth that of the atmosphere, about the same as the difference in atmospheric pressure between a couple of floors, nowhere enough to make your ears pop. But a shockwave sweeps past you in a fraction of a second so your ears experience a very rapid rise and fall in air pressure. It is the rapid change in pressure that is responsible for the sonic boom However, even though it is audible, a shockwave is not a wave because no vibrations are involved, i.e. the air between you and the source does not vibrate. A better term for the phenomenon is "shock front" (Fig. 6.10).

Sonic Booms

Shells are not the only objects that can travel faster than sound. Day and night meteoroids plunge earthward at speeds of several kilometres per second and the larger ones are accompanied by powerful shock fronts. Indeed at these hypersonic speeds the extremely high pressure within the shock front raises

---

[75] The shadowgraph method used by Mach is known as Schlieren photography and invented by August Toepler (1836–1912) in 1864.

[76] Mach, E. (1898) On Some Phenomena Attending the Flight of Projectiles. In: Mach, E. (1898) Popular Scientific Lectures. Open Court Publishing Co., p 309–38.

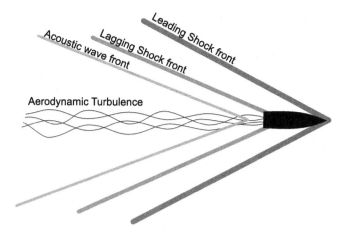

Fig. 6.10 Shock fronts produced by a bullet travelling supersonically. The acoustic wavefront is due to aerodynamic turbulence in the bullet's wake

the temperature of the air to such a degree that it ablates the surface of the meteor and causes the resulting vapour to glow brightly, a glow that makes the meteoroid visible. On those occasions when these powerful shock fronts reach the ground they are heard as a sonic boom. So long before meteoroids were known to have an extraterrestrial origin, the evidence for which was first collected and evaluated by Ernst Chladni in 1794, witnesses of particularly bright meteoroids sometimes reported that they also heard loud sounds.[77] Of course, they did not know the real cause of these sounds and assumed that they were due to explosions or distant thunderstorms.

These days people associate sonic booms with jet planes, though they were heard on hundreds of occasions several years before there were aircraft capable exceeding the speed of sound. The very first sonic boom due to a man-made vehicle was heard across London on 8th September, 1944 when a V2 rocket exploded on reaching the ground in Chiswick, West London, killing three people and injuring seventeen.[78] The explosion was followed by double boom resembling a thunderclap caused by the shock front catching up with the rocket, which had been travelling at four times the speed of sound.[79] The first sonic boom due to a manned vehicle was supposedly heard on the 14th

---

[77] Chladni, E. (1794) Über den Ursprung der von Pallas gefundenen und anderer ihr änlicher Eisenmassen und über einige damit in Verbindung stehende Naturerscheinungen. Riga, J.F. Hartknoch. [On the Origin of the Iron Masses Found by Pallas and Others Similar to it, and on Some Associated Natural Phenomena].

[78] I have not come across any evidence that German scientists and engineers heard sonic booms when they were test-firing V2s.

[79] Some 3000 V2s were launched (1400 at London and SE England and 1600 at Antwerp).

of October, 1947 when Chuck Yeager piloted the Bell X-1 rocket plane. It reached a speed of 1,070 kph or Mach 1.05 at 13,000 m, which it maintained for just 20 s.[80] People on the ground said that they heard a faint sonic boom, though doubt has been cast on the claim because atmospheric conditions on the day would have prevented the shock waves from the tiny rocket plane reaching the ground.[81] It is likely that they were recalling sonic booms produced during later test flights.

The idea that a sonic boom signals the moment an aircraft exceeds the speed of sound is incorrect. Rather it is the audible effect of an expanding shock front, and will be heard only as long as the aircraft is travelling faster than the speed of sound. Moreover, the shock wave trails the aircraft and forms what is known as a "sonic boom carpet" when it reaches the ground. Depending on the altitude of the plane, the sonic boom carpet can be tens of kilometres across. Everyone on the periphery of the envelope of the shock front will hear the sonic boom as a single, momentary sound. The fact the sonic boom is heard only once by someone on the ground as the shock wave sweeps past them while the aircraft remains visible as it streaks across the sky is why they assume that the boom is a single event that marks the moment when the aircraft breaks the so-called sound barrier. In fact, a second sonic boom often accompanies the first one because a body travelling supersonically is accompanied by two shock fronts, one formed in front of it and the other at its rear. Hence the double sonic boom that was often heard after the explosive warhead in the V2 had detonated.

Those outside the envelope of the sonic boom carpet are unlikely to hear anything unless the shock front is reflected back into the atmosphere by the ground and subsequently refracted back to earth by a temperature inversion, giving rise to what is known as a secondary sonic boom.

Nor is sound the only evidence of the shock front. The spike in pressure due to a supersonic plane high in the sky is unlikely to be large enough by the time it reaches the ground to break the glass in a window though it may make the frame rattle. This, along with the accompanying sonic boom, is why aircraft are allowed to fly supersonically over land only in exceptional circumstances such as fighters scrambled to intercept encroaching enemy planes or

---

[80] Although Mach introduced the concept of Mach number as the ratio of the speed of a moving object to the speed of sound, the term itself was coined by J. Ackeret (1898–1981) in 1928. An object travelling at the speed of sound is said to travel at Mach 1. Mach 1 is also known as the sound barrier.

[81] Benson, L.R. (2019) Quieting the Boom: The Shaped Sonic Boom Demonstrator and the Quest for quiet Supersonic Flight. NASA, p 29, note 8.

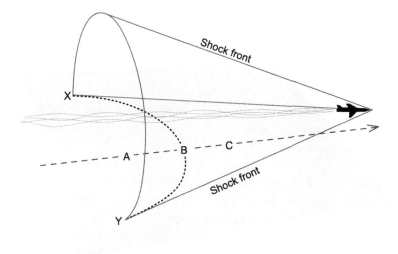

Fig. 6.11 Sonic boom carpet. The dotted curve XY marks the intersection between the cone of the shock front trailing the aircraft and the ground. The person at A has just heard the sonic boom, the person at B is just hearing the sonic boom and the person at C has yet to hear the sonic boom

unidentified commercial aircraft. It was that prohibition that fatally undermined the commercial viability of Concorde. It is also why sonic booms are rarely heard these days (Fig. 6.11).

Bullwhips

The crack of a whip is a weak shock wave produced by the supersonic motion of the tip. This is achieved by the gradual reduction of the diameter of the whip along its length. As the energy of initial flick of the hand holding the handle moves along the tapered whip, the speed at which it moves along the whip increases until by the time it reaches the braids at the tip (known as a *popper*), it is travelling supersonically and producing a loud crack. That sound is not due to a collision between the braids but the shock wave when their speed momentarily exceeds that of sound. The same explanation applies to "snapping" a towel.

Explosions

An explosion can be the source of a shock front as long as the initial expansion of the gases produced by the reaction between the chemicals of the explosive is supersonic. This compresses the air before it, resulting in a shock front that initially travels away from the site of the explosion at more than the speed of sound, i.e. supersonically. The increase in the density of the air within the shock front changes the refractive index of the air and alters the direction of light passing through it, which makes the shock front visible as an expanding, hemispherical, optical discontinuity that stands out from the background if the explosion is filmed in slow motion.[82] There are any number of websites on which you can find videos and photographs of shock fronts due to heavy calibre guns and explosions. Close to the explosion the rapid rise and fall in pressure can be large enough to bring down buildings, destroy vehicles and kill people. And as with a sonic boom, it is the brief spike in pressure as the shock front sweeps past you that is the cause of the sound you associate with the explosion.

There are several natural sources of shock fronts due to explosions. We saw in chapter four that sudden uplift of the ground or the sea at the moment seismic waves reach the surface can sometimes produce an audible shock front. Volcanoes that erupt explosively may also produce powerful shock fronts. A memorable example captured on video was the eruption of Mount Tavurvur on August 29th, 2014.[83]

The bang you hear when an inflated rubber balloon bursts is also due to a shock wave because the speed with which the rubber membrane ruptures can exceed the speed of sound, allowing the high pressure air within it to expand supersonically for a brief moment, creating a weak shock wave that quickly decays into a pressure spike travelling through the surrounding air at the speed of sound. You hear a bang as that spike travels past your ears.

Thunder

You will know from your own experience of thunderstorms that the sound of thunder can vary from a sharp crack to a long, low rumble that may persist for several seconds. The source of that sound is a shock front created by the

---

[82] Video of "huge shockwave captured at high-speed": https://www.military.com/video/ammunition-and-explosives/explosives/shockwave-captured-on-high-speed/763995636001 (accessed 19/07/2021).

[83] Video of "Eruption of Mount Tavurvur" https://www.youtube.com/watch?v=oMxIlXW56cQ (accessed 19/07/2021).

almost instantaneous superheating of the entire length of the channel of air through which the electrical discharge passes. About 99% of this energy is dissipated as heat and is responsible for the flash of light. The remaining 1% is converted into a shock front within the air surrounding the stroke and that is the source of thunder.

The actual sound you hear depends on several factors, three of which are particularly important. The first is the length of the discharge channel and its spatial orientation with respect to the observer. The other is its attenuation as it is scattered, absorbed and refracted within the atmosphere.

A lightning stroke never travels in a straight line for any distance. It continually twists and turns, following the path of least electrical resistance. The resulting shock front propagates perpendicularly to the channel, so most of its energy also propagates in this direction. This is why a single stroke will be heard as a series of claps and rumbles. Generally speaking you will hear a sharp clap from that part of a lightning channel that is perpendicular to your line of sight. A rumble is produced by a discharge that is moving either towards or away from you. At the same time, those sounds fade with distance as their high frequencies are absorbed, altering their timbre to a low growl. And if there is a strong temperature inversion, those sounds will be refracted away from the ground and will be inaudible to someone several kilometres from the thunderstorm.

Armed with this knowledge you can discover quite a lot about a lightning stroke merely by listening to it. To estimate how far away the nearest branch of the lightning stroke is from you, multiply the time in seconds between the flash and the moment you first hear the thunder by the speed of sound in air (340 m/s). The time between the flash and the final rumble will enable you to calculate the minimum length of the lightning channel. And, of course, by listening to the rumbles and claps, you can make an informed guess as to the path that the stroke has taken.

A Final Clap

Unless you are an aficionado of flamenco you may not have given much thought to the art of the handclap. To produce the sharp, loud snap that characterises flamenco clapping it is necessary that either the fingers or the ridge just below the fingers of one hand strike the slightly cupped palm of the other hand. Cupping the palm increases the area of contact it makes with the fingers of the other hand, resulting in a louder sound. A palm-on-palm clap makes a softer, lower pitched sound. The quietest claps are those where both hands are rigid.

The source of the sound in flamenco style clapping is, surprisingly, a weak shock wave because air between the colliding fingers and palms forces out air from the space between them at supersonic speeds. If the palms are cupped, however, the air in the space between them acts as a resonating cavity, the pitch of which depends on the degree to which the palms are cupped. If both hands are rigid, the area of contact when they collide is small, as is the resulting sound.[84]

Clearly, flamenco-style clapping is the most effective way to summon and investigate echoes. Perhaps it's time to get practicing.

---

[84] Fletcher, N.H. (2013) Shock Waves And The Sound Of A Hand-Clap — A Simple Model. Acoustics Australia, Vol. 41, No. 2, p 165–8.

Index

'n' indicates footnote page numbers.

© The Editor(s) (if applicable) and The Author(s), under exclusive
license to Springer Nature Switzerland AG 2021
J. Naylor, *Now Hear This*,
https://doi.org/10.1007/978-3-030-89877-9

Printed in the United States
by Baker & Taylor Publisher Services